PHILOSOPHY, RHETORIC,

AND THE END OF KNOWLEDGE

RHETORIC OF THE HUMAN SCIENCES

General Editors

John Lyne
Donald N. McCloskey
John S. Nelson

Philosophy, Rhetoric, and the End of Knowledge

THE COMING OF SCIENCE AND

TECHNOLOGY STUDIES

STEVE FULLER

THE UNIVERSITY OF WISCONSIN PRESS

The University of Wisconsin Press
114 North Murray Street
Madison, Wisconsin 53715

3 Henrietta Street
London WC2E 8LU, England

Copyright © 1993
The Board of Regents of the University of Wisconsin System
All rights reserved

5   4   3   2   1

Printed in the United States of America

Library of Congress Cataloging-in-Publication Data

Fuller, Steve.
    Philosophy, rhetoric, and the end of knowledge:
the coming of science and technology studies / Steve W. Fuller.
      444 pp.   cm. – (Rhetoric of the human sciences)
      Includes bibliographical references and index.
      ISBN 0-299-13770-8   ISBN 0-299-13774-0 (pbk.)
      1. Science-Philosophy.   2. Science-Social Aspects.
3. Knowledge, theory of.   4. Rhetoric-Philosophy.   5. Social
sciences-Philosophy.   I. Title.
Q175. F926   1993
501-dc20   92-34689

CONTENTS

Acknowledgments ix

Introduction xi

PART ONE THE PLAYERS AND THE POSITION

Chapter 1 The Players: STS, Rhetoric, Social Epistemology 3

HPS as the prehistory of STS 3
The turn to sociology and STS 9
Rhetoric: The theory behind the practice 17
Enter the social epistemologist 24

Chapter 2 The Position: Interdisciplinarity as Interpenetration 33

The terms of the argument 33
The perils of pluralism 38
Interpenetration's interlopers 44
The pressure points for interpenetration 48
The task ahead (and the enemy within) 56
Here I stand 64

PART TWO INTERPENETRATION AT WORK

Chapter 3 Incorporation, or Epistemology Emergent 69

Tycho on the run 70
Hegel to the rescue 84
Building the better naturalist 94
Conclusion: Naturalism's trial by fire 100

Contents

Chapter 4 Reflexion, or the Missing Mirror
          of the Social Sciences   102

          Why the scientific study of science might just
              show that there is no science to study   106
          The elusive search for science in the social
              sciences: Deconstructing the five
              canonical historiographies   114
          How economists defeated political scientists
              at their own game   126
          Conclusion: The rhetoric that is science   133

Chapter 5 Sublimation, or Some Hints on How
          to be Cognitively Revolting   139

          Introduction: Of rhetorical impasses
              and forced choices   140
          Some impasses in the artificial
              intelligence debates   142
          Drawing the battle lines   144
          AI as PC-positivism   146
          How my enemy's enemy became my friend   151
          But now that the coast is clear   156
          Three attempts to clarify the cognitive   168
          AI's strange bedfellows: Actants   179

Chapter 6 Excavation, or the Withering Away of History and
          Philosophy of Science and the Brave New World
          of Science and Technology Studies   186

          Positioning social epistemology in the transition
              from HPS to STS   187
          The price of humanism in historical scholarship   192
          A symmetry principle for historicism   200
          Historicism's version of the cold war   203
          Under- and overdetermining history   210
          When in doubt, experiment   213
          STS as the posthistory of HPS   220

Contents

PART THREE	OF POLICY AND POLITICS
Chapter 7	Knowledge Policy: Where's the Playing Field? 227

Science policy: The very idea 229
An aside on science journalism 234
Managing the unmanageable 237
The social construction of society 251
The constructive rhetoric of
 knowledge policy 255
Armed for policy: Fact-laden values
 and hypothetical imperatives 262
Machiavelli redux? 270
A recap on values as a prelude to politics 275

Chapter 8 Knowledge Politics: What Position Shall I Play? 277

Philosophy as protopolitics 277
Have science and democracy outgrown
 each other? 281
Back from postmodernism and into the
 public sphere 290
Beyond academic indifference 300
Postscript: The social epistemologist at the
 bargaining table 307

PART FOUR	SOME WORTHY OPPONENTS
Chapter 9	Opposing the Relativist 319

The Socratic legacy to relativism 319
The sociology of knowledge debates:
 Will the real relativist please stand up? 321
Interlude I: An inventory of relativisms 324
Interlude II: Mannheim's realistic relativism 326
Is relativism obsolete? 328
Counterrelativist models of
 knowledge production 335

Chapter 10   Opposing the Antitheorist   347

A Fish story about theory   347
What exactly does "Theory has no consequences" mean?   352
Fish's positivistic theory of "theory"   354
Toward a more self-critical positivist theory of "theory"   358
The universality, abstractness, and foolproofness of theory   360
Convention, autonomy, and Fish's "paper radicalism"   363
Consequential theory: An account of presumption   367

Postscript   The World of Tomorrow, as Opposed to the World of Today   377

Appendix   Course Outlines For Science and Technology Studies in a Rhetorical Key   383

References   393

Index   415

ACKNOWLEDGMENTS

If one believes, as I do, that knowledge is the property of a distributed network of exchanges, then an author's acknowledgments are no less substantive than anything else that might appear in a book. Starting with the economic infrastructure, then, I would like to thank the National Science Foundation and the National Endowment for the Humanities for personal and collective grants that provided me the time and space to interact with many of the parties discussed in this book during the period 1988-91. Also, my thinking about "knowledge policy" and "knowledge politics" was stimulated by the assemblage of academics and policymakers at a 1989 conference I organized with Will Shadish, entitled 'The Mutual Relevance of Science Studies and Science Policy," which was jointly sponsored by Virginia Tech and Memphis State Universities.

As faculty member in the Center for the Study of Science in Society at Virginia Tech, I have been able to help forge the identity of Science and Technology Studies, the emergence of which frames the issues raised in this book. Virginia Tech has awarded the first doctoral degrees in this field, which officially prepares students to think critically about the production of knowledge in both academic and policy contexts. This promise is easier said than satisfied, not only for the reasons raised in this book but also for more local reasons, ones which make the metaphor of the *crucible* an apt one for our day-to-day existence here. In this regard, I want to thank Gary Downey and Skip Fuhrman for supporting me during the extraordinarily difficult period in which this book was written. In the midst of the turmoil, a singularly bright moment occurred when Sujatha Raman and I were married. We anticipate a lifelong partnership in the never-ending battle to keep the academy wedded to public life.

I have benefited immensely from the insight and goodwill of colleagues all over the world. I would especially like to single out the STS communities of the Netherlands, Finland, Sweden, and the United Kingdom for their generous invitations to lecture–in particular, Hans Harbers, Dick Pels, Hans Radder, Matti Sintonen, Erkki Kaukonen, Aant Elzinga, Thomas Brante, David Edge, Barry Barnes, David Bloor,

Acknowledgments

Steve Woolgar, and William Outhwaite. I would also like to thank Bruno Latour and Steve Woolgar–who remain the two most probing thinkers in any field–for their critical support. Thanks, too, to Joseph Rouse, who gave lots of good advice and is most responsible for having me take feminist critiques of science seriously. Ron Giere, Paul Roth, Jim Maffie, and Larry Laudan showed that analytic philosophers can be good guys, too. For editorial advice and support when some of the chapters were mere articles (they have since been substantially changed), I would like to thank Nick Jardine (*Studies in the History & Philosophy of Science*), Ian Jarvie (*Philosophy of Social Sciences*), David Shumway (*Poetics Today*), and Stephen Turner.

My interest in rhetoric is long-standing, going back to my graduate student days in Pittsburgh. (The Speech Communication Department was located on the floor above the History and Philosophy of Science Department.) My oldest friend here is Charles Willard. In 1989, the Project on the Rhetoric of Inquiry at the University of Iowa provided me with a unique opportunity to experience what I call in this book "interpenetrative interdisciplinarity." This book is, in part, repayment to Don McCloskey, John Nelson, John Lyne, Allan Megill, Bob Boynton, Kate Neckerman, Lorna Olson, and Jay Semel for an especially fruitful and pleasant stay. Many of these can also be thanked for acquainting me with the Rhetoric of the Human Sciences series at the University of Wisconsin Press, whose director, Allen Fitchen, has been singularly accommodating in getting this book to press. Finally, a special debt of gratitude must be extended to the ever vehement Bill Keith of the University of Louisville Communication Department, for pushing me (probably in directions that he would not altogether approve of) to clarify my views on rhetoric. I eagerly await his own opus in this area.

Exceptional students continue to come my way. They are the people with whom I have the most sustained intellectual contact. The great virtue of students, aside from open-mindedness, is that their natural orientation toward the future forces you to think normatively: What should I teach that will be of use to them once they become professional researchers? My former student Steve Downes exemplifies the new breed of STS practitioner that this book hopes to promote. Other students whose work and talk have helped me think through rhetoric's place in the future of STS are Garrit Curfs, Ranjan Chaudhuri, and especially Jim Collier, whose textbook proposal was a sustaining inspiration for what rhetoric of science *in practice* can be. I dedicate this book to another former student, Bill Lynch, the only person I know who will someday be foolish enough to write a book like this.

INTRODUCTION

It must be the end of knowledge. Never before have both philosophy *and* rhetoric had to endure such hard times in the academy. Usually, one has had to suffer at the expense of the other. Yet today philosophy is typically seen as having exhausted itself, with whatever good it contained being transported to the more fertile fields of the special sciences. Rhetoric, on the other hand, has expanded its domain to fill the available space, and in so doing has lost much of the incisiveness that traditionally put claimants to epistemic authority on their guard. What is on the verge of being lost is what I seek to recover–namely, language as a force that collectively enables people to transform their world. In particular, I want us again to take seriously the proposition that theorizing is a politically significant practice (cf. Skinner 1987).

To see theorizing as political is to recognize it as having consequences that go beyond its intended audiences. Much of the gloom generated by the "end of philosophy" scenarios proposed from Nietzsche to Rorty turns on a misapprehension of philosophy's real effects. These effects are typically much stronger on those who actually follow some philosophical advice and leave the discipline than on those who stay to repeat the advice to the next generation of students. For example, to take logical positivism seriously is not to mimic the positivists' verbal gestures in philosophy classes, but rather to take up a special science for the purpose of logically analyzing its conceptual foundations and empirically testing its most challenging claims. Paradoxically, like most of the distinctive philosophical movements of this century, logical positivism prescribed the withering away of the social conditions that made its emergence possible. Thus, insofar as positivism continues to flourish, and sometimes even to be dominant, in philosophy departments, it has failed at its mission. A similar conclusion was reached by the historian of science Derek de Solla Price (1986). He discovered that the perceived "hardness" of a science within the academy corresponded to the perceived efficacy of the science outside the academy. Price speculated that the link was that academic practitioners of the efficacious discipline devote less time and energy to replacing themselves

Introduction

than to spreading their ideas and techniques to people who will eventually work outside the academy.

However, we must be careful here about what to count as "efficacy." As the fate of rhetoric in recent years demonstrates, widespread acceptance is not a reliable indicator of significant impact. The realization that "everything is rhetorical" has much the same ring as learning that one has been speaking prose all of one's life. Nothing much changes, except that one now feels a little more comfortable with what one has already been saying. In that respect, today's rhetorical turn in the human sciences has tended to valorize the very features of rhetoric–its ornamental, expressive side–that by the seventeenth century were taken to be symptomatic of rhetoric's decadence (Vickers 1987). Nowadays, this value reversal is most evident in the fact that rhetorical criticism is more literary analysis than sociology of reception. That is, rhetoricians often spend more time unearthing the "implied reader" of a text than tracking down the text's real readers and seeing what, if any, relation they have to the implied reader projected in the text. Unfortunately, this literary approach to rhetoric fails to capture the distinctiveness of rhetoric as a practice that has periodically interrupted, as well as represented, the normal flow of discourse. Socrates is, of course, the archetypal practitioner of such a rhetoric. As a result of his sophistic interventions, a tradition of questioning and revising the structure of knowledge and society was launched, one which is closely associated with philosophy at its best.

I write as a philosopher of science who has been promoting the idea of *social epistemology* (Fuller 1988a, 1989a, and the journal by that name) in order to foster closer cooperation between humanists and social scientists in the emerging interdisciplinary complex known as Science and Technology Studies (STS), a field that is capable, I believe, of not only redrawing disciplinary boundaries within the academy but ultimately, and more important, of making the academy more permeable to the rest of society (Pickering 1992 is a representative recent anthology). The trick is that STS practitioners employ methods that enable them to fathom both the "inner workings" and the "outer character" of science without having to be expert in the fields they study. The success of such a practice bodes well for extending science's sphere of accountability, presumably toward a greater democratization of the scientific decision-making process. These concerns are also shared by the assemblage of people who travel under the rubric of *rhetoric of science*. The success of this book, then, should be measured in terms of its ability to persuade philosophers, theoretical humanists and social scientists, STS practitioners, and rhetoricians of science to see each

Introduction

other as engaged in a common enterprise. In order for the reader to appreciate the dimensions of this enterprise, let me briefly recount the origins of its component parts.

Social epistemology arose in the 1980s as an attempt to overcome the false dilemmas that marked the birth of STS, namely, between "normative" philosophical approaches concerned with what science ought to be and "empirical" sociological approaches concerned with what science actually is. While these dilemmas are still discussed in the literature, they are gradually losing their polemical interest, largely because a middle-of-the-road consensus of philosophers and sociologists has been forged–often with historians acting as midwives–which admits that the actual conduct of science evinces more philosophically respectable qualities than the sociologists originally maintained (even though it would be an exaggeration to say that these qualities amount to something that a philosopher would dignify as a "theory of scientific rationality"). But in opposition to this "era of good feeling" is emerging a more numerous and decidedly grassroots force within STS. Challenging the "High Church" of historians, philosophers, and sociologists who share a largely intellectual interest in the hold that scientific knowledge has over society is a "Low Church" of such heterogeneous elements as policymakers, feminists, political activists, schoolteachers, journalists, as well as administratively oriented political scientists and economists and civic-minded natural scientists and engineers–all of whom are more concerned with the problems that science has caused and solved in modern society than with the special epistemic status that science enjoys vis-à-vis other forms of knowledge.

Two indicators capture the difference between High and Low Church STS. First, both sects like to trace their origins to the 1960s. But whereas the High Church points to Thomas Kuhn's *The Structure of Scientific Revolutions* (1970) as the watershed STS text, the Low Church portrays STS as a response to the disturbing symbiosis that had developed between scientific research and the military establishment during the Vietnam War. (Fuller 1989b and Cutcliffe 1989 provide brief histories of the High and Low Churches, respectively.) The second indicator pertains to the different senses in which STS can become more "radical." High Church radicalism heads toward "reflexivity," an inward turn whereby STS practitioners apply to their own work the very principles which have enabled them to deconstruct the epistemic authority of the scientists they study. In this phase, STS research is revealed to offer no overarching lessons about the nature of science, but rather specific points that vary significantly across contexts in which

Introduction

STS might be practiced. High Church radicalism tends to undercut Low Church radicalism, which is basically a version of the "emancipatory" politics associated with Western socialist parties. Here the STS practitioner invokes her own initially privileged "standpoint" on science, whose emancipatory capacity is tested by the extent to which it can be made available to the entire citizenry (Harding 1986, 1991). Thus, science is put squarely in the service of humanity, perhaps even to the point of "downsizing" science so that more people can participate in its conduct and evaluation.

My characterization of STS in Chapter 1 is meant to transcend the distinction between High and Low Churches, especially in their radical forms. If the old false dilemma of STS was "normative" versus "empirical" approaches to the study of science, I suspect that a new false dilemma is emerging between the High Church's "theoretically" informed perspective and the Low Church's "practically" oriented one. This tension is evident in the pages of each issue of *Science, Technology, and Human Values*: once a favorite literary haunt for the Harvard-MIT science policy set, but now the official organ of the Society for Social Studies of Science (4S), to which most of the High Church belongs. A relevant Low Church organ is the *Bulletin of Science, Technology, and Society*, published by the National Association for Science and Technology in Society (NASTS). With regard to this new tension, social epistemology holds that, for all its haughtiness, the High Church makes sense only as doing more or less what the Low Church is doing, but at a more abstract, but no less socially relevant, level. What are different, of course, are the audiences—which brings us back to rhetoric.

Rhetoric of science is in an ideal position to heal any rifts that may be opening up between the two STS sects, as its own constituency reflects the natural interpenetration of theory and practice. Perhaps the most visible group to identify openly with the rhetoric of science is the Project on the Rhetoric of Inquiry (POROI) at the University of Iowa, which also houses one of America's leading rhetoric departments. From POROI has come such landmarks as Donald McCloskey's *The Rhetoric of Economics* (1985) and the anthology *The Rhetoric of the Human Sciences* (Nelson, Megill, and McCloskey 1987), which abundantly illustrate how distinguished humanists and social scientists use the resources of rhetoric to stem the tide of disciplinary fragmentation and the academy's growing irrelevance to public debate. Professional rhetoricians have increasingly adopted this agenda, some deepening the debate over the resolution of expert disagreement in a democratic forum (e.g., Willard 1983, 1990, 1991, 1992), and others reaching back

Introduction

into classical rhetoric for *topoi*, or argumentative frameworks that commonly arise in the legitimation of scientific claims (e.g., Prelli 1989; A. Gross 1990). Most recently, programs in technical writing have evolved from their humble origins as required "composition" courses in English departments to the site of some of the most promising research on the reading and writing conventions in the academy and the liberal professions (e.g., Bazerman 1988, 1989a, 1989b; Collier and Toomey 1994; a good history is Russell 1991).

It goes without saying that the social epistemologist needs to establish credibility with both academics and policymakers. This is quintessentially a problem in rhetoric, namely, one of cultivating *ethos*. Specifically, the social epistemologist must overcome the classical stereotypes of *both* the philosopher (as Platonist) and the rhetorician (as Sophist). The stereotypical philosopher invokes norms as an excuse for distancing herself from the people, who (so says the philosopher) willfully fail to meet her lofty standards. For her part, the stereotypical rhetorician abandons norms for gimmicks that can secure short-term success for her client (often in willful disregard of more long term and less tangible benefits). The social epistemologist's way out of these stereotypes is to realize that the normative is constitutively rhetorical: that is, no prescription can have force if the people for whom it is intended have neither the ability nor the wish to follow it. This point implies two *principles of epistemic justice* (à la Rawls 1972) that I propose as procedural constraints on normative transactions. I call these *humility* and *reusability*.

The turn to political philosophy here is quite deliberate. Philosophy's public service is to promote "Enlightenment," an idea that first reached self-consciousness in the eighteenth century, when the efficiency of the capitalist mode of production freed up enough people's time from the material necessities of life that a relatively widespread discussion of societal ends could be conducted on a sustained basis: Where are we going? Should we be heading there? If so, how should we get there? Who should be doing what in the meanwhile? Even though many twentieth-century theorists have questioned the Enlightenment's emphasis on managed talk over directed action, the project has nevertheless been unique in examining tradition for the sake of transforming it, rather than simply continuing it (Wuthnow 1989: pt. 2). The most inspiring case in point is the United States Constitution, my best example of rhetorically effective theorizing in the Enlightenment spirit, whose full realization requires the participation of all the members of a society.

Introduction

The U.S. Constitution is sometimes described as the only successful instance of "philosophically designed order," in marked contrast to the failed instances that make up the entire history of totalitarian politics. (Has there ever been a form of totalitarianism that was *not* philosophically inspired?) However, the highlighted turn of phrase misleadingly leaves the impression that the Constitution involved the "application" or "implementation" of a particular philosophical theory, when, in fact, I would claim that the Constitution is itself an example of philosophical theorizing fully actualized (or "rendered self-conscious," as Hegel might say). For the Constitution does exactly what every philosophical theory—especially the ones that have gone by the name of "metaphysics"—has aspired to do, namely, to provide a procedural language for articulating a variety of distinct perspectives on the world. The worth of such a theory is measured by the transformations of perspective that it enables: Are the perspectives simply given the opportunity to pursue what they have already identified as their own interests, or are they constrained to take into account the interests of others in such a way that they reach positions better able to address the standing hopes and fears of the day? To put the point as an insight of social epistemology, philosophical theories are diffuse social movements. If law and politics actualize philosophical theories, then metaphysics and epistemology, respectively, as commonly understood (that is, as the study of reality and its modes of access) are what result when such theorizing *fails* to be actualized.

The power of the great philosophical theories of the twentieth century—Marxism, pragmatism, logical positivism, existentialism, and structuralism—lay not in the truth of their specific doctrines but in the ability of their procedural languages—what is often disparagingly called their "jargons"—to get people from quite different walks of life to engage in projects of mutual interest. Such collaboration was made possible by the several registers in which each of these languages could be articulated. Thus, to restrict logical positivism to Rudolf Carnap, Hans Reichenbach, and a handful of Euro-American academic philosophers adept in formal logic and conversant with cutting-edge scientific research would be to ignore logical positivism's more lasting significance as a social movement (Proctor 1991: pt. 3). Here we need to look to the constituencies for works like A. J. Ayer's *Language, Truth, and Logic*, I. A. Richards' *Practical Criticism*, Count Korzybski's *Science and Sanity*, S. I. Hayakawa's *Language, Thought, and Action*, Stuart Chase's *The Tyranny of Words*, and even Samuel Beckett's novel *Watt*. Each extended representation in the "Positivist Constitution" to such Low Church outposts as psychiatry, political science, education, com-

Introduction

munication studies, literary criticism, and–dare I say–the general reading public.

In our own time, Jürgen Habermas (especially since Habermas 1985) has singularly excelled at theorizing in a way that not only draws into his discourse variously interested intellectuals but also intervenes in the public affairs of his native Germany (Forrester 1985 is the best survey of similar impacts outside Germany). But as Habermas (1987) himself has rightly observed, the biggest threat to rhetorically effective theorizing in the late twentieth century is *postmodernism*, with its refusal to believe in the possibility of the sort of constitution that I have been describing (e.g., Lyotard 1983), that is, a form of talk that sublimates without entirely eliminating the deep divergences that exist in contemporary society. More so than Habermas, social epistemology accepts the facts which inspire postmodernists but not their skeptical normative conclusions.

In response to the skeptical postmodernist, I would ask whether a constitution really requires a meeting of minds, or simply a confluence of behaviors. Following a convergence of opinion within both analytic and continental philosophy (e.g., Quine 1960 and Derrida 1976), I believe that only a philosophical conceit, backed by a dubious mental ontology, makes agreement on meanings, values, and beliefs a necessary condition for coordinated action. Instead, parties simply need to realize that they must serve the interests of others as a means of serving their own. That is, their diverse perspectives are causally entangled in a common fate. Much thinking about public policy reifies zero-sum gamesmanship, and, as a result, illicitly presumes that opposing interests require opposing courses of action that eventuate in one side's succeeding at the expense of the other. Such thinking is compelling only if one imagines that parties are fixed in their positions, a situation that will obtain only if the parties do not communicate with one another. It is not that communication breeds consensus. Rather, communication may cause all concerned to change their positions such that their still quite different goals can be pursued in harmony and perhaps even to the benefit of others who are not directly involved. In any case, in the long term, both sides to a dispute will either win or lose together. Contrary to a recent trend in STS thinking (following Latour 1987), I believe that the relevant model here is the ecology rather than the battlefield. Competition should be between different coalition formation strategies, each designed to implicate all parties in a common fate.

The traditional strategy for instilling this sense of mutually implicated inquiry has been to engage in a *rhetoric of truth*, whereby inquirers are led to believe (usually with the help of a philosophical the-

Introduction

ory) that they are all *already* heading in a common direction, fixated on a common end, and that all subsequent discussion should be devoted to finding the most efficient means toward that end. The historical persuasiveness of this strategy is revealed by the extent to which the subject matter of epistemology and the philosophy of science has been traditionally defined as "methodology" (a search for means) rather than as "axiology" (a search for ends). My own strategy is to develop a rhetoric that does not, in the name of "truth," preempt the articulation of significant disagreements over the ends of inquiry. This is the *rhetoric of interpenetrability*, which I deploy in four cases of interdisciplinary negotiation to which I have been party in recent years, and which remain open for others to explore. In large part, this book is a report of my practice as a theorist moving within the academy, as well as between the academy and the rest of society. Before surveying the contents of this book, a little lesson in political theory may go some way toward elucidating my social epistemology.

Adapting a distinction drawn in recent democratic theory (e.g., Held 1987), we may speak of the *plebiscitarianization* and the *proletarianization* of knowledge production. For short, call these *plebiscience* and *prolescience*. Plebiscience argues that there should be only as much public involvement in knowledge production as will allow the process to flow smoothly. Scientific research in the United States approximates plebiscientism in that the public normally ends up being involved in decisions about scientific research only when that research has potentially adverse consequences for a particular community, and then the scope of public involvement is restricted to that community. Prolescience reverses the priorities by arguing that knowledge production should proceed only insofar as public involvement is possible. In a prolescientific state, research agendas and funding requests would have to be justified to a board of nonexperts, not simply to a panel of scientific peers. Although critics of prolescience typically read it as a veiled form of antiscience, it would be more instructive to regard prolescience as an implicit challenge to many of the elitist assumptions of plebiscience. In economic terms, this elitism appears in plebiscience's strong distinction between the production (*by* experts) and the distribution (*to* nonexperts) of knowledge. The mutation of representative democracy known as *corporatism* has done the most to embody this distinction in recent times.

The corporatist reverses the democratic impulse by making the people beholden to their representative—in this case, a scientist. Corporatists suppose that as the world becomes a more complex place, people *ought* to lose interest in managing more of their lives—in fact, all

Introduction

but the most locally effective aspects of their lives. The extent to which the corporatists have successfully cultivated this ethos may be seen in how "abstract" or "remote" people come to regard, say, the workings of foreign policy or scientific research vis-à-vis their own daily concerns. Yet the felt abstractness or remoteness of these activities, which the corporatist promotes as grounds for rule by experts, does not necessarily reflect any causal detachment from everyday life. After all, the price of consumer goods at home, as well as their access, could easily be affected by either a breakdown in international relations or a scientific redefinition of product safety standards. What *is* detached from everyday life, however, is a rhetoric for talking about the causal connections, and hence there is only a limited potential for a variety of constituencies to realize that they have a stake in the conduct of affairs that take place outside their neighborhoods.

One way of diagnosing the deficiency brought on by such rhetorical detachment (a diagnosis inspired by Piaget's child development experiments) is that people no longer have the opportunity to make the sorts of decisions that would force them to appreciate the complexity of the human condition–decisions, in other words, that would get them to break out of the simplistic schema within which they normally make political judgments (Rosenberg 1988). But this is not the only diagnosis. It may be, instead, that the experts are not provided enough opportunity to account for themselves in ways that would force them to reduce the complexity of their own cognitive situation. This is the STS diagnosis, as I see it. In this context, it is worth recalling an observation that Bruno Latour (1981) made about the discourse of lab biologists: the biologists can modulate their speech and writing patterns, depending on whether they need to justify themselves to an audience of like-minded researchers or to a committee of scientifically illiterate members of Congress. I agree with Latour that probably little of scientific importance is lost in the translation, since, if need be, the scientists can descend from abstract formulae to simple drawings in order to explain a point. The scientists may prefer to concentrate the expression of their knowledge claims in dense jargon rather than to diffuse it through a cognitively permeable ensemble of words, pictures, artifacts, and ambience. But that is a guild privilege that we can ill afford scientists to enjoy. Thus, every time someone like Latour enters the lab, STS strikes a blow for proleschience by demonstrating that much of what would otherwise be considered "external" to science quickly becomes "internal" once scientists need to answer to a wider audience.

As for what follows in the pages ahead, Part I lays out the basic position of the book which is that the emerging field of Science and

Introduction

Technology Studies has the potential to be an emancipatory practice, given its dual mission of dissolving disciplinary boundaries and democratizing knowledge production. However, a properly renovated sense of philosophy and rhetoric is needed for the normative project of STS to get off the ground. After locating the roots of STS in nineteenth-century concerns about the proliferation of rival epistemic authorities, Chapter 1 outlines the major contemporary STS orientations and discusses why normative questions have been generally given the silent treatment. An account of rhetoric is then given that is designed to empower the STS practitioner with an empirically responsive normative sensibility. The account is based on the idea that norms are prescribed to compensate for already existing tendencies in order to reach some mutually desirable goal. Finally, the standpoint and scope of my brand of social epistemology is introduced. I advance a "shallow" conception of science, one that locates the authoritative character of science, not in an esoteric set of skills or a special understanding of reality, but in the appeals to its form of knowledge that *others* feel they must make in order to legitimate their own activities. In this way, rhetoric goes to the very heart of science.

In Chapter 2, I argue that the desultory character of most interdisciplinary research and the lack of epistemic standards that cut across the disciplines are really two sides of the same problem. The scent of banality that often seems to accompany calls to interdisciplinary scholarship arises from a failure to take to heart the (merely) conventional character of the differences separating academic disciplines, as well as between the academy and society at large. This point is repeatedly driven home by STS research. It implies that interdisciplinary exchanges have the potential for significantly transforming the work that disciplines do, especially by constructing new epistemic standards to which several disciplines agree to hold themselves accountable. However, a "knowledge policy" initiative of this sort requires a special rhetoric, which I dub *interpenetrative*. I present several pressure points for interpenetration in the academy, while at the same time distancing this rhetoric from both a blandly tolerant humanism and a maniacally technocratic enthusiasm.

Part II characterizes four cases of interdisciplinary interpenetration in which I have participated, mainly with regard to rhetorical strategies available and used, as well as their socio-epistemic implications. Common to the four types–incorporation, reflexion, sublimation, excavation–is the suggestion that many, if not most, of the "philosophically deep" problems generated by the sciences are the function of unreflective, often downright bad, communication habits. Entrenchment

Introduction

thus is mistaken for profundity. However, I do not mean to imply that these problems can be easily resolved. Part III is meant to show that these problems have a deep political and economic character that cannot be dealt with apart from all the other issues involved in governing a polity. If nothing else, science is *not* autonomous. Taken on its own terms, however, Part Two provides a fairly comprehensive sense of the state of play in the epistemological debates that currently dominate the social and cognitive sciences, as well as much of the humanities. The status of STS as a player in this game is the subject of Chapter 6.

Chapter 7 elaborates the sensibility that social epistemology brings to knowledge policy, namely, that the knowledge system may have problems even if nobody is complaining. Indeed, the institutional inertia that governs most science policy is itself the biggest problem. After showing how both independent and advocacy journalism obscure this problem, I suggest strategies for constructing normative considerations in a policy setting. Finally, I consider objections that "knowledge policy" would have to be Machiavellian in order to succeed.

Chapter 8 moves from the systemic to the political, suggesting a continuity between philosophy's classic normative mission and "knowledge politics." Basically, philosophers are in the business of questioning standards and achievements that are normally found exemplary. But practically speaking, the union of Big Science and Big Democracy currently provides no public forum for conducting the politics of knowledge. I then consider the distinct possibility that "science" and "democracy" have outgrown each other, given that none of the old normative models seem to have much purchase on the sorts of activities that pass under those names today. Here, inspired by work in economic sociology and mass media law, I propose a principle of *epistemic fungibility* to cut Big Science down to a democratically manageable size. I conclude by considering the rhetorical indifference with which academics conduct their professional lives, which prevents them from appreciating the political character of their work, as well as preventing policymakers from using that work in the most appropriate manner. This chapter contains an appendix that discusses the negotiating style of the social epistemologist as interdisciplinary mediator.

Part 4 tackles the two main foes of the knowledge policy maker: the relativist (in many guises) and the antitheorist (in the person of literary critic Stanley Fish). As it happens, these two foes are weakest where they advertise themselves as strongest: the relativist operates with an obsolete conception of society, while the antitheorist has a rather unrhetorical, positivistic conception of theory. In place of these

Introduction

inadequacies I offer, respectively, some conceptions of society compatible with social epistemology and a conception of presumption as a legal or scientific norm (an embedded theory, if you will) that counteracts a community's acknowledged worst tendencies. The book ends with a utopian postscript–so readers can get a vivid sense of what difference my position would make to the way we think about knowledge in the world–and a practical appendix on teaching STS rhetorically.

Now, for some advice on reading this book. Of course, I would like you to read all of it, but the chapters are sufficiently self-contained that you can devise your own reading plan. Everybody should read Part I (Chapters 1-2) as well as the Postscript and Appendix. After that, here are my time-saving tips:

> Philosophers (other than philosophers of science):
> Chapters 3, 8-10;
> Philosophers of science: Parts II and IV;
> Social scientists: Parts II and III, and Chapter 9;
> Humanists (especially rhetoricians): Chapters 4-8, 10;
> Politicos and policymakers: Part III.

Finally, to readers initially skeptical of this enterprise, please keep in mind that if the pursuit of knowledge policy or the satisfaction of normative impulses seems inherently authoritarian, that is only because not enough people are doing it. In *The Open Society and Its Enemies* (1950), Karl Popper first complained about the "transcendental" viewpoint of Marxists and Freudians who thought it better to meet an objection with a metalevel diagnosis of the objector's (faulty) state of mind than with a straightforward counterargument. In a sense, postmodernists who are reluctant to engage in the normative enterprise that follows in these pages have drawn a perverse lesson from Popper's complaint. After all, a theoretical language is not born transcendental, but it can be unwittingly elevated to that status if the audience feels that the theory must be either accepted whole cloth or rejected in its entirety. True believers do the former, postmodernists the latter. Either way, transcendence is rhetorically accomplished. A better strategy would be to resist the theoretical language from within–the pages that follow.

PART ONE

THE PLAYERS AND THE POSITION

1 THE PLAYERS: STS, RHETORIC,

SOCIAL EPISTEMOLOGY

HPS as the Prehistory of STS

Although the seventeenth century is touted as the intellectual and institutional source of "science" as a distinctive form of life, the idea of a "Scientific Revolution" was itself a product of the late eighteenth and early nineteenth centuries (I. B. Cohen 1985). It is during this period that the modern issues associated with science as a social problem first get raised (Restivo 1988). After Kant, it became acceptable to replace the question of whether science can come to know ultimate reality with whether science can come to know itself. Specifically, by what means are we to study and evaluate the proliferation of jargons, techniques, and artifacts passing for authoritative knowledge in a rapidly changing society? And how should all this bear on the future pursuit of knowledge? Science and Technology Studies (STS) is the contemporary descendent of this tradition of thinking about the sciences. Although STS is primarily an empirical field devoted to describing and explaining the scientific enterprise, I will be concerned here with the genealogy of the field's normative-prescriptive wing, *social epistemology*.

Most nineteenth-century theorists of science are classed today as "philosophers," although virtually all had scientific training and a historical orientation. British theorists tended to be concerned with the popular reception of science and the role of scientific reasoning in democratic decision-making processes; French ones were often interested in science and technology as extensions of the state and instruments of social progress; and German theorists were preoccupied with the division of academic labor in the emerging structure of the research univer-

One: The Players and the Position

sity (Ben-David 1984: chaps. 5-7). Ultimately, however, the cognitive exigencies of the modern world dictated the uses to which these theorists would put science and its history. And, for the most part, the uses have been highly "rhetorical," in that theorists sought ways of expressing scientific claims that would move appropriately educated audiences to support emergent scientific institutions over their competitors for cognitive authority—religion, craft guilds, folk wisdom, as well as explicitly pseudo- and antiscientific movements—many of which were housed within the academy. The task was by no means easy nor its performance completely evenhanded. Science's strongest suit was its claim to derive knowledge by experimental observation, yet the preferred rhetorical strategy—the enumeration of "demarcation criteria" that science could alone meet—was persuasive insofar as it inclined the public not to scrutinize but rather to trust the scientists' observational powers, on the basis of verbal accounts that enabled the audience to "virtually witness" what scientists had done (Ezrahi 1990: chap. 3).

The first self-proclaimed "positivist," Auguste Comte, initiated the task of demarcating science from nonscience in order to identify theories worthy of further pursuit without having to precommit significant intellectual and material resources. To insure the economic viability of this presorting process, philosophers tried to read epistemic merit off the relatively superficial verbal features of theories that one might find in a student's textbook. It was supposed that something about the way a theory is presented can indicate whether it is likely to push back the frontiers of knowledge. By the time logical positivism caught the philosophical imagination in the 1930s, it was thought that scientific theories ought to wear their logical structures and operational definitions on their rhetorical sleeves.

The history of science was used in an equally rhetorical fashion, unashamedly Whiggish by today's standards. In 1840, the Cambridge geologist and cleric William Whewell both coined the term "scientist" and opened the field called "History and Philosophy of Science," or HPS (Fisch and Schaffer 1991 collect recent scholarship on Whewell). HPS explicitly sought what was best to believe about the past in order to construct a desirable future. This project entailed a twofold strategy of (1) eliciting principles of epistemic growth that could be transferred across disciplines—and thus could potentially be made the possession of all inquirers—and (2) favoring the study of certain revolutionary periods in which the process of major epistemic change was more evident. In the case of (1), it is important to keep in mind that in the nineteenth century physics was regarded as a discipline that had largely run its course, and whose methodological vitality was thus better placed in

The Players

the more exciting developing proto-sciences of life, mind, and society. This explains, in the case of (2), the bias toward focusing on great showdowns over theory choice and agenda setting—which continues in philosophical writing about science to this day—at the expense of studying the workaday methods of the most advanced sciences of the day (as STS nowadays tends to do).

The sense of "normative" that I pursue under the rubric of *social epistemology* returns to this nineteenth-century idea of philosophers intervening to improve the course of knowledge production. Nineteenth-century philosophical interventions ran the entire gamut of prescriptive activities. Whewell advised Faraday and Darwin on the conception and interpretation of their theories. Comte and John Stuart Mill laid down the steps by which the fledgling social and psychological disciplines might become truly scientific. Ernst Mach used the history of physics as a critical wedge in contemporary debates by recovering minority dissents to which the dominant Newtonian paradigm had failed to respond adequately. (Einstein later credited Mach's critical appeal to history with having prodded his own thinking in a relativistic direction. In our own time, Paul Feyerabend made himself the master of this form of history.) Pierre Duhem normalized science's relations with a traditional cultural authority like the Roman Catholic church by stressing continuity between science and religion on some occasions (e.g., in the medieval origins of modern physical concepts) and a division of labor between the two institutions on others (e.g., in the search for ultimate truths). For his part, John Herschel normalized science's relations with the emerging reading public of Victorian Britain by portraying scientific reasoning as an extension and formalization of common sense. John Dewey's influence in graduate schools of education enabled him to play largely the same role in early-twentieth-century America.

In retrospect, the most distinctive normative contribution of these theorists of science was their attempt to streamline the knowledge production process by isolating a lingua franca, a procedural language that would enable all the sciences to develop toward greater methodological unity and, hence, greater public accountability. "Positivism" is still the term normally used to capture this project in both its nineteenth- and twentieth-century forms. The project of social epistemology is sympathetic with positivism's instinctive question: *How do we cope with an increasingly diversified social and cognitive order?* However, the possibilities that we pursue in response to this question are mediated by recent developments in STS, which have veered considerably off the course of philosophical positivism. But before suggesting where posi-

One: The Players and the Position

tivism went wrong, it is important to point out that not all twentieth-century philosophers of science have relinquished the robust normative perspective of the previous century's theorists. In this regard, Karl Popper and Paul Feyerabend come most readily to mind as precursors of social epistemology. Although Popper and Feyerabend would probably object to seeing their philosophies as ways of accommodating the nineteenth-century posture to the exigencies of today's "Big Science," this is in fact a good way to explain their projects.

Popper and Feyerabend intervene not in the activities that most scientists would consider the core of their practice (e.g., what they do in the laboratory), but rather in the shadows of policy forums where research is initially stimulated and ultimately evaluated. Thus, these philosophers have stressed that science needs to be evaluated in terms more of consequences than of conception. Also, in their writings is the theme that, given the increasing access to resources that science commands, research has become—if it hadn't been already—both an investment opportunity and a public trust. It needs to be acted upon as such. To put the point in the signature Popperian way, science must be supported as an "open society" that will serve as a model for all of society. Social epistemology embraces the spirit of this enterprise, yet it cannot help but notice that Popper and Feyerabend—both armed with sharp philosophical tools but only a philosopher's understanding of the history of science and society—come relatively ill prepared to do the normative job they set out to do (cf. Gellner 1979: chap. 10, on Feyerabend's particular lack of preparation). To be sure, Popper and Feyerabend allow a much more cosmopolitan range of considerations to enter their thinking about science than did the logical positivists or their descendants in analytic epistemology and the philosophy of science. But it is still not enough. Here is where STS makes a decisive difference. Before specifying this difference, however, we should appreciate just what an improvement Popper and Feyerabend are over the fate of HPS in the twentieth century.

Whewell's original project withered under the weight of its own disciplinization. The interventionist projects just sketched were gradually subserved to projects pursued for their own sake. Thus, today HPS contains people who continue to fuss over the logical form of scientific theories and the great moments in the history of science, as if such things were worth pursuing as ends in themselves. The demise of HPS came slowly, however. As philosophical and scientific training started to drift apart into two recognizably different expertises, philosophers saw this as a sign that science had come of age and could therefore be left to regulate itself. The burden was then shifted to philosophers try-

The Players

ing to maintain the autonomy of their own enterprise (Rorty 1979). This institutional development was coupled with an interesting conceptual move that I trace back to the early-twentieth-century Italian political economist Vilfredo Pareto (especially 1935). The move is responsible for the so-called *internal history of science* the putative object of HPS inquiry (cf. Fuller 1989a: chap. 1).

To the progressive nineteenth-century mind, it was only natural to suppose that if science gave us the most comprehensive grasp of the world, then the most comprehensive grasp of science could be gotten by studying science scientifically. However, Pareto gave this line of reasoning a particular spin. His idea was not so much to study the actual practice of science by scientific means (as STS would eventually do), but rather to treat scientific practice as if it were like the world as represented by our best scientific theories, namely, an idealized mechanics closed under a system of rational principles operating on the inputs of nature, but frequently subject to extraneous ("external") influences. Thus, internalists tried to deliver on Hume's promise to provide a "mental mechanics" that paralleled Newton's physical mechanics. More generally, *science was taken to have the qualities of the things that science studies*. We can see this trend as a kind of "homeopathic" theorizing that bears an uncanny resemblance to the idea of the individual as the microcosm of nature or of the species. The resemblance was further strengthened in twentieth-century philosophy of science, when not only was science seen as reproducing in its own structure the structure of the world it represented, but this process was itself also seen as potentially transpiring in the mind of a single individual, namely, the philosopher of science. Comte already had anticipated this micro-internalism when he justified his hierarchy of the sciences in terms of its enabling him to reenact the history of science in his mind. In our own time, this "rational reconstructionist" position has been represented by a host of positivist (Reichenbach), Popperian (Lakatos), and historicist (Shapere) philosophers of science.

In 1962 Thomas Kuhn began, perhaps unwittingly, to undo HPS. His major breakthrough, in *The Structure of Scientific Revolutions* (1970), was to show that it was possible to account for the history of science as internally driven without concluding that it was being driven anywhere in particular. A veteran instructor in the Harvard general education program, Kuhn reminded his readers that the memorableness of the sequence of episodes in internalist histories of science was due to these episodes having been canonized in science textbooks, which are, in turn, the vehicles by which the "normal science" of a paradigm is transmitted. However, the specific episodes in the sequence will vary

One: The Players and the Position

from paradigm to paradigm, thereby relativizing any conclusions about "progress" and the "ends of knowledge" that internalists might want to draw from the ordering.

It would be hard to exaggerate the blow that Kuhn dealt to philosophers of science. Some (mistakenly, I believe) have even taken his book to mark the revenge of the humanists against the positivists. After all, with Kuhn's sequence of paradigm-anomalies-crisis-revolution-new paradigm, cyclical history would seem to have finally made a major inroad into the last bastion of linear progress, science. Although few philosophers officially conceded any ground to Kuhn, it is nevertheless true that, since 1965, scientific progress has been philosophically defended on grounds unrelated to any substantive ends that science might be presently pursuing (e.g., L. Laudan 1977, only to be lamented in L. Laudan 1984). Rival conceptions of "verisimilitude" and "increased empirical adequacy" are contested on such purely formal grounds that, even if agreement were reached on one of these notions, philosophers would still be in no position to evaluate, let alone improve on, the degree of progress enjoyed by the standing research programs of today. This has given the subsequent debate a scholastic cast, as philosophers retreat from explicit historical appeals to quasi transcendental arguments about the "nature" of science: "How would science be possible at all without a certain conception of progress?" Questions of this sort were wisely passed over by Kuhn in silence (cf. Leplin 1984; Fuller 1988a: chap. 3).

With the increasing internalization of HPS's field of inquiry has come a more restricted normative sensibility. Thus, today HPS seems to be conducted more in the spirit of a schoolmaster giving marks than of the policymaker trying to improve the conduct of inquiry. Philosophers of science know that it was good to choose Copernicus over Ptolemy by Galileo's day, and that it would have been better to have made the choice sooner, but they have precious little to say about what line of research we ought to pursue *now* (an exception is Nickles 1989). One wonders what HPS practitioners would say if they realized just how close their current schoolmasterly mode places them to literary criticism and art connoisseurship, two disciplines whose practices have become increasingly alienated from their putative objects of evaluation. Contrary to nineteenth-century hopes, the judgments of critics typically do not feedback into the creation of better art or even better publics for the reception of art. What is produced, instead, is a self-sustaining body of scholarly literature. Any positive impact of critics on the course of art in this century has been fortuitous, much like the impact of philosophy on the course of science today.

The Turn to Sociology and STS

The fate of HPS notwithstanding, Kuhn's overall impact on the academy has been more liberating than inhibiting. Although Kuhn betrayed little knowledge of sociology in *Structure*, nevertheless his own example suggested to sociologists (especially Barnes 1982) that it was possible to explain most of what was interesting about science without having to make reference to such philosophical categories as "truth," "objectivity," "rationality," or even "method"–categories which had traditionally led sociologists to enforce a double standard in the way they studied science vis-à-vis the way they studied other social practices. Indeed, this double standard is operative in the work of the founder of the sociology of knowledge, Karl Mannheim (especially 1936), and his distinguished American successors Robert Merton (especially 1973) and Joseph Ben-David (1984). To varying degrees (Mannheim more so than Merton), these early sociologists unwittingly diminished the public accountability of science–if not contributed to its outright mystification–by refusing to scrutinize science by its own principles. In practice, this meant that the sociologists typically drew conclusions about science based on the authoritative testimony of the great philosophers and scientists, or anecdotal evidence from great episodes in the history of science. Such prescientific sources of knowledge would not have been tolerated in the study of other social phenomena, so why should methodological standards be lowered for what is supposedly society's premier cognitive institution?

And so, inspired by Kuhn's work, it came to pass in the 1970s that the first school in STS was founded. The Strong Programme in the Sociology of Scientific Knowledge (Barnes 1974; Bloor 1976), or the "Edinburgh School," rejected the double standard in the sociological study of science by laying down what I will dub:

> *The Fundamental Mandate of STS.* Science should be studied as one would any other social phenomenon, which is to say, scientifically (and not by relying uncritically on authoritative testimony, anecdotal evidence, and the like).

Major contributors to STS over the last fifteen years include Barry Barnes, Wiebe Bijker, David Bloor, Michel Callon, Daryl Chubin, Harry Collins, Susan Cozzens, David Edge, Aant Elzinga, Thomas Gieryn, Donna Haraway, Sandra Harding, Evelyn Fox Keller, Karin

One: The Players and the Position

Knorr-Cetina, Bruno Latour, John Law, Donald MacKenzie, Michael Mulkay, Andy Pickering, Trevor Pinch, Sal Restivo, Arie Rip, Simon Schaffer, Steven Shapin, Peter Weingart, Richard Whitley, Steve Woolgar. Their work is featured in such journals as *Social Studies of Science, Science, Technology, and Human Values* and, more recently, *Social Epistemology* and *Science in Context*. Also, the development of STS can be traced through a series of watershed anthologies (Barnes and Edge 1982; Knorr-Cetina and Mulkay 1983; Stehr and Meja 1984; Law 1986; Bijker et al. 1987; Cozzens and Gieryn 1990; Pickering 1992). Although surprisingly few of these researchers are actually trained as sociologists–and in fact they often betray the influences of anthropology, psychology, economics, politics, and literary criticism–they can all be broadly identified as "sociological" in the sense of denying an "internal" history of science that is distinguished in its categories and methods from the history of the rest of society. And despite the mix of methods that these researchers have used to study science, it is fair to say that analogies from, allusions to, and even actual instances of ethnographic practice enjoy epistemic privilege in the field, largely for enabling STS researchers to "observe on site" the divergence between the words and deeds of scientists without having to adopt an explicit normative stance toward that finding; hence the much ballyhooed "relativism" of STS research (Fuller 1992a, and Traweek 1992 offer alternative views of the strengths and weaknesses of this tendency).

Contrary to what one might think, the target in this sociological dressing down of science is not the scientists themselves, who are generally portrayed in STS research as modest toilers trying to make the most of difficult situations in which expectations are high but resources are often embarrassingly low. Rather, the real foes are many of the philosophers mentioned earlier in this discussion. They have fostered these unwarranted expectations by making it seem as though science works by a "method" that manifests a "rationality" quite unlike anything else that society could offer. Positivist philosophers may be the ultimate source of this sentiment, but it continues to be found in popular accounts of science, which speak the language of "hypothesis generation," "theory testing," and "falsifiability"–words that sound right *only* if one is speaking of science. In that regard, the demarcationist rhetoric practiced by Comte and his successors proved all too effective. For when one actually steps into the labs and the other workplaces where science is done, a variety of quite ordinary, and often inconsistent, activities that could be said to fall under these fine rubrics are observed. Thus we arrive at the normative crossroads facing STS:

The Players

How should STS conduct itself in light of what it learns about science?

This question may be subject to various elaborations. For example, should STS advise the public to abandon its faith in science? Should STS scrutinize science more but expect less (or vice versa)? Or rather, should STS let the scientists go about their business, and simply put an end to all this mystifying talk about "method" and "rationality"? Of course, these questions hit still closer to home, as STS must decide whether its own practices need to be changed. This is the *reflexive dimension* of the normative question, which in the history of philosophy has been most strongly associated with the Hegelian tradition. In the case of STS, we might wonder, if science is, indeed, the product of sociohistorical contingency, how is it that only now (and here) do we come to learn this, and, more specifically, how should this knowledge be allowed to affect our subsequent practice? The answers to this important question have been far from uniform. Some argue for minimal effect, an epistemic "business-as-usual" attitude, whereby the STSer pursues her inquiries alongside those of the sciences they study (e.g., Collins 1985 and the more orthodox ethnographers). Others suggest that STS should purge this newfound contingency from its own practice and thereby become more scientific than the scientists themselves (e.g., probably the original intent of the Strong Programme). Still others argue that STS should incorporate contingency into the content of its own findings so as to lend a more partisan and political flavor to its research (e.g., roughly speaking, my own and other critical-theoretic approaches). Finally, still others recommend that STS adopt a self-deconstructive style of writing that reveals the contingent character of distinguishing "factual" from "fictional" accounts of science (e.g., Woolgar 1988a and the radical ethnographers).

The Achilles heel of STS has so far been its reluctance to *argue* about the relative merits of these reflexive postures. Instead STS researchers have tended to resolve these matters silently in their practice (though Pickering 1992 makes a promising start at engagement). The problem, of course, with the silent treatment is that it leaves perilously open the question of *what is the point of closely studying the knowledge system?* Social epistemology is designed to provide a forum for such normative considerations, which will serve to bring STS closer to both the most abstract and the most concrete of students of the knowledge enterprise–epistemologists and science policy analysts–as well as to critical social theorists. Perhaps the best way to begin to identify the desired forum is by distinguishing two general attitudes toward science that can be found among STS practitioners. On the one hand,

One: The Players and the Position

some seem relatively satisfied that the training of scientists ensures that they know what they are doing and should continue doing, largely without the misguided commentary of philosophers and other outside scrutinizers. Other STS practitioners are less satisfied, taking their own success in penetrating the inner workings of science to imply that nonspecialists should have more of a say about which science is done, and how. Let us call the first viewpoint *Deep Science*, and the second *Shallow Science*.

As a first pass at this distinction, we might think of the two viewpoints as alternative answers to the question: *Where does one find knowledge in society?* The Deep Science inquirer locates knowledge in the skills that scientists display in their workplaces, which are taken to be intimately connected with the things they produce, which are then "applied," for better or worse, throughout society. This is not so different from the way we ordinarily think about science. However, no such distinction between knowledge and its applications is presumed by the Shallow Science inquirer, who sees knowledge as something distributed across the network of authority and credibility with which a particular piece of scientific work–especially a text–is associated. Thus, whereas the Deep Scientist (i.e., the scientist studied by the Deep Science inquirer) has knowledge in virtue of her unique powers of mind and body, the Shallow Scientist has knowledge in virtue of others' letting her exercise discretion. As will become increasingly clear, my brand of social epistemology is linked to Shallow Science.

Deep Science is a largely non-verbal craft, or "tacit knowledge," that requires acculturation into long-standing disciplinary traditions and is best studied by a detailed phenomenology of scientific practice. Opposed to this image is that of Shallow Science, a largely verbal craft that consists of the ability to negotiate the science-society boundary to one's own advantage in a variety of settings; it is best studied by deconstructing the seamless rhetoric of scientists so as to reveal the clutter of activities–the positivist's "context of discovery"–that such rhetoric masks. Typical students of Deep Science are many of the historians of experiment, who follow Michael Polanyi (1957, 1969) in devaluing the role of theorizing–and the use of language, more generally–in everyday scientific practice (e.g., Gooding et al. 1989). Students of Shallow Science include most social constructivists, discourse analysts, and actor-network theorists. Despite my sympathies with this camp, to whom I assign the generic label *constructivist* throughout this book, I depart from many of its members in believing that a robustly normative approach to science is compatible with–indeed, even facilitated by–the Shallow Science perspective. For, in their own inimitable attempts

The Players

to isolate one all-purpose methodological trick, philosophers of science originated the Shallow Science perspective so as to enable non-scientists such as themselves to hold science accountable for its activities. In this way the classical philosophical focus on the context of *justification* has metamorphosed into a sociological interest in science's mode of *legitimation*. By contrast, students of Deep Science tend to be purely descriptive in their aspirations, tacitly presuming that science works well as long as the scientists do not complain (cf. Fuller 1992a). Is it any surprise, then, that Deep Science tends to be concentrated in labs, whereas Shallow Science is spread diffusely across society?

The Deep and Shallow images define polar attitudes toward the cognitive powers of the individual scientist. At the Deep end is the idea that scientists are especially well suited, in virtue of their training, to represent the nature of reality. The practices of scientists, however disparate their origins, have fused into a "form of life" with its own natural integrity. At the Shallow end is the idea that scientists are no better suited than laypeople to represent reality, an idea that is rarely appreciated not only because scientists share with laypeople basic limitations in their ability to scrutinize their own practices, but also because the epistemic cost of admitting the fallibility of scientific judgment is especially dear: How would engineering be possible if the judgments of physicists were not well grounded? Yet it is precisely this easy relation between science and technology that the Shallow Science perspective has endeavored to challenge (e.g., Bijker et al. 1987). The basic problem with Deep Science here is that its conception of the social is unbecoming to anyone who wishes to hold science accountable to someone other than the scientists themselves. It is a conception that provincializes society into jurisdictions of "local knowledge," the authority of which is meant to be taken on trust, regardless of the potential consequences for those outside a given jurisdiction. On this basis, most partisans of Deep Science claim to be relativists. Indeed, generally speaking, it is easy to be a relativist, if you presume that your utterances have effect only on the ears of the intended audiences in your community. However, if you believe that language enhances, diminishes, or reverses existing social orders as it is appropriated by others outside the original context of utterance, then the well-defined jurisdictions of the relativist will be impossible to maintain. The methodology of actor-network theory, which tracks the alignment of interests—both scientific and nonscientific—that have a stake in the fate of a piece of research (Callon, Law, and Rip 1986, as popularized in Latour 1987), makes this point quite vividly.

Someone with a Shallow Science perspective clearly refuses to take

One: The Players and the Position

a term like "tacit knowledge" at face value. Rather than presuming that the term has a positive referent, namely, a scientist's unarticulated craft ability, the Shallow Science perspective treats appeals to the tacit dimension as rhetorical indicators of when one should stop asking scientists to account for their activities. (A fascinating social history could be told about the shifting boundary between the "tacit" and the "articulate" in accounts of science. Such a history would search for the sorts of things that scientists and their epistemological mouthpieces have identified as the "proper objects" of intuition or immediate experience, which "as such" can be transmitted only by personal contact. I would guess that the more items contained in a society's inventory of the tacit dimension, the more successful the scientists were at staving off the bureaucrats.) From this perspective, Deep Science historians treat tacit knowledge somewhat naively by drawing a spurious distinction between the transience of explicit formal theories and the persistence of tacit laboratory practices. As the Shallow Science partisan sees it, this distinction may be due less to an absolute difference between theory and practice than to a difference between a practice that is legitimated by verbal means and one legitimated by nonverbal means. For a practice that passes muster by saying (or measuring) certain things can be subjected to a finer-grained level of analysis–and hence of criticism and directed change–than a practice that requires simply that it appear (to the relevant audience) to be proceeding smoothly. The tacit practice may vary historically just as much as the verbal one, but it would be harder to detect those variations in the tacit practice, let alone whether they add up to improvements on the practice. Admittedly, matters quickly become complicated once it recalled that uttering the right words at the right time is itself routinely treated as a kind of silencing practice (or a "display of competence," as the Polanyiites would say) that absolves the speaker from any further scrutiny of her position.

How can Deep Science be brought around to the normative perspective of Shallow Science? Simply put, Deep Science needs to "thicken" its conception of language use. Instead of the Deep Science partisan's sense of language as a pale abstraction of an ineffably rich world, the Shallow Science partisan presents language as a construction that sharply focuses an otherwise indeterminate reality. The thickener is rhetoric. If I may be allowed the philosophical indulgence of reconstructing history for my own purposes, the first stage in the thickening process takes us back to the Sophist Protagoras' invention of language as something that could be standardized and controlled, specifically,

The Players

by shifting the paradigm of usage from sincere speech to grammatical writing (Billig 1987: chap. 3). This shift from an aurally to a visually based communicative medium—or "externalization"—was accompanied by a creation of scarce conditions for access to this medium, a sure sign of language's materiality (cf. McLuhan 1962, 1964). Thus, people were shown to have differential access to communicative skills, the remediation of which required training in the verbal arts of rhetoric and dialectic. The final step that Protagoras took was to charge for his services, thereby converting a scarce resource into a marketable good. It was this last move that enabled Socrates to launch one of the earliest attacks on the capitalist spirit. After all, as Socrates portrayed it, the Sophists were proposing to alienate the client from his soul and then reacquaint him with it at a cost. The Sophists failed to meet the Socratic challenge because the ease with which they flaunted their dialectical prowess, in both serious and playful settings, only served to undermine the idea that the good they were peddling was truly scarce.

At that point, Plato pushed Socrates offstage and concluded that right-minded speech was not scarce at all, and indeed, was universally available—except that certain people, the ones whose activities the Sophists fostered, unjustifiably tried to restrict access to such speech by eloquence, obfuscation, and threats. Plato's step here undid the thickening of language that Protagoras had begun. Had the thickening process continued, the Sophists would have supplemented their *embodiment* speech in grammar with an account of grammar's *embeddedness* in the material context of utterance. In other words, as their conception of language became thicker, rhetoric would have yielded to the sociology of knowledge and political economy. Similar conclusions have been drawn by cognitive scientists and sociologists (cf. Shrager and Langley 1990: 15-19; Block 1990: chap. 3) as well as rhetoricians (cf. McGee and Lyne 1987).

If Deep Science is wedded to a "thin" conception of language (as a kind of transparent representation of the world) and Shallow Science is wedded to a "thick" conception (as one fortified with rhetoric), then the natural question to ask is how does one thicken the thin? Let me suggest here two translation strategies that capture the moments of embodying and embedding language. The idea behind the two strategies is that to thicken language is to give it spatiotemporal bearings. The boundaries of language so thickened constitute an "economy," by which I mean here simply the metaphysical notion that not everything that is possible can be realized in the same time and place, and that therefore every realization involves a trade-off of one set of

One: The Players and the Position

possibilities against another. Embodiment and embeddedness address, respectively, the temporal and spatial dimensions of the thickening process. Thus, using "speech" to refer to a unit of discursive action, we have the following definitions:

> *Embodiment (Temporalization):* Language is embodied insofar as the goal of speech is manifested in the manner in which the speaker conducts herself during the time that she is speaking.
>
> *Embeddedness (Spatialization):* Language is embedded insofar as a speech is treated not as an instance of a universally attributable type which everyone in the speech community possesses to the same extent, but rather as part of an object the possession of which is finitely distributed among the speech community's members.

By way of exemplifying embodiment, consider the sorts of activities that are said to be done "for their own sake" or as "ends in themselves." Such Kantian talk signals that the consequences of pursuing these activities will not figure in their evaluation. Not surprisingly, Kantian talk is most effective when the activities in question have undetectable or diffuse consequences, as knowledge production is typically said to have. As we become more accustomed to planning our epistemic practices, and hence to monitoring their social consequences, this Kantian talk will probably lose currency. However, the so-called ultimate ends–such as peace, survival, happiness, and (yes) even *truth*–refer not to radical value choices for which no justification can be given, but rather to constraints on the manner in which other instrumentally justifiable ends are pursued. Thus, happiness in life is achieved not by reaching a certain endpoint, but by acquiring a certain attitude as one pursues other ends. A related point applies to the pursuit of truth. "Serious inquirers" comport themselves in a way that, over time, reinforces in others the idea that they have caught the scent of the truth. Admittedly, there is considerable disagreement over the exact identity of the relevant traits (e.g., how respectful of tradition?), but few doubt that there are such traits. Verbal attitudes that are incongruous with one's avowed aim do not wear well over time, and are likely to be dismissed as inauthentic "mere rhetoric," failing to manifest "methodological rigor."

If, in terms of our metaphysical economy, embodiment is a measure of "return on investment" (i.e., whether my manner tends to diminish or enhance the audience's sense of my purpose), embeddedness is a measure of "currency flow," which is related to what Michel Foucault (1975) called the "rarity" of an utterance. Embeddedness is intimately tied to the social epistemologist's problem of determining what gives knowledge its "value," a point to which I will return at the end of this chapter. The basic idea is that whenever someone speaks effectively, either she increases the effectiveness of what is said by decreasing the ability of others to follow suit or she decreases the effectiveness of what is said by increasing the ability of others to follow suit. Thus, the "currency" of what is said is either strengthened through restriction or weakened through inflation (cf. Klapp 1991). For example, magicians for centuries have passed down their lore through a highly guarded process of apprenticeship. This process ensures restricted access to the lore, which is integral to the "success" of magic on lay audiences. However, once a professional magician like the Amazing Randi breaks rank and divulges the secrets of his craft, magic loses much of its effect, devalued to simply another performing art or form of entertainment. (Would something similar happen if a band of Nobel Prize laureates publicly endorsed the Shallow Science perspective, admitting how it perfectly explained their own careers as scientists?) A related strategy, which will be prominent in Chapter 5, is to destabilize the power relations embedded in restricting the applicability of value terms, for example, applying "rationality" or "intelligence" exclusively applying to human beings, and then really only to certain human beings in certain settings. By developing semantic conventions (metaphoric extensions, if you will) for applying, say, "rationality" to non-human entities or atypical humans, we make it harder to take politically significant action on the basis of that term. In this way, "rationality" is neutralized as a source of power.

Having now begun to lay some of my rhetorical cards on the table, I had best confront the most vexed player in this field, *rhetoric*.

Rhetoric: The Theory Behind the Practice

Using the word *rhetoric* is hardly the most rhetorically effective way of referring to anything, let alone something that might be properly called "rhetoric"! Moreover, it is not clear whether the friends or the foes of rhetoric are more to blame. Thus, I am com-

One: The Players and the Position

pelled to shroud the guilty in a cloak of anonymity. Friends of rhetoric tend to overemphasize the community-building functions of well-chosen language, often harboring some fairly nostalgic (if not downright mythical) views about the degree of common ground that is achievable, or desirable, between people. Yes, even desirability may be questioned, insofar as a community where people are always *pleased* to listen to each other probably will learn little from whatever is said. Where in such a place is rhetoric's potential for *re*configuring the ways in which people relate to each other and to the world? For their part, the foes of rhetoric have got that part of the story right. At the same time, however, their stress on the demystifying, divisive, and otherwise debasing character of rhetoric presupposes a trumped up (if not downright paranoid) view of rhetoric's pervasive and corrosive powers. Are all adept rhetors such sinister sirens? (Only your advertising agent knows for sure!) What, then, could we want to preserve from these vexed conceptions of rhetoric?

Perhaps the best way to capture rhetoric's place in my approach is to say that it overcomes the antinomies that plague STS thinking today, which have so far inhibited the development of the field toward social epistemology. These antinomies largely result from STS's having decisively discredited certain philosophical conceptions of science without leaving anything in their place. For openers, consider the following, very basic antinomy:

> (T+) Philosophers have claimed that language stands apart from the natural order it passively represents. Language thus functions as a "mirror of nature."
>
> (T-) STSers have shown that language is part of the natural order, with just as much capacity to move and be moved as anything else. Indeed, language is much of the stuff out of which "nature" is actually constructed.

Although swords appear to be crossed over the nature of language, if we follow the long line of Western thinkers from Aristotle to Habermas who believe that linguistic ability is the mark of the human, then the antinomy may be more profitably seen as covertly expressing the dispute between determinism (T+) and free will (T-). Rhetoric operates between these two extremes by proposing a sphere of "freedom within limits," an expression which harkens back to Kant and Hegel. It involves drawing a distinction between rational and irrational freedom,

The Players

which corresponds to the distinction between there being limits and no limits on the range of options that one is in a position to consider. Rhetoric is, then, the exercise of rational freedom. I can act rationally, in the sense of deliberating over alternatives, only if my options are limited and thereby surveyable. The truly free being, God, always sets her own limits, but the rest of us usually make do with being thrown into limited situations not of our own creation. This is what rhetoricians have traditionally called *exigence*, the feature of the world which brings forth the occasion for rhetorical invention (Bitzer 1968). Now, the horizon of this inquiry can be broadened to include the conditions under which exigences are reproduced time and again–why it seems that we have only a limited set of options for dealing with certain recurrent situations (cf. McGee and Lyne 1987). This would involve a study of conventions ultimately grounded in an analysis of the power structure that maintains them. Understood as a systematic enterprise, STS is largely oriented toward this goal. The social epistemologist enters the picture to locate exigences that enable the destabilization of this power structure.

The reader is perhaps beginning to see how rhetoric fits in. To reinforce this perspective, let us consider rhetoric's role in resolving related antinomies. I have marked rhetoric's resolution as (T'):

> (T+) Philosophers have claimed that rational language use conceptually presupposes that a discourse could be understood by any other language user, regardless of her particular interests.
>
> (T-) STSers have shown that rational language use is relative to the standards of particular linguistic communities, whose differing interests may render their discourses mutually incomprehensible.
>
> (T') Rhetoricians have ways of helping disparately interested parties overcome their language differences in order to join in common cause.

Thus, a priori normative claims to a universal audience are met with a posteriori empirical claims to incommensurable worldviews, only to be resolved by a posteriori normative claims to what, in the next chapter, I call *interpenetrable* discourses. Another version of this antinomy highlights the distinctiveness in how the rhetorician begins her

One: The Players and the Position

inquiry:

> (T+) Philosophers erase the past and begin from scratch, much as God would have ideally designed the universe: first things first. This enables the philosopher to operate with maximum freedom, constrained only by the principles that she herself has already laid down.
>
> (T-) STSers begin in medias res on the same ontological plane as the people they study, constrained only by whatever the people under study have let constrain their own practices.
>
> (T') Rhetoricians also begin in medias res but then design strategies for transforming recognized exigencies into normatively acceptable action.

The importance of this last antinomy for demarcating rhetoric from other disciplines cannot be overestimated. For example, philosophers typically propose normative theories of action that satisfy their colleagues but rarely the people (say, actual scientists, in the case of philosophers of science) whose actions would be judged or governed by those theories. The same may be said of the models of rationality proposed by neoclassical economists. Consequently, as Laymon (1991) and others have observed, these theories are idealizations without being approximations of the phenomena they model. In other words, as such theories are supplemented with more realistic assumptions–about, say, human psychology, sociology, and the decision-making environment– their ability either to predict or to prescribe behavior does not improve accordingly. Rather, if the theory is to work at all, the normative theorist must supply *un*realistic auxiliary assumptions about human beings (the path of fictionalism), blame reality for its failure to conform (the path of moralism), or try to force reality into the mold of the theory (the path of coercion). While the rhetorician would not deny the occasional efficacy of these approaches, she would argue that the normative project may be pursued more effectively by factoring in more realistic assumptions about the intended audience at the outset, namely, by respecting the fact that people are not blank slates at the beginning of normative inquiry just waiting for the pronouncements of philosophers to give their lives direction. Rather, people in search of guidance already come with certain concerns, habits of mind, and situations in which they are prepared to act. Any normative proposal must

The Players

therefore take the form of advice that *complements* this state of affairs. In the jargon of cognitive psychology, such advice must function as a "heuristic" that strategically compensates for biases and processing limitations that already exist in the target knowledge system. In terms more familiar to rhetoricians, norms are proposed in the spirit of shifting the burden of proof in a direction that enables more fruitful arguments to be made.

Among the various branches of philosophy, the rhetorician would find more kindred spirits in ethics than in epistemology. Traditionally, the standard of knowledge presupposed by epistemologists has been omniscience, God's point of view. Opinions that thrive on anything less—no matter how methodologically scrupulous they may be—are susceptible to the illusions of Descartes' Evil Demon. As a result, when epistemic norms are proposed, relatively little attention is paid to whether they would actually improve the conduct of inquiry if they were in place. Instead, one is told that inquiry would improve in an ideal setting, but unfortunately, given the ever-present possibility of the Demon, it is unlikely that the real world of inquiry is such a setting. By contrast, ethicists do not typically aim to provide a set of moral principles that would always enable its adherents to resist the temptations of Satan. On the contrary, a point is often made of saying that ethics would not be needed if there were "moral saints," because in that case there would be no need to give advice on how to *improve* one's conduct. Ethics presupposes moral imperfection but also its corrigibility. Whereas epistemologists have only recently turned to cognitive science in order to grasp the psychological backdrop against which epistemic norms operate, moral psychology has been an integral part of ethical inquiry from Plato and Aristotle onward. Moral principles, such as Kantianism and utilitarianism, have been proposed in the spirit of disciplining or mitigating features of "human nature" that are already present when the ethicist begins her inquiry (Baier 1985). The exact consequences that these principles have for conduct will depend on the conception of human nature that the audience brings to the ethical forum. A utilitarian confident in her understanding of the world will take "the greatest good for the greatest number" as an injunction to engage in projects of deferred gratification that promise big long-term payoffs. A utilitarian of a somewhat more skeptical disposition will interpret the slogan as a call for incremental policy and reversible decisions. Similarly, a confident Kantian will be relentless in her dutifulness, ignoring consequences completely, whereas a less confident adherent to the categorical imperative will harbor a guilty conscience as she

One: The Players and the Position

wonders whether she is, indeed, steadfast in her duty.

The closest that epistemology gets to this sort of spirit is Popper's falsification principle, which was designed to counteract our predisposition toward finding evidence that supports our own opinions. Popper (e.g., 1959) repeatedly complained that by setting a superhuman standard for knowledge, epistemologists had fostered two sorts of overreactions, either of which was sufficient to undercut any motivation for doing science. On the one hand, those who were confident in their fundamental beliefs wanted everyone to share them. On the other hand, those skeptical of their beliefs did not leave open the prospect that another set of beliefs might mitigate their skepticism. Indeed, upon considering the reaction of philosophers and STSers to Popper's principle, it would be easy to conclude that Popper had a rhetorical sensibility (cf. Orr 1990). Consider the following three opinions on the viability of falsificationism, which correspond to the philosophers' (T+), STSers' (T-), and Popper's own (T'):

> (T+) Since it is easy to find counterinstances to any hypothesis, strict adherence to falsificationism would not allow hypotheses enough time to be developed before being tested.

> (T-) People are psychologically ill disposed to falsificationism, which explains why the principle has been rarely applied–despite claims to the contrary by philosophers and scientists alike.

> (T') Although it is easy to find counterinstances to any hypothesis, precisely *because* people are psychologically ill disposed to falsificationism, advising scientists to apply the principle will issue in an optimal turnover of hypotheses. The scientists' native resistance to falsificationism will cause them to fortify their hypotheses from attack, so that only developed versions will ever be decisively falsified.

Arch-rationalist that he is, Popper would probably be the last to want to identify his approach with that of the rhetorician's. Yet, it would seem that, unlike most philosophical pieces of advice, his is of the sort that might actually lead to better results the *closer* one moved toward a realistic understanding of human beings (cf. Gorman et al. 1984). By contrast, consider a formula such as the ever-popular Bayes theorem, a mathematical equation that determines the most plausible of a set of

The Players

rival hypotheses by comparing their probabilities before and after a test has been run. The idea behind the formula, what philosophers after Peirce call "abductive" reasoning, is impeccable (Salmon 1967). Yet this very precise guide to scientific reasoning fares poorly when addressed to human beings, whose computational powers are severely strained very quickly, even when they are well disposed to using formal methods (cf. Faust 1985, Cherniak 1986). In an entirely serious vein, then, Clark Glymour (1987) has argued that such formal models of rationality are really suited to computer androids. Rhetorically speaking, the positivists who developed and promoted these models had a radically mistaken sense of audience, as they had failed to realize that their proposals could make sense only to machines that had yet to be invented! The history of formal reasoning as a philosophical institution prior to the computer revolution testifies to this point. With the exception of elementary logic exercises and cutting-edge logic research, formal models have functioned less as tools for the actual conduct of reasoning and more as yardsticks or templates for the evaluation of informally expressed arguments (cf. Toulmin 1958).

A historically salient feature of rhetoric that is responsible both for its virtues and for its ambivalent place in the academy is its self-image as primarily a practice, from which a body of doctrine may ex post facto be derived and taught. The pecking order implied here is quite the reverse of the one normally found in the academy. Whereas conventional academic disciplines tend to regard practice–with more or less contempt–as an application of theory-driven research, rhetoricians have been inclined to see matters the other way around, with academically certified knowledge being the ultimate safe haven for the failed practitioner. Those whose theories of rhetoric are confined to the classroom never meet the test of the marketplace. Those who can't do, teach. In terms of its epistemic prejudices, then, rhetoric is the cousin of the liberal professions, such as law, medicine, and engineering. I would argue that rhetoricians make good models for how STS practitioners ought to conduct themselves, given their understanding of the nature of knowledge production.

Like practitioners of the liberal professions, rhetoricians are alive to the fact that the classroom and the textbook represent a limited range of communicative possibilities, in terms of what speakers and audiences can do and where and how they can do it. Thus, rhetoricians are expert in constructing the occasions and sites that call for certain forms of argument and persuasion. The kindred professional strategy is to create a universally felt "need" to see a doctor or lawyer when various

One: The Players and the Position

personal and social exigences arise. STSers need to craft such a need–namely, that of addressing the ongoing problem of epistemic economy: the questions that arise from the production, distribution, and consumption of knowledge in society. However, as it stands, STS practitioners share with other academics a rather unimaginative sense of how to make use of their space and time. Where are the attempts to mix media, engage different audiences at different registers? Perhaps academics interested in STS should be taught not only public address (as I require in my own seminars: see the Appendix to this book), but also the performing arts and architecture to refine their spatiotemporal sensibilities (cf. Soja 1988, who represents a school of "postmodern geographers" who urge this point in all seriousness). In any case, it is not enough to continue to write the same sorts of articles and books to the same audiences–except that one now says that the fact-fiction distinction is being "blurred" or "crossed." If the communicative environment remains largely unchanged, these "new literary forms," as they are sometimes called (e.g., Clifford and Marcus 1986; Woolgar 1988a; Ashmore 1989), will simply have poured old wine into new caskets–the thin conception of language, yet again, whereby *only* the words have changed but not the social relations in which they are embedded.

Enter the Social Epistemologist

My version of social epistemology begins by reading the findings of STS research through a Shallow Science perspective. This generates three presumptions that inform the strategies and positions adopted in this book. In particular, they motivate the alliance between rhetoric and STS that I wish to forge, as well as encapsulate the issues raised up to this point:

> *The Dialectical Presumption*: The scientific study of science will probably serve to alter the conduct of science in the long run, insofar as science has reached its current state largely in the absence of such reflexive scrutiny.
>
> *The Conventionality Presumption*: Research methodologies and disciplinary differences continue to be maintained only because

The Players

no concerted effort is made to change them–not because they are underwritten by the laws of reason or nature.

The Democratic Presumption: The fact that science can be studied scientifically by people who are themselves not credentialed in the science they study suggests that science can be scrutinized and evaluated by an appropriately informed lay public.

These presumptions, in turn, generate certain semantic consequences that have been implicit in my past work, but which I now make explicit so readers are not misled by what follows. These consequences consist of the following collapsed binaries:

Reasons = Causes: This follows in the wake of the Dialectical Presumption. Both supporters and critics of science typically capitalize on the distinction between these two terms to quite opposite effects. Supporters use it to ground the difference between an autonomously driven knowledge enterprise (governed by "reasons") and one driven by external social factors (swayed by "causes"), while critics use it to separate the ideology that scientists invoke to legitimate their activities (mere "reasons") from the true account of why they do what they do (real "causes"). The possibility of drawing this distinction–and the internal/external histories of science that it breeds–diminishes as scientists come to justify their activities in the sorts of terms that best explain them. That a distinction between reasons and causes continues to exist is a measure of the extent to which knowledge generated by STS has yet to feed back into the conduct of the inquiry (cf. Fuller 1988a: Appendix B, 1989a: chap. 1).

Natural = Social: This follows in the wake of the Conventionality Presumption. I typically mean "science" in the generic, German sense of *Wissenschaft*, a systematic body of knowledge closed under a canonical set of methods and a technical vocabulary. "Discipline" is the best one word English translation. Unless otherwise indicated, it is meant to refer indifferently to the natural and the social (human) sciences. I do this, not simply because I am pitching my claims at a level of abstraction where such a distinction no longer makes a difference

One: The Players and the Position

> (certainly, that would accord with the "epistemological" character of social epistemology), but more important because, from the STS perspective, the natural sciences consist of certain strategies for mobilizing societal resources. Indeed, as will become clear in my discussion of the rhetoric of science policy, natural scientific research indirectly tests hypotheses about social organization and political economy. The success or failure of those strategies and hypotheses determines the longevity of a given science—and much else of societal import.
>
> *Public = Policymaker*: This follows in the wake of the Democratic Presumption. If the promise of STS is delivered, and the workings of science can be understood by nonexperts, then each person currently identified as a "knowledge policymaker"—a government bureaucrat, say—will have the status of primus inter pares, that is, someone whose role as policymaker is potentially interchangeable with that of any other concerned and informed citizen. This projected state of affairs will be brought about, not by everyone acquiring formal training in all the sciences, but by scientists learning to account for their activities to larger audiences, which will thereby enable everyone to assume a stake in the outcome of research. Thus, a high-priority item for social epistemology is the design of rhetorics for channeling policy-relevant discussions in which everyone potentially can participate.

The larger context in which social epistemology is situated is the profound ambivalence that Western philosophers have had toward the equation of knowledge and power. Admittedly, this ambivalence has become increasingly obscured in the twentieth century, as epistemology (including philosophy of science) and ethics (including social and political philosophy) have evolved into separate specialties, especially in the Anglo-American analytic tradition. However, the problem is easily recovered, once we see the Western tradition as having been fixated on the problems of *producing knowledge* but *distributing power*. Consequently, epistemology has tended to concentrate on practices with the highest levels of epistemic productivity ("science"), regardless of their access to society at large, while ethics has focused on schemes for equitable distribution, without considering the costs of (re)producing the institutions needed for implementing those schemes. Thus, social epistemology is born with an "essential tension" (Roth 1991): How to

The Players

balance Machiavellian and democratic impulses?

The Machiavellian impulse is toward maximizing the production of knowledge-and-power, even if the means of production are concentrated in an elite cadre of "epistemocrats." By "epistemocrats" I mean those whose superior knowledge of people (and what is good for them) enables them to mask their own interest in bringing the world into alignment with their normative model. The ultimate source here is Plato. Unlike the Aristotelian *phronesis* approach to politics, in which the rulers are no smarter than the ruled except in their ability to represent several constituencies at once, the Platonic *episteme* approach involves the ruler in strategic overclarification and illusion in order to guide the populace toward a normatively acceptable end. As I will show in Chapter 4, economists have been especially skillful in converting "purity" to "power" in this manner (Proctor 1991). By contrast, the democratic impulse aims to maximize the distribution of knowledge-and-power, even if this serves to undermine the autonomy and integrity of current scientific practices. Its modes of persuasion are entirely open-faced: If I can't justify my knowledge claims to you, then you have no reason to believe them.

Social epistemology's relevance to rhetoric and argumentation lies in its stress on the integral role that communication, both its facilitation and impedance, plays in contemporary thinking about knowledge and power. The most distinctive contributors to social epistemology in our time—Karl Popper, Thomas Kuhn, Michel Foucault, and Jürgen Habermas—can be best understood in terms of the type of communication that they take to be realizable in today's world. A useful way of configuring these four disparate thinkers is in terms of the following chain of ideas: Free access to the communicative process breeds increased accountability, which, in turn, forces aspiring authorities to couch their claims to knowledge in terms that can be understood by the largest number of people. This leveling of terminology conveys the idea that we all live in the same world. Any apparent differences in the access we have to that world is attributed to epistemic artifice—"ideology," if you will—which typically masquerades as ontological differences, or "incommensurable worlds." These world-differences restrict the number of eligible critics of one's claims, namely, to the class of people known as "experts" or "natives." In *Social Epistemology*, I argued that this chain of ideas implies that communication breakdown is the leading cause of cultural difference, the diachronic version of which is conceptual change (Fuller 1988a: xiii).

Popper's "open society" account of knowledge production articulates

One: The Players and the Position

the positive relation between cognitive democracy and one world suggested in the scenario above, whereas Kuhn's "paradigm" picture of the scientific enterprise asserts the negative one between cognitive authoritarianism and a plurality of discrete worlds. However, I see both Kuhn and Popper as talking mainly about the implications of opening or closing discourse for one's own pursuits. In contrast, Foucault and Habermas are more concerned with the implications that these possibilities have for what *others* do. Foucault teaches that the power associated with claims to superior knowledge accrues to those who can suppress alternative voices, or, in Kuhnian terms, can consign others to worlds incommensurable with one's own. In the case of scientific authority, this suppression is best studied in terms of the presumptions that aspiring revolutionaries need to overturn before being granted a complete hearing (Fuller 1988a: chap. 4). Habermas, however, wants each inquirer to submit her claims to a series of validity checks that exert a measure of self-restraint, which, in turn, gives others a chance to stake their own claims. If Foucault is an other-directed Kuhn, Habermas is an other-directed Popper—at least from the social epistemologist's vantage point. The result is shown in Figure 1. The particular philosophical lesson about the knowledge-power nexus that social epistemology teaches from this configuration of Foucault, Kuhn, Habermas, and Popper is that *knowledge differences become reality differences when it becomes impossible to communicate across those differences.*

As a positive research program, social epistemology proposes inquiries into the maintenance of the sort of institutional inertia that has made social epistemology's three presumptions radical rather than commonplace. Why don't research priorities change more often and more radically? Why do problems arise in certain contexts and not others, especially, why is there more competition for resources within a discipline than between disciplines? A sensitivity to latent incommensurabilities turns out to *help*, not hinder, this sort of critical knowledge policy. First, armed with the tools of the STS trade, the social epistemologist can isolate the quite heterogeneous set of interest groups that derive enough benefits, in their own distinctive ways, from the status quo that they have little incentive to change. The strategy, then, would be to periodically restructure the environments in which researchers compete for resources. The terms of this restructuring may be quite subtle, e.g., providing incentives to reanalyze data gathered by earlier researchers. Less subtly, researchers may be put in direct competition with one another where they previously were not. Moreover, they may be required to incorporate the interests of another discipline, including that discipline's practitioners, in order to receive adequate

The Players

Knowledge Politics / Implications For	COGNITIVE DEMOCRACY (Levelled Playing Field)	COGNITIVE AUTHORITARIANISM (Multiple Jurisdictions)
SELF INTEREST	Popper's Open Society	Kuhn's Paradigms
TREATMENT OF OTHERS	Habermas' Ideal Speech Situation	Foucault's Suppressed Voices

Figure 1. Social Epistemology's Universe of Discourse

funding. Finally, researchers may be forced to account for their findings, not only to their own discipline's practitioners, but also to the practitioners of other disciplines and maybe even the lay public. While I see this last step as a long-term goal, it is nevertheless essential to social epistemology's project of locating the *value* of knowledge (Fuller 1988a: chap. 11). Insofar as the value of knowledge has been discussed at all in the philosophy of science, it has been in one of two ways, mirroring a dichotomy already present in theories of value available in economics (cf. Mirowski 1989).

On the one hand, there is a kind of *labor* theory of epistemic value, which locates the value of knowledge in the difficulty or improbability of extracting knowledge from the world. Knowledge itself is natural stuff (brains, books, etc.) that has been substantially transformed by a scientist's labor. There are High Church and Low Church prototypes for this view. The High Church evokes Francis Bacon's view of clever experiments as the means by which humans overcome their own ignorance and nature's resistance in yielding its secrets. The Low Church evokes diligence, testing one's mettle, and "hard thought" as educational virtues.

On the other hand, there is a kind of *utility* theory of epistemic

One: The Players and the Position

value, which points to the capacity of knowledge for organizing a wide variety of phenomena, which can, in turn, be used to realize a wide variety of ends. Knowledge itself is a field of rival means-ends relations (or if-then statements) that pull the scientist in different directions to different degrees. On the High Church side lies Newtonian mechanics as a model of parsimonious explanatory theory for all the sciences; hence, the ultimate means to every scientist's ends. On the Low Church side lie the consumer technologies that enable large numbers of people to satisfy their wants with ease.

As Joseph Agassi (1985) has observed in another context, these two classical views–the labor theory associated with basic research and the utility theory with applied research–are fundamentally opposed, the former tending toward challenging the most by the least and the latter toward accommodating the most by the least. They coexist only as a result of a hard-won exchange forged in the academy, whereby basic researchers exchanged some of their prestige and allowed applied researchers to work alongside them, in return for a piece of the applied researcher's credentialing process, which assured a steady stream of students for the pure sciences. This grafting of labor onto utility theories of epistemic value is reflected in every curriculum that requires engineers to study branches of physics and mathematics, or that requires medical practitioners to study branches of biology and chemistry. In many cases, the study of basic research diverts from, if not outright impedes, the mastery of the relevant applied techniques.

As against both the labor and utility views, I propose that the value of knowledge lies in the ability of its possessor to influence the subsequent course of its production. Thus, the physicist's knowledge of physics is worth more than, say, a popularized account of quantum mechanics or even an introductory college course in the subject, not because of its inherent profundity or its ability to ease the lives of the physicist and others but because of the relative ease with which the trained physicist can intervene in the production of physical knowledge. The most obvious advantage of my view is that it brings under one rubric the epistemic idea of *demonstration* and the political idea of *empowerment*: that is, competence is judged in terms of an appropriate alteration of the tradition rather than a simple reenactment of it.

However, a subtler advantage of this view is that it calls into question the value of being a mere possessor, or "consumer," of knowledge– which, in turn, has implications for how one thinks about the ends of education (cf. Fuller 1991). Thus, epistemic value is gauged not only in terms of certain products, but more important in terms of certain *productive capacities* that are ideally distributed through the knowledge system. In that case, education can itself serve to devalue the currency of knowledge, if students come to "understand," say, the nature of scientific research or democratic government, without being provided the opportunity to affect the course of these institutions. Feminists have been especially sensitive to this point, insofar as women much more quickly gained access to seats in college classrooms than to places at the lecture podium (cf. Hartman and Messer-Davidow 1991). In the first half of this century, courses in "civics" in American public schools aimed to address this problem by instructing students on the political mechanisms at their disposal. Nothing comparable has yet been done for science education, so that, at best, schools produce "pure consumers" of science, who regard scientific research and its technological extensions as being as normal as any of their own daily activities, even though they are unlikely ever to be in a position to influence the direction that science and technology take. Education of this sort, for all its distribution of facts and figures, is akin to indulging in a high-calorie diet without vigorous physical exercise: the citizenry's epistemic energy is converted to an acquiescent adiposity!

By helping to reconfigure the variables of knowledge production, the social epistemologist can ensure that disciplinary boundaries do not solidify into "natural kinds" and that the scientific community does not acquire rigidly defined class interests. Such reconfigurations will go a long way toward keeping the channels of communication open between sectors of society that seem increasingly susceptible to incommensurability. Indeed, this strategy would even alter the character of the knowledge produced, including perhaps what we take something to be when we call it "knowledge." In all this, social epistemology needs to be a thoroughly rhetorical enterprise. Consider the two different contexts of persuasion that are implicated in the above discussion. First, there is the need to motivate scientists to restructure their research agendas in light of more general concerns about the ends that their knowledge serves. Second, there is the need to motivate the public to see their fate as tied to the support of one or another research program. It is not that there are no norms whatsoever currently governing these two contexts. However, as long as the rhetorical transactions underlying a set of norms in force remain unexamined, people will tend to behave in ways that were perhaps appropriate for those who participated in instituting the norms, but which now fail to serve the interests of those who continue to abide by them. As a result, the norms fail to receive any explicit consent of the governed: inertial producers matched with inert consumers. Thus, the social epistemologist recognizes the essentially rhetorical character of normative action, to wit:

One: The Players and the Position

> *A necessary (though not sufficient) condition for the appropriateness of a norm is that the people to whom the norm would apply find it in their interest to abide by the norm.*

The standpoint of *interpenetrative interdisciplinarity* will consider who these people are and what their interests might be.

2

THE POSITION: INTERDISCIPLINARITY

AS INTERPENETRATION

The Terms of the Argument

Interdisciplinarity can be understood as either a fact or an ideology (cf. Klein 1990). I endorse both. The fact is simply that certain sorts of problems–increasingly those of general public interest–are not adequately addressed by the resources of particular disciplines, but rather require that practitioners of several such disciplines organize themselves in novel settings and adopt new ways of regarding their work and co-workers. As a simple fact, interdisciplinarity responds to the failure of expertise to live up to its own hype. Assessing the overall significance of this fact, however, can easily acquire an ideological character. I am an ideologue of interdisciplinarity because I believe that, left to their own devices, academic disciplines follow trajectories that isolate them increasingly from one another and from the most interesting intellectual and social issues of our time. The problem is only masked by dignifying such a trajectory with the label "progress." Thus, I want to move away from the common idea that interdisciplinary pursuits draw their strength from building on the methods and findings of established fields. Instead, my goal is to present models of interdisciplinary research that call into question the differences between the disciplines involved, and thereby serve as forums for the renegotiation of disciplinary boundaries. This is perhaps the most vital epistemological function for rhetoric to perform in the academy, the need for which has become clear only with the emergence of STS.

An interesting, and probably unintended, consequence of the increasing disciplinization of knowledge is that the problem of interdisciplinarity is drawn closer to the general problem of *knowledge policy*, that is, the role of knowledge production in a democratic society. In the

One: The Players and the Position

first place, as disciplines become more specialized, each disciplinary practitioner, or "expert," is reduced to lay status on an expanding range of issues. Yet, specialization serves to heighten the incommensurability among the ends that the different disciplines set for themselves, which, in turn, decreases the likelihood that the experts will amongst themselves be able to coordinate their activities in ways that benefit more than just their respective disciplinary constituencies. The increasingly strategic roles that deans, provosts, and other trans-departmental university administrators play in shaping the future of departments testify to this general tendency to assimilate the problem of interdisciplinary negotiation to the general problem of knowledge policy (e.g., Bok 1982). A complementary trend is the erosion of the distinction between academic and nonacademic contexts of research. Nowadays, corporations not only subsidize academic research but also often pay for the university buildings in which the research takes place. Be it through government initiatives, venture capitalism, or the lure of the mass media, the nonacademic public is potentially capable of diverting any narrowly focused disciplinary trajectories. Social epistemology's contribution to these tendencies, one might say, is to make such initiatives intellectually respectable. The key is to cultivate *the rhetoric of interpenetrability*. Although the technofeminist Donna Haraway (especially 1989) has recently revived the idea behind "interpenetration" (to produce "cyborgs," techno-organisms that interpenetrate the nature/culture distinction), the term probably still carries enough of the old Marxist baggage to merit unpacking.

"The interpenetration of opposites," also known as "the unity and conflict of opposites," is one of the three laws of dialectics identified by Friedrich Engels in his 1880 work on the philosophy of science, *The Dialectics of Nature*, which has since become a staple of orthodox Marxism. To put it metaphysically, the idea is that stability of form—the property that philosophers have traditionally associated with a thing's identity—really inheres in parts whose tendencies to move in opposing directions have been temporarily suppressed. Perhaps the most familiar application of this idea appears in Marx's concept of *structural contradiction*, which purports to explain the lack of class conflict between the workers and the bourgeoisie by saying that the workers unwittingly buy into capitalist ideology and hence fail to identify themselves as a class with interests opposed to those of the bourgeoisie. The Italian humanist Marxist Antonio Gramsci popularized the term "hegemony" to capture the resulting ideological harmony, which leads workers to blame themselves for their lowly status (cf. Bocock 1986). However, armed with the Marxist critique of political economy, the

The Position

workers can raise this latent contradiction to the level of explicit class warfare, for once they identify exclusively with each other, they are in a position to destroy the stability of the capitalist system. Now consider a rhetorical example that makes the same point. Whereas philosophers since Plato have supposed that communication involves speaker and audience partaking of a common form of thought having its own natural integrity, rhetoricians have taken the more interpenetrative view that any apparent meeting of minds is really an instance of strategically suppressed disagreement that enables an audience to move temporarily in a common direction.

One of the least likely places for this last point to apply, yet where it applies with a vengeance, is in the *history of tolerance*, a concept worthy of the rhetorician's conquest by division. First, there is what might be called *passive tolerance*, the ultimate target of sophisticated forms of censorship, yet still unrecognized by philosophers as a legitimate epistemological phenomenon. In the 1950s Carl Hovland and his Yale associates captured it experimentally as "the sleeper effect." It is the tendency for subjects to become better disposed to a message after repeated exposure over time, even if they were originally ill disposed because of the source of the message (Hovland et al. 1965). Thus, even conservatives may start to express sympathy for a social program originally proposed by a liberal, once they get used to hearing it and forget that a liberal first proposed it. At least, the burden of proof starts to shift in their minds, so that now they might want to hear arguments for why the program should *not* be funded. Managing this form of tolerance, especially in a democracy whose mass media are dedicated to the equal-time doctrine, is a rhetorical and epistemological challenge of the first order, especially as the proliferation of messages serves only to increase the amount of passive tolerance in society. The trick then is to "activate" tolerance without thwarting it.

Thus, there is *active tolerance*, the sort for which one openly campaigns. In theory, it aims to empower groups by channeling their attention toward one another. In practice, it often turns out to be a version of "my enemy's enemy is my friend," whereby otherwise squabbling factions agree to cease hostilities to fend off a still greater and mutual foe. In the case of John Locke's *Letter on Toleration*, which was influential in the establishment of religious tolerance in the American colonies, the common enemy was an ominously defined band of "atheists" who had no place in a Christian commonwealth. The logic of interpenetration can work in this environment, if the threat posed by the foe forces the factions beyond mere peaceful coexistence to active

One: The Players and the Position

cooperation in combating the foe (cf. Serres 1982, on the strategy of removing a "parasite"). For once the foe has been removed and all the factions are able to go their own way, they will have been substantially transformed as a result of their previous collaboration.

The rhetoric of interpenetrability aims to recast disciplinary boundaries as artificial barriers to the transaction of knowledge claims. Such boundaries are necessary evils that become more evil the more they are perceived as necessary. The rhetoric that I urge works by showing that one discipline already takes for granted a position that contradicts, challenges, or in some way overlaps a position taken by another discipline. As a dialectical device, interpenetrability goes against the grain of the current academic division of labor, which typically gives the impression that issues resolved in one discipline leave untouched the fate of cognate issues in other disciplines. For example, it is routine to think that whatever psychological findings are reached in laboratory settings have no necessary bearing on the psychological makeup of the sort of ordinary "situated" reasoners that historians and other humanists study. That is to say, no mutual challenge is posed by the juxtaposition of laboratory cognizers and historical cognizers, and hence any interaction between the two types will be purely a matter of the inquirer's discretion. It is in this context that advocates of interdisciplinarity, especially the cultural anthropologist Clifford Geertz (1983), have traditionally spoken of social scientific theories as "interpretive frameworks" that can be applied and discarded as the inquirer sees fit–but never strictly tested.

In stressing applicability over testability, Geertz and other interdisciplinarians are, in part, reacting–perhaps overreacting–to positivist academic rhetoric, which culminated in Popper's falsificationist methodology, with its explicit aim of *eliminating* false hypotheses. The finality of such eliminationist rhetoric made one close follower of Popper, Imre Lakatos, squirm over the possibility of preemptively squashing fledgling research programs, and ultimately drove another of Popper's famous students, Paul Feyerabend, to espouse the anarchistic doctrine of letting a thousand flowers bloom. Moreover, even as a simple fact about the history of science, eliminationism is hard to justify. For better or worse, once articulated, theories tend to linger and periodically reemerge in ways that make "half-life" an apt unit of analysis.

Unfortunately, the explicitly nonconfrontational strategy of Geertz and his cohort plays in the worst way to the exigencies of our cognitive condition. There is little need to belabor the point that, for any field, more theories are generated than can ever be given a proper hearing.

The Position

How, then, does one decide on which theories to attend to, and which to ignore? Testability conditions of the sort Popper offered under the rubric of falsifiability constitute one possible strategy. For example, a theory may challenge enough of the current orthodoxy that the orthodoxy would be significantly overturned if the theory were corroborated. This is the sort of theory that Popper would test. However, if inquirers are allowed complete discretion on how they import theories into their research, then it is likely that they will capitalize on their initial conceptions as much as possible and ignore—not test—the theories that implicitly challenge those conceptions. Thus, in the long term, the non-confrontational approach would probably lead to the withering away of subversive theories that could be accommodated into standing research programs only with great difficulty. My point here is not that inquirers will converge on dogma if they are not required to confront each other critically; rather, unless otherwise prevented, inquirers will diverge in ways, mostly involving the elaboration of incommensurable technical discourses, that will make critical engagement increasingly difficult.

I believe that much of the sting of Popper's rhetoric could be avoided if testing were seen more in the spirit of a Hegelian *Aufhebung*, that is, the incorporation and elimination of opposites in a more inclusive formulation. Concretely, I am suggesting that when disciplines (or their proper parts, such as theories or methods) interpenetrate, the "test" is a mutual one that transforms all parties concerned. It is not simply a matter of one discipline being tested against the standards of its epistemic superior, or even of both disciplines being evaluated in terms of some neutral repository of cognitive criteria (as might be provided by a philosopher of science). Rather, *the two disciplines are evaluated by criteria that are themselves brought into being only in the act of interpenetration*. And while these criteria will undoubtedly draw on the settlements reached in earlier interdisciplinary disputes, the exact precedent that they set will depend on the analogies that the current disputants negotiate between these prior exchanges and their own.

Consider, once again, historians and psychologists confronting each other's explanations of scientific behavior. To put the difference in starkest terms, whereas humanistic historians think that scientists strive to emulate the geniuses in their fields, cognitive social psychologists are more inclined to believe that scientists are motivated to take each other down in order to make room for themselves. In both cases, the scientist is portrayed as "rational," albeit on the basis of divergent conceptions of rationality, which presuppose a strong difference of

One: The Players and the Position

opinion on whether intelligence is intrinsic to the individual or emergent from the group. Now suppose that after protracted discussion historians came to abolish the term "genius" from their lexicons to reflect a revision in their estimates of intellectual merit in light of the psychological evidence, yet the psychologists remained unmoved by anything in the historians' original explanatory strategy. Although the psychologists would have thereby gained the upper hand in dialogue with the historians, that would not be the end of the story. In particular, onlookers in other disciplines also have a stake in whether the knowledge system as a whole has benefited from this capitulation of history to psychology. Their opinions will count significantly with regard to the long-term viability and generalizability of the psychologists' victory. However, if there were sustained discourse between history and psychology, then both disciplines would probably be transformed in somewhat different ways that together help enable them to see each other as engaged in a common enterprise. This would be especially true if practitioners of the two disciplines considered how their respective methodologies functioned as means toward a common end, say, an understanding of the scientific mind. In this regard, the historian might be able to show the psychologist that the sort of laboratory findings that she would normally attribute to inherent properties of the subject are in fact emergent features of the social situation co-produced in the lab by experimenter and subject (cf. Billig 1987: chap. 4; Danziger 1990). In that case, the roles of experimenter authority and subject compliance are highlighted in a way that renders "laboratory life" more political, and hence more reformable, without necessarily invalidating the experimental method as such. It is this prospect that calls for the resources of the epistemically interested rhetorician.

The Perils of Pluralism

While the three presumptions that social epistemology takes from STS—dialectical, conventionality, democratic—make me a natural enemy of "traditionalists" in the academy (e.g., Bloom 1987), my comments in the last section are meant to throw down the gauntlet to many of the so-called *pluralists* (e.g., Booth 1979) who normally oppose the traditionalists. In spite of their vocal support of interdisciplinary research, pluralists nevertheless tend to assume that, left to

The Position

their own devices and absent any overarching institutional constraint, the practitioners of different disciplines will spontaneously criticize one another in the course of borrowing facts and ideas for their own purposes. If Popper's Open Society were indeed a by-product of such a pluralistic academic environment, the social epistemologist would not need to cultivate interventionist impulses. However, I believe that, like any activity which is clearly beneficial if everyone does it but is potentially dangerous if only a few do so, criticism requires special external incentives. Otherwise, each discipline will tend to politely till its own fields, every now and then quietly pilfering a fruit from its neighbor's garden but never suggesting that the tree should be replanted in a more mutually convenient location. My view here rests on the observation that criticism flourishes in the academy—insofar as it does—only within the confines of disciplinary boundaries (say, in journal referee reports) and erupts into symbolic violence when it spans such boundaries. Given this state of affairs, the "tolerance" that is much revered by pluralists turns out to be the consolation prize for those who are unwilling to face their differences.

In terms of the idea of active tolerance raised initially in this discussion, there are two directions in which a tolerant community may go at this point. On the one hand, it may take advantage of the opportunity provided by realizing that "my enemy's enemy" is really "my friend" and foster an interpenetrative intellectual environment. On the other hand, it may foster just the reverse, perhaps out of fear that voiced disagreements would allow the enemy to reappear. As the "tolerant" Christian commonwealth would have it, interdenominational strife is Satan's calling card. In a more secular vein, it has been quite common in the history of academic politics for rival schools of thought to cease fire whenever it seemed that a more powerful "third party," usually a government agency, was in a position to use the disagreement to gain advantage over the feuding parties, often by discrediting the knowledge produced in such a fractious setting. For example, Proctor (1991: chap. 8) has argued that sociologists in early-twentieth-century Germany became preoccupied with appearing as "value-neutral" inquirers, just at the point it became clear that, from within their ranks, an assortment of conflicting normative programs were being advanced on the basis of scholarly research. By suppressing these deep disagreements, the sociologists believed (with mixed results) that they could counter government suspicions that the classroom had become the breeding ground for alternative ideologies, and thereby salvage the "autonomy" of their inquiries. (Furner 1975 offers the American analogue to this story.) From the standpoint of social epistemology, a bet-

One: The Players and the Position

ter strategy would have been for the sociologists to argue openly about what they really cared about–the normative programs that they wanted their research to legitimate–and to enroll various government agencies as allies in the ensuing debate, thereby dissipating whatever leverage the state could exercise in its official capacity as "external," "neutral," and, most important, *united*.

We see, then, that tolerance works homeopathically: in small doses, it provides the initial opportunity for airing differences of opinion, which will hopefully lead to an engagement of those differences; however, in large doses, tolerance replaces engagement with provincialism, thereby producing Robert Frost's policy of "good fences make good neighbors," and the veiled sense of mutual contempt that it implies. The unconditional protection of individual expression not only fails to contribute to the kind of collaborative inquiry that sustains the growth of knowledge (cf. Elgin 1988, on a related contrast between the pursuits of "intelligence" and "knowledge"), but also fails to foster healthy social relations among inquirers. In particular, individual expression instills an ethic of *learning for oneself* at the expense of *learning from others*. This accounts for the tendency of interdisciplinarians to become "disciplines unto themselves," increasingly fragmented sects that unwittingly proliferate old insights in new jargons that are often more alienating than those of the disciplines from which they escaped.

My complaint here is not that interdisciplinary discourses tend to mutate into autonomous fields, but rather that they mutate *without replacing some already existing fields*. Thus, they merely amplify, not resolve, the level of babble in the academy. Given the exigencies of our epistemic situation, pluralists hardly help matters by magnanimously asserting that anyone can enter the epistemic arena who is willing to abide by a few procedural rules of argument that enable rival perspectives to remain intact and mutually respectful at the end of the day. (After all, isn't the security of this outcome what separates the interdisciplinary environment of the academy from the rough-and tumble world of politics?) In practice, this gesture amounts to one of the following equally unsavory possibilities:

> 1. Everybody gets a little less attention paid to her own claims in order to make room for the newcomer.
>
> 2. The newcomer starts to adopt the disciplinary perspective of the dominant discussants, and consequently is seen as not adding to the level of academic babble.

3. Given that the newcomer starts late in the discussion, her claims never really make it to the center of attention.

Newcomers, of course, fear that (3) is the inevitable outcome, though the path of cooptation presented in (2) does not inspire confidence, either. As a result, newcomers have been known to force themselves on the discussion by attempting to "deconstruct" the dominant discussants, which is to say, to call into question the extent to which the discussants are really so different from one another, especially in a world where there are still many other voices yet to be heard. Aren't they all *men?* Aren't they all *white?* Aren't they all *bourgeois?* Aren't they all *normal scientists?* The suggestion here is that if the discussants are "really" all the same, then they can easily make room for the genuine difference in perspective offered by the newcomers. Clearly, the deconstructive newcomers are trying to totalize or subsume all who have come before them, which gives their discourse a decidedly *theoretical* cast. Critics of untrammeled tolerance and pluralism have observed that pluralists become extremely uncomfortable in the face of this theoretical cast of mind, regardless of whether the source of the theory is Marxism (Wolff et al. 1969), feminism (Rooney 1991)–or positivism, for that matter. (Kindred suspicions have surrounded "synthetic" works in history, which, while not especially theoretical, nevertheless juxtapose pieces of scholarship in ways other than what their authors originally intended; cf. Proctor 1991: chap. 6). After all, the deconstructors have turned the pluralist's procedural rules into topics in their own right. No longer neutral givens, the rules themselves now become the bone of contention, as they appear to foster a spurious sense of diversity that, in fact, excludes the most challenging alternatives. I will return to this topic under the rubric of "knowledge politics" in Chapter 8.

People aside, the basic problem with pluralist forms of interdisciplinarity is that they reinforce the differences between disciplines by altering the *products* of research, while leaving intact research *procedures*. A good piece of interdisciplinary research is supposed to abide by the local standards of all the disciplines drawn upon–this, despite the fact that most disciplines are born of methodological innovations that, in turn, reflect deep philosophical dissatisfaction with existing methods. Given such a historical backdrop, it is hard to imagine that research simply combining the methods of several disciplines–say, a study of attitude change that wedded historical narrative to phenomenological reports to factor analysis–would constitute an improvement on the rigorous deployment of just one of the methods. To think that such a combination would automatically constitute an improve-

One: The Players and the Position

ment is to commit the *fallacy of eclecticism*, the belief that many partial methods add up to a complete picture of the phenomenon studied (rather than simply to a microcosm of cross-disciplinary struggles to colonize the phenomenon). The fallacy is often undetected in practice because interdisciplinarians deftly contain the reach of any one method so as to harmonize it with other methods that together "triangulate" around the author's preferred account of the phenomenon. Readers, of course, are free to infer that one method was brought in to compensate for the inadequacies of another, but the nature and potential scope of the inadequacies are passed over by the author in tactful silence.

In this regard, it is worth considering the favorable light in which triangulation is regarded in the social science methods literature (e.g., Denzin 1970; Webb et al. 1981). Here triangulation appears as a means to ensure that the inherently partial and reductive nature of a given research tool does not obscure the underlying complex reality that the researcher is trying to capture. Not surprisingly, then, discussions of triangulation tend to focus on the need for multiple methods in order to achieve a balanced picture of reality—not on the more basic fact that the biases introduced by divergent methods persistently reemerge across virtually all research contexts. Thus, triangulation simply defers an airing of these differences to another day—or perhaps another forum, philosophy of social science, where the results of deliberations are less likely to felt by research practitioners. (It is for this reason that ethnomethodologists have been especially insistent on letting these metascientific concerns interrupt and shape their research practices; cf. Button 1991: especially chaps. 5-6).

Consider, by way of illustration, the oft-cited exemplar of triangulation in the sociological literature, James Coleman's *The Adolescent Society* (1961), which revealed the existence of a high school subculture in the United States, whose value structure is oriented more toward athletics and extracurriculars than toward academics. Coleman studied students, parents, and school personnel, using data collected from questionnaires, interviews, and school records. Although Coleman says that all the methods were administered to all three groups, in fact *Adolescent Society* reports most of its findings about students from questionnaires and most of its findings about the adults from interviews. Coleman (1961: viii) even remarks on this asymmetry, but makes little of it. From a narrative standpoint, the combination of student questionnaires and adult interviews biases the reader toward Coleman's view that the world of adolescents is so detached from that of adults that interpersonally based methods may provide unreliable access to the adolescent mind. No doubt, too, it was easier to administer

The Position

questionnaires to students captive in the classroom, for whom the discipline of exam taking is part of their everyday lives, than to parents and school personnel, who are allowed the luxury of circumventing the exercise for more convenient means of expressing their views. These features of the social backdrop against which the triangulation occurred served only to reinforce Coleman's expressed (and influential) desire to demonstrate uniformity among adolescents across a wide range of schools, so as to embolden adults to design educational policy for keeping students within striking distance of the normative mainstream.

Another sort of triangulation is prominent among humanists who attempt to "blur genres," in Clifford Geertz's (1980) memorable phrase. Geertz (especially 1973) himself is among the most masterful of these eclectics. A discussion ostensibly devoted to understanding the practices of some non-Western culture will draw upon a variety of Western interpretive frameworks that sit well together just as long as they do not sit for too long. For example, an allusion to the plot of a Shakespearean tragedy may be juxtaposed with Max Weber's concept of rationalization to make sense of something that happens routinely in Southeast Asia. The juxtaposition is vivid in the way a classical rhetorician would have it, namely, as a novel combination of familiar tropes. In fact, the brilliance of the novelty may cause the reader to forget that it is meant to illuminate how a non-Western culture actually is, rather than how a Western culture might possibly be. But most important, Geertz's eclecticism caters, perhaps unwittingly, to what the structural Marxist Louis Althusser (1989) astutely called the *spontaneous philosophy of the scientists*. By this Althusser meant the tendency for an inquirer to understand her own practice in terms of her discipline's standing with respect to other disciplines, which is usually as part of a sensitive and closely monitored balance of power. Goldenberg's (1989) survey of scientists' attitudes toward science–to be discussed at the end of this chapter–illustrates nicely the way in which the philosophical self-images of the various sciences reinforce one another. Of special interest here is the fact that this reinforcement takes place, *regardless* of whether the sciences in question respect or loathe one another. In both cases, interdisciplinary differences are merely affirmed without being resolved, which, to follow Althusser, disarms the critical impulse that has traditionally enabled the discipline of philosophy–and now social epistemology–to force the sciences to see the deep problems that arise, in part, from the fact that they treat each other as "separate but equal."

It is understandable why an eclectic author would want to make it

One: The Players and the Position

seem as though the mere juxtaposition of methods establishes common epistemological ground. The impulse is a strategic one that also caters to readers' interests. After all, if you accept the validity of any of the methods used in an eclectic study, you can incorporate the study into your own research. Such a study is thus very "user-friendly" to the normal scientist. By contrast, revolutionary theorists have refused to ignore the problematic status of common epistemological ground. Their answers have typically involved an interpenetration that leaves the constitutive methods or disciplines permanently transformed. New presumptions are instituted for the threshold of epistemic adequacy, which, in practice, means that new people with new training are needed for the evaluation of knowledge claims. Consider these uncontroversial cases of successful revolutionary theorizing. After Newton's *Principia Mathematica*, astronomy could no longer just yield accurate predictions, but also had to be physically realizable. After Darwin's *Origin of Species*, no account of life could dispense with either the "nature" or the "nurture" side of the issue. After Marx's *Capital*, no study of the material forces of production would be complete without a study of the social relations of production–a point that was rhetorically conceded even by Marx's opponents, who then started designating their asocial (i.e., "neoclassical") economics a "formal" science. After Freud's *Interpretation of Dreams*, any psychology based primarily on conscious introspection would be dismissed as at least naive (and at most spurious, à la behaviorism's response to cognitivism).

Interpenetration's Interlopers

Equipped with her rhetorical skills, the social epistemologist can facilitate revolutionary theorizing and its attendant mutations in our epistemic institutions. Whereas classical epistemologists and philosophers of science normally evaluate revolutionary theories in terms of their adequacy to the phenomena they purport to explain, the social epistemologist wants to unearth the implicit principles by which the revolutionary theorist managed to translate the concerns of several fields into an overarching program of research. In the days of logical positivism, this project would have been seen as involving the design of the "metalanguage" that enables the revolutionary theory to subsume disparate data domains. However, the social epistemologist

The Position

regards translation as very much a bottom-up affair, one in which the concerns of different disciplines are first brought to bear on a particular case—be it historical, experimental, hypothetical, or anecdotal—and then bootstrapped up to higher levels of conceptual synthesis. In that case, the relevant linguistic model is borrowed, not from metamathematics, set theory, and symbolic logic, but from the evolution of a trade language, or pidgin, into a community's first language, or creole, which over time may become a full-fledged, grammatically independent language. The positivists erred, not in thinking that there could be global principles of knowledge production, but in thinking that those principles could be legislated a priori from the top down rather than inferred inductively as inquirers pool their epistemic resources in order to reconstitute their world.

At this point, let me distance what I have in mind from a related idea that has become popular in rhetorical studies of physics and economics, namely, the *trading zone*, an idea most closely associated with the historian of twentieth-century physics Peter Galison (1992), and economist Donald McCloskey (1991). McCloskey offers the most succinct formulation of the idea, one that goes back to Adam Smith's *Wealth of Nations*. As a society becomes larger and more complex, people realize that they cannot produce everything they need. Consequently, each person specializes in producing a particular good that will attract a large number of customers, who will, in exchange, offer goods that the person needs. Thus, one specializes in order to trade. McCloskey believes that this principle applies just as much to the knowledge enterprise as it does to any other market-based activity. Galison's version of the trading zone draws more directly from the emergence of pidgins mentioned in the previous paragraph. His account has the virtue of being grounded in a highly informed analysis of the terms in which collaborative research has been done in Big Science-style physics. For example, determining the viability of the early nuclear bombs required a way of pooling the expertise of pure and applied mathematicians, physicists, industrial chemists, fluid dynamicists, and meteorologists. The pidgin that evolved from this joint effort was the Monte Carlo, a special random number generator designed to simulate stochastic processes too complex to calculate, such as the processes involved in estimating the decay rate of various subatomic particles. Nowadays, the Monte Carlo is a body of research in its own right, to which practitioners of many disciplines contribute, now long detached from its early nuclear origins (cf. Galison 1992: chap. 7). There are two questions to ask about the models that McCloskey and Galison propose:

One: The Players and the Position

> 1. Are they really the same? In other words, is Galison's history of the Monte Carlo trade language properly seen as a zone for "trading" in McCloskey's strict economic sense?
>
> 2. To what extent does the trading-zone idea capture what is or ought to be the case about the way the knowledge enterprise works?

The short answer to (1) is no. McCloskey is talking about an activity in which the goods do not change their identities as they change hands. The anticipated outcome of McCloskey's trading zone is that each person ends up with a greater number and variety of goods than when she began. The process is essentially one of redistribution, not transformation. In this way, Galison's trading zone is closer to the idea of interpenetration. The Monte Carlo simulation is an emergent property of a network of interdisciplinary transactions. It is not just that, say, the applied mathematicians learn something about industrial chemistry that they did not know before, but rather that the interaction itself produces a knowledge product to which neither had access previously. By contrast, then, McCloskey's idea perhaps captures the eclecticism of the human sciences in the postmodern era, which, to answer (2), calls its desirability into question. Interestingly, another economist, Kenneth Boulding (1968: 145-47), has already offered some considerations that explain why "Specialize in order to trade!" is not likely to become a norm of today's knowledge enterprises–though it perhaps should be. Boulding points out that in order to enforce Smith's imperative in the sciences, one would need two institutions, one that was functionally equivalent to a common currency (e.g., a methodological standard that enabled the practitioner of any discipline to judge the validity, reliability, and scope of a given knowledge claim) and one to an advertising agency (e.g., brokers whose job it would be to persuade the practitioners of different disciplines of the mutual relevance of each other's work). Short of these two institutions, the value of knowledge products would continue to accrue by producers' hoarding them (i.e., exerting tight control over their appropriate use) and making it difficult for new producers to enter their markets.

However, Galison's trading zone has its own problems from the standpoint of interpenetration to be promoted here. It does a fine job of showing how a concrete project in a specific place and time can generate a domain of inquiry whose abstractness enables it to be pursued subsequently in a wide variety of disciplinary contexts. In this way, he partly overcomes a limitation in McCloskey's trading zone; namely, he shows that the trade can have consequences–that is, costs and benefits–that go beyond the producers directly involved in a transaction. But Galison does not consider the *long-term* consequences of pursuing a particular trade language. Not only does a pidgin tend to evolve into an independent language, as in Galison's own Monte Carlo example, but it also tends to do so at the expense of at least one of the languages from which it is composed. Either that, or one of the source languages reabsorbs the developed pidgin in a process of "decreolization." In any case, there is just no practical way of arresting language change, short of segregating entire populations (cf. Aitchison 1981: especially pt. 4).

This empirical point about the evolution of pidgins may carry some normative payoff, insofar as the mere invention of new languages does not clarify the knowledge enterprise, if old ones are not at the same time being displaced. Since we are ultimately talking about scientists whose energies are distributed over a finite amount of space and time, cartographic metaphors for knowledge prove appropriate. You cannot carve out a new duchy without taking land away from neighboring realms–even if the populations of these realms are steadily growing. The strategy of interpenetrability that I support is, ultimately, a program for rearranging disciplinary boundaries. It presumes that the knowledge enterprise is most creative *not* when there are either rigid boundaries or no boundaries whatsoever; nor is creativity necessarily linked with the simple addition or elimination of boundaries; rather, creativity results from moving boundaries around as a result of constructive border engagements.

The social epistemologist thus imagines the texts of, say, Marx or Freud as such border engagements, the conduct of cross-disciplinary communication by proxy. They implicitly represent the costs and benefits that members of the respective disciplines would incur from the revolutionary interpenetration proposed by the theorist. For example, in the case of *Capital*, the social epistemologist asks what would an economist have to gain by seeing commodity exchange as the means by which money is pursued rather than vice versa, as the classical political economists maintained. Under what circumstances would it be worth the cost? Such questions are answered by examining how the acceptance of Marx's viewpoint would enhance or restrict the economist's jurisdiction vis-à-vis other professional knowledge producers and the lay public. More specifically, we would have to look at the audiences that took the judgment of economists seriously (for whatever reason) at the

One: The Players and the Position

outset, Marx's potential for affecting those audiences (i.e., his access to the relevant means of communication), and the probable consequences of audiences' acting on Marx's proposal. And while this configuration of *Capital*'s audience would undoubtedly do much to facilitate understanding the reception and evolution of Marxism, the social epistemologist is aiming at a larger goal, namely, the generalizability of the judgments that Marx made about translating distinct bodies of knowledge into a common framework: What was his strategy for removing interdisciplinary barriers? How did he decide when a key concept in political economy was really bad metaphysics in disguise, and hence replaceable by some suitably Hegelized variant? How did he decide when a Hegelian abstraction failed to touch base with the conception of material reality put forth in classical political economy? Is there anything we can learn from Marx's decisions for future interdisciplinary interpenetrations? So often we marvel at the panoramic sweep of revolutionary thought, when in fact we would learn more about revolutionary thinking by examining what was left on the cutting-room floor.

The practice of the social epistemologist would thus differ from that of mainstream hermeneuticians and literary critics in emphasizing the *transferability* of Marx's implicit principles to other potentially revolutionary interdisciplinary settings. However, none of these possibilities can be realized without experimental intervention, specifically, the writing of new texts that will, in turn, forge new audiences, whose members will establish the new terms for negotiation that will convert current differences into strategies for productive collaboration. The three presumptions that social epistemology derives from STS are meant to render explicit what revolutionary theorists have tacitly supposed about the nature of the knowledge enterprise.

The Pressure Points for Interpenetration

In terms of the original three STS presumptions, the kind of pressure point I want is the unit that best epitomizes the Conventionality Presumption. As it turns out, a survey of the various sociological units in which the knowledge enterprise can be analyzed reveals that the most conventional are disciplines, which correspond more exactly to technical languages and university departments than to

The Position

sets of skills or even distinct subject matters (Fisher 1990). For example, some skills are common to several disciplines, and other skills may be combined across disciplines with potentially fruitful results. However, the institutional character of disciplinary differences encourages inquirers to forgo these points of contact and to concentrate, instead, on meeting local standards of evaluation. This, in turn, perpetuates the misapprehension that disciplines carve up a primary reality, a domain of objects (cf. Shapere 1984), whereas interdisciplinary research carves up something more derivative. Indeed, sometimes in the effort to shore up their autonomy, disciplines will retreat to their signature topics, which are highly stylized (or idealized) versions of the phenomena they purport to study. Thus, when political science wants to demonstrate that it is a science, its practitioners will retreat from its programmatic aspirations of wanting to explain life in the *polis*, and instead point to the track record of empirical studies on voting behavior, as if the full complexity of political life could be constructed from a concatenation of such studies (J. Nelson 1987). If special steps are not taken to stem this tide of gaining more control over less reality, it is by no means clear that the situation will remedy itself (cf. Campbell 1969; Schaefer 1984; Fuller 1988a: chap. 12). On this basis, we can specify two sets of tensions–conveniently labeled *spatial* and *temporal*–that make disciplines especially good pressure points for interpenetration.

In terms of the spatial tension, disciplines are defined by two forces–*the university* and *the profession*–that are largely at odds with one another, although much of the conflict remains at the implicit level of structural contradiction. A university occupies a set of buildings and grounds in (more or less) one place, and each discipline a department in that place. The limits of university expansion are dictated by a budget, from which each department draws and to which each contributes. The idea of "budget" should be understood here liberally, to include not only operating funds, but also course assignments and space allocation (cf. Stinchcombe 1990: chap. 9). Of course, universities expand, but the interests of particular departments are always subserved to that of the whole. The brutest way of making this point is to recall the overhead costs that researchers receiving government grants must turn over to their universities for general operating purposes. But, in more subtle ways, the particularity of departments comes out in how curricular responsibilities are distributed among disciplines in different universities. The intellectual rigor or epistemic merit of a discipline may count for little in determining the corresponding department's fate in the realm of university politics.

One: The Players and the Position

In moving from the university department to the professional association, we see that the latter has indefinite horizons that stretch across the globe and determine the networks within which practitioners do and share their work. Such an association is more readily identified with technical languages and their ever-expanding publication outlets than with fixed ratios of money, courses, or space. Indeed, much of the information explosion that makes the access to pertinent knowledge increasingly difficult may be traced to the fact that most professional associations view the relentless promotion of their activities to be an unmitigated good (cf. Abbott 1988: chap. 6). The spatial tension between universities and professions is recognizable in many sociodynamic guises. Sociologists, following Alvin Gouldner (1957), see university versus profession as a case of "local" versus "cosmopolitan" allegiances. Political theorists interested in designing a "Republic of Science" may see a couple of familiar options for representing the disciplined character of knowledge: the subordination of professional to university interests, on the one hand, and the subordination of university to professional interests, on the other. The former is analogous to representation by geographical region, whereby the republic is conceptualized as a self-contained whole divided into departments; the latter resembles representation by classes, whereby a given republic is simply one site for managing the interplay of universally conflicting class interests. One might expect the teaching-oriented faculty to prefer regional representation, while research-oriented ones prefer the more corporatist model. But perhaps the most suggestive way of presenting the structural contradiction in disciplined knowledge is in terms of Immanuel Wallerstein's (e.g., 1991) world-system model, which attempts to explain the course of modern history as temporary resolutions of the ongoing tension between the proliferation of capitalist markets across the world (most recently in the guise of transnational corporations) and the attempts by nation-states to maintain and consolidate their power base (most recently in terms of high-tech military systems).

But how close is the analogy between, on the one hand, capital and professional expansion or, on the other, national and university consolidation? To consider just the first analogy, the sociologist Irving Louis Horowitz (1986) has argued that transnational publishing houses have been decisive in the proliferation of professional specialties by making it easier to start journals than to publish books, as the latter tend to attract a larger and more interdisciplinary audience but in a one-shot fashion that generates much smaller revenues. (This has to do with the

The Position

traditionally transient character of most interdisciplinary endeavors: once the specific interdisciplinary project is complete, the parties return to their home disciplines.) Beyond this rather literal case of professionalization as a form of capital expansion, a fruitful site for investigation is intellectual property law. Here the explicit treatment of knowledge as a material, specifically economic, good forces professional bodies to think of themselves as companies and universities to think of themselves as states. As the economic consequences of embodied forms of knowledge become more apparent (especially as the difference between "basic" and "applied" science vanishes), universities are claiming proprietary rights to knowledge products and processes that would otherwise be more naturally identified with the professional skills of its creator. Will there come a point in which a widely distributed technology is more closely associated with the name of a university than of its creator's profession? How literally should we take the nickname of the first patented genetically engineered animal, "The Harvard Mouse"?

In presenting the spatial tension surrounding a discipline, I may have given the impression that, on balance, professional interests are more "progressive" than university-based ones. I happen to think that this is generally true, if by "progress" is meant the tendency to make the academy more permeable to the public. (I say this in the full knowledge that professionalism has much the same self-serving motive as capitalism's own reduction of indigenous social barriers—namely, to increase mobility of the labor force and the number of paying customers.) Yet, at the same time, professionalism left to its own devices will reify itself into perpetuity, a tendency that this book is largely designed to combat—that is, the tendency of professional associations to cast themselves as having special access to distinct realms of being. Here the university functions as an effective foil, as budgetary constraints naturally curb ontological pretensions. It is a mistake to think that knowledge is best served by maximizing the pool of funds available. At most, an ample budget will enable all to continue on their current trajectories as they see fit. However, it is an open question whether the undisturbed course of "normal science" will likely lead to genuine epistemic growth, since there would then be little incentive to engage in interdisciplinary interpenetration. Tight budgets, by contrast, provide an incentive for interpenetration by forcing a discipline to distinguish essential from nonessential aspects of its research program, and to recognize situations where some of those aspects may be more efficiently done in collaboration with, if not turned over to, researchers in other

One: The Players and the Position

disciplines. Nevertheless, the emancipatory character of budgetary constraints is often obscured because of the bad rhetoric that accompanies talk of "eliminating programs," which forces departments to think that some of them will benefit only at the expense of others.

Fatalism is superimposed on this image of fatalities by a version of the fallacy of division that I will dub *The Dean's Razor*, namely, the inference that because interdisciplinary programs consist of people trained in regular disciplines, it follows that nothing essential to the knowledge production process will be lost by eliminating the programs (and keeping the original disciplines) when times are tough. Instead of a razor, a better instrument for the Dean to wield would be what economists call "zero-based budgeting," whereby each discipline would have to make its case for resources from scratch each year. I would go further: In the university's accounting procedure, while members of the faculty would continue to be treated as university employees, they would no longer be considered the exclusive properties or representatives of particular departments. Their specific departmental affiliations would therefore have to be negotiated with each academic year. Departments would take on the character of political parties pushing particular (research) programs, probably at the behest of professional associations, but allowing also for some locally generated interdisciplinary alliances, to which faculty will need to be recruited from the available pool each year. While, in practice, few faculty members would probably want to shift departmental affiliation very often, such a set-up would nevertheless loosen the grip that professional associations often have on the constitution of departments, as departments would have to come up with ways to attract particular personnel who might also be desired by competing departments within the university. There are more epistemic consequences to budgetary practices, specifically at a national level. I will turn to these after discussing the temporal tension that defines a discipline.

A discipline's temporal tension can be analyzed in terms of two countervailing forces: the *prospective* judgment required to legitimate the pursuit of a research program and the *retrospective* judgment that figures in explaining the research program's accomplishments. Our earlier example of the fate of political science makes the point nicely. The original promise of the discipline, repeatedly stressed by its most innovative theorists, was to explain the totality of political life by mechanisms of power, ideology, and the like, whose ontological purchase would cut across existing disciplinary divisions in the social sciences.

However, when forced to speak to the field's empirical successes, political scientists fall back on, say, the many studies of voting behavior, which display the virtuoso use of such discipline-specific techniques as cross-national questionnaires (cf. Almond and Verba 1963) but which make little direct contribution to the larger interdisciplinary project.

Reflected in this tension are the two sorts of strategies that philosophers have used to account for the "success" of science. *Realists* put the emphasis on the prospective judgments, which are often expressed as quests for a desired set of mechanisms or laws able to bring disparate phenomena under a single theory. Realists see the scientific enterprise as continuing indefinitely, anticipate many corrections and even radical reversals of the current knowledge base, and typically regard the current division of disciplinary labor, at best, as a necessary evil and, sometimes, as a diversion from the path to unity. By contrast, *instrumentalists* stress retrospective judgments of scientific success, which turn on identifying specific empirical regularities that have remained robust in repeated testings under a variety of conditions. These regularities continue to hold up long after theories explaining them have come and gone (cf. Hacking 1983: pt. 2). Indeed, any new theory is born bearing the burden of "saving" these phenomena. Quite unlike the realist, the instrumentalist welcomes the increased division of disciplinary labor as issuing in a finer grained level of empirical analysis and control.

Symptomatic of the atemporal way in which philosophers think about this matter is their failure to see that the relative plausibility of realism and instrumentalism depends on the historical perspective on science that one adopts. From the standpoint of the present, the realist is someone who projects an ideal future in which the original promise of her research program is fully realized, whereas the instrumentalist is someone who reconstructs an ideal past in which the actual products of her research turn out to be what she had really wanted all along. Both perspectives are combined in the history of science that philosophers—both realists and instrumentalists—have typically told since the advent of positivism: The Greeks started by asking about the nature of the cosmic order, and today we have answers that, in part, complain about the ill-formedness of their original questions and, in part, specify empirical regularities by which we can elicit more "order" (properly redefined) than the Greeks could have ever imagined. It is quite common for philosophers in this context to claim that, insofar as the early Greeks were "seriously" inquiring into the nature of

One: The Players and the Position

things, they would recognize our accomplishments as substantial steps in that direction. The difference here between the Greeks looking forward to us and our looking backward at them reflects an underlying psychodynamic tension. Generally speaking, the history of disciplines presents a spectacle of research programs whose actual products are much more modest than what their original promise would suggest–if not actually tangential to that promise; yet, those products would probably not have been generated had inquirers not been motivated by a more comprehensive project. Thus, one doubts that any of the special sciences would have inspired much initial enthusiasm if its proponents promised merely to produce a set of empirical correlations, the reliability of which could be guaranteed only for highly controlled settings. Such prescience on the proponents' part would have doomed their project at the outset!

The psychodynamics between the realist and instrumentalist orientations may provide a neat explanation for what Hegel and Marx called "the cunning of reason" in history. But from the standpoint of social epistemology, this psychodynamics has more immediately pressing implications. Consider the most recent comprehensive statement by the US government on research funding and evaluation: *Federally Funded Research: Decisions for a Decade* (Chubin 1991). This report, prepared for Congress by the Office of Technology Assessment, drew attention to the fact that research funding increasingly goes to glamorous and expensive "megaprojects," such as the Human Genome Project, the Orbiting Space Station, and the Superconducting Supercollider. These megaprojects promise major breakthroughs across several disciplines and many spinoffs for society at large. However, a megaproject is rarely evaluated by its original lofty goals. Rather, its continued support typically depends on a series of solid empirical findings, the significance of which, however, is probably too limited to justify (in retrospect) the amount of money that was spent to obtain them. Nevertheless, these findings are typically couched as "just the start" toward delivering on the original promises. If the history of science policy is a good inductive guide, however, the odds that this is an accurate prognosis of the project's research trajectory are low. But that does not stop policymakers from being suckered into supporting projects that can only be counted on to deliver diminishing returns on continued investment. As suggested above, the interactive effects of the policymaker's prospective and retrospective judgments on research make any solution to this problem complicated. On the one hand, it might be reasonably argued that even findings of limited scope would not have been made had scientists not aspired to more. On the other hand, such a judgment itself becomes clearer as it seems less feasible to divert funding from that line of research. This should give us pause.

The political theorist Jon Elster (1979, 1983) has a striking way of characterizing our quandary. The realist vision of a megaproject is necessary to "precommit" policymakers to a funding pattern that they would otherwise find very risky. In that sense, realism girds the policymaker against a weakness of the fiscal will. But evaluating the products of a megaproject by the instrumentalist criteria of particular disciplines makes the policymaker prone to develop a version of "sour grapes" called "sweet lemons," an exaggerated sense of the project's accomplishments that results from deflating "what can now be seen" as the project's original pretensions, which no one could have been expected to meet. But does sour grapes do anything more than pervert precommitment? In whose moral psychology is self-deception an adequate solution to weakness of the will?

My point here is not to dump the idea of megaprojects, largely because I do not (yet) have a substitute for the motivational role that the realist vision has played in scientific research throughout the ages. However, if delusions of grandeur are unavoidable at the planning stage of a megaproject, it does not follow they must also dominate the evaluation stage. In particular, policymakers should be able to separate out their interest in sustaining the vision that informs the megaproject from whatever interest they might have in supporting the specific research team that first proposed it. Sour grapes may be seen to result from too closely associating the project's potential with the actual research results, which ends up leading policymakers to indefinitely support the team behind the results, regardless of whether that team is *now* in the best position to take the next step toward realizing the project's full potential. Thus, one way to address this problem is to carefully distinguish the processes of *rewarding* and *reinforcing* scientists for their work. To prevent the scientists who first staked out a megaproject from indefinitely capitalizing on their original research investment, they should be rewarded for their pioneering work but not expected to continue in their original trajectory. Incentives may be set in place–say, in terms of the grant sizes made available–to encourage the research team to break up and recombine with members of other teams in other projects, with the megaproject's own future being placed in the hands of another team (or at least a significantly altered version of the

One: The Players and the Position

original one).

The Task Ahead (and the Enemy Within)

Whether one approves or disapproves of the current state of knowledge production, there is a tendency to see "science" as a unitary system, a *universitas* in the original medieval sense, which emphasizes the departmental over the professional character of disciplines. This suggests that the disciplines see themselves as part of the same team, engaged in relations of mutual respect, if not outright cooperation. In that case, criticisms of the knowledge enterprise should appear as rather generic attacks on academic practices, not as cross-disciplinary skirmishes. Indeed, this is a fair characterization of the scope of science evaluation, ranging from science policy advisors to popular critics of science. Not since C. P. Snow's famous 1959 Rede Lecture on "two cultures" has anyone systematically raised the social epistemological consequences of disciplines' refusal to engage issues of common and public concern because they suspect *one another's* methods and motives (Snow 1964; cf. Sorell 1991: chap. 5). The rhetoric of interpenetration is meant to address this most open of secrets in the academy. The Canadian sociologist Sheldon Goldenberg (1989) has recently performed an invaluable service by surveying both social and natural scientists about their attitudes toward the knowledge enterprise: What books influenced how they think about the pursuit of knowledge? Can work in other disciplines be evaluated by the same standards used to evaluate work in their own? If not, is the difference to be explained by the character of the discipline or of its practitioners? Before proceeding to my own specific interdisciplinary incursions, it might be useful to get a sense of the dimensions of the task ahead for the social epistemologist interested in having disciplines deal with each other in good faith. For Goldenberg enables us to map *the structure of academic contempt*.

Telescoping Goldenberg's data somewhat, we can discern three general attitudes to the knowledge enterprise that are in sharp tension with one another. These attitudes are associated with *natural scientists, social scientists,* and *philosophers of science*. Natural scientists tend to think that something called "the scientific method" can be applied across the board, but that social scientists typically fail to do so

The Position

because they let incompetence, politics, or sloth get in the way. In this portrayal, social scientists suffer from weakness of the will, while natural scientists persevere toward the truth. Not surprisingly, social scientists see the matter much differently. They portray themselves as reflective, self-critical inquirers, who are not so easily fooled by the idea of a unitary scientific method bringing us closer to the truth. Natural scientists appear, in this picture, to be naive and self-deceived, mistaking big grants and political attention for epistemic virtues. Philosophers of science occupy a curious position in all this. On the one hand, social scientists are more likely to read the philosophical literature than natural scientists, but, on the other hand, they are also more likely to disagree with it, insofar as philosophers tend to believe that science does, indeed, work if applied diligently. Thus, social scientists often regard philosophers as dangerous ideologues who encourage natural scientists in their worst tendencies, whereas philosophers regard the natural scientists as spontaneously vindicating philosophical theses in their daily practices. As the philosopher sees it, her job is to raise the efficacious aspects of scientific practice to self-consciousness, as scientists themselves tend not to have the broad historical and theoretical sweep needed to distinguish what is essential from what is nonessential to the growth of knowledge. In that sense, philosophers and social scientists are in agreement that natural scientists are typically ignorant of the principles that govern their practice–the difference between the two camps being that philosophers also tend to believe that science works *in spite of* that ignorance, as if it were governed by an invisible (philosophical) hand.

The rhetoric needed to perform social epistemology in this environment consists of a two-phase "argumentation practice" (Keith 1992). This practice may be illustrated by the following exchange between "you" and "me." Before I am likely to be receptive to the idea that I must change my current practices, I must be convinced that you have my best interests at heart. Here the persuasive skills of the Sophist come into play as you try to establish "common ground" with me. The extent of this ground can vary significantly. At one extreme, you may simply need to point out that we are materially interlocked in a common fate, however else our beliefs and values may differ. At the other, you may claim to be giving clearer expression to views that I already hold. In either case, once common ground has been established, I am ready for the second, more Socratic side of the process. I am now mentally (and socially) prepared to have my views criticized without feeling that my status as an equal party to the dialogue is being undermined. Ideally,

One: The Players and the Position

this two-step strategy works a Hegelian miracle, the mutual cancellation of the Sophist's manipulative tendencies and Socrates' intellectually coercive ones. For persuasion arises in preparation of an open encounter (and so no spurious agreement results), while criticism arises only after the way has been paved for it to be taken seriously (and so no fruitless resistance is generated).

The argumentation practice of classical epistemology is distinguished from that of social epistemology by its elimination of the first phase. Instead of establishing common ground between "you" and "me," the classical epistemologist simply takes common ground for granted. As a result, any failure on my part to respond adequately to the second phase, criticism, is diagnosed as a deep conceptual problem, not as the consequence of a bad rhetorical habit, namely, your failure to gauge the assumptions I bring to our exchange prior to your beginning to address me. This diagnosis of classical epistemology is supported by the following rhetorical construction of how the problem of knowledge is currently posed by analytic philosophers.

In order to understand this "modern problem of knowledge," we must first realize that it is a technical problem of definition, most of which has already been solved. This explains the narrowness of the debate over the "missing term." All parties to the debate seem to follow (more or less) Plato, Descartes, and Brentano in granting that knowledge is *at least* "justified true belief." The putative advance that has been made since World War II (according to a standard textbook, Chisholm 1977) is to realize that there is a little bit more to the story–but what? A major breakthrough was staged in a three page article by Edmund Gettier (1963), who independently restated a point that was neglected when Bertrand Russell first raised it, fifty years earlier. The breakthrough consisted of some thought experiments designed to isolate the missing term. In brief, the "Gettier Problem" is the possibility that we could have a justified true belief which ends up being mistaken for knowledge, because the belief is grounded on a false assumption that is never made explicit. For example, outside my house are parked two cars, about which I have a justified true belief that one belongs to John and the other to Mary. Thus, when asked for the whereabouts of one of the vehicles, I rightly say, "John's car is outside." Unfortunately, John and Mary traded cars with each other earlier that morning, and so the car that I thought was John's now turns out to be Mary's. And, if my interlocutor does not ask which car is John's, my ignorance will remain undetected as a false assumption. There is a tendency for people outside of epistemology to dismiss the Gettier Problem as simply more of the idle scholasticism for which they have come to fear and loathe philosophers. However, the unprecedented extent to which Gettier has focused the efforts of epistemologists over the last thirty years testifies to the rhetorical appeal of the problem bearing his name (cf. Shope 1983). A brief look at the social dynamics presupposed in the problem should, therefore, reveal something telling about the susceptibility of philosophers to persuasion.

Let us start by taking the Gettier Problem as a purely linguistic transaction, or speech act. I am asked two questions by you, my didactic interlocutor. In response to the first, I correctly say that John's car is outside; in response to the second, I incorrectly say that Mary's car is John's. You frame this sequence of questions as occurring in a context that changes sufficiently little to allow you to claim that our second exchange is an attempt at deepening the inquiry begun in the first exchange. As a piece of social dynamics, this "deepening" is none other than your ability to persuade me that your evaluation of my second response should be used as a standard against which to judge my first response, which, prior to your asking the second question, seemed to be unproblematic. But why should I assent to this shifting of the evaluative ground? The reason seems to be that I accept the idea that my second response had already been implicit in the first response when I made it, and in that sense constitutes the deep structure of the first response. As the "essence" of the first response, the second response existed *in potentia* all along. If nothing else, this linguistic transaction defines the social conditions for attributing the possession of a concept to someone: to wit, I have a concept, if you can get me to follow up an initial response with an exchange that you deem appropriate to the situation.

Now, this ontologically loaded view of language as replete with hidden essences and deep structures–concepts, to say the least–recalls the Socratic rhetoric of *anamnesis*, the recovery of lost memories. However, as against all this, it may be said in a more constructivist vein that reality normally transpires at a coarser grain of analysis than our language is capable of giving it, which implies that if all talk has some purchase on reality, then it is only because talk can bring into being situations and practices that did not exist prior to their appearance in discourse. In terms of the Gettier Problem, why should we suppose that, under normal circumstances, I would have something definite to say about which car is John's prior to your actual request? Moreover, why should we suppose that the answer I give to your request has some retroactive purchase on my answer to your previous query, instead of simply being a new answer to a new question posed in a new context?

One: The Players and the Position

The constructivist view that I make up new levels of analysis as my interlocutor demands them of me, and then back-substitute those levels for earlier ones, puts a new spin on the verificationist motto that all conceptual (or linguistic) distinctions should make an empirical (or "real-world") difference. The Gettier Problem shows that instead of eliminating conceptual distinctions that make no empirical difference, the epistemologist, in her role as my interlocutor, can produce empirical differences in my response based on the conceptual distinctions raised in her questions. To follow a line of reasoning initiated by social psychologist Michael Billig (1987: chap. 8), here the epistemologist proves herself a master dialectician, as she manufactures a world which I am willing to adopt as my own, even at the (unwitting) expense of relinquishing my old one.

If the reader detects perversity in the epistemologist's strategy of manufacturing occasions that enable her talk to acquire a significance that it would not have otherwise, then you have just demonstrated some rhetorical scruples. Joseph Wenzel (1989) has observed that a good way of telling the "rhetoricians" from the "dialecticians" (or philosophers) among the Sophists was that the former engaged arguments only as part of a general plan to motivate action, whereas the latter argued so as to reach agreement on a proposition. What philosophers have traditionally derided as "mere persuasion" is simply the idea that talk only goes so far toward getting people to act appropriately. But the truly heretical thesis implied in this simple idea is that from the standpoint of appropriate action, it may make no difference whether everyone agrees on a given proposition or whether they instead diverge significantly, perhaps even misunderstanding each other's point of view. Contrary to what many philosophers continue to believe, rhetoricians realize that consensus is not a prerequisite for collaboration—in fact, consensus may often prove an obstacle, if, say, a classical epistemologist has convinced the practitioners of different disciplines that they must agree on all the fundamentals of their inquiry before proceeding on a joint venture. In that case, the convinced parties would have simply allowed the epistemologist to insert her project ahead of their own without increasing the likelihood that theirs will ever be carried out. The *social* epistemologist promises not to make that mistake!

While the social epistemologist cannot be expected to resolve incongruous, contempt-breeding cross-disciplinary perspectives immediately, she may begin by identifying modes of interpenetration appropriate to situations where several disciplines already have common

The Position

concerns but no effective rhetoric to articulate those concerns *as* common. Four such modes are examined in the first part of this book. They vary along two dimensions. The first dimension concerns the difference between *persuasion* (P) and *dialectic* (D): rhetoric that on the one hand aims to minimize the differences between two disciplines, and on the other aims to highlight those differences. In terms of a pervasive stereotype, persuasion is the Sophist's art, dialectic the Socratic one. The former seeks common ground, the latter opposes spurious consensus. The second dimension concerns the direction of cognitive transference, so to speak. Does a discipline engage in persuasion or dialectic in order to import ideas from another discipline (I) or to export ideas to that discipline (E)? This distinction corresponds to the two principal functions of metaphor (Greek for "transference") in science: respectively, to test ideas in one domain against those in another ("negative" analogy), and to apply ideas from one domain to another ("positive" analogy). Together the two dimensions present the following four interpenetrative possibilities. Each possibility is epitomized by a current interdisciplinary exchange in which I have been a participant. In the elaborations that follow in the next four chapters, I do not pretend that these exchanges represent "pure" types. However, for analytical purposes, we may identify four distinct processes:

> (P + I) *Incorporation*: Naturalized epistemologists claim that epistemology can itself be no better grounded than the most successful sciences. Classical epistemologists counter that naturalists presuppose a standard for successful knowledge practices that is logically prior to, and hence must be grounded independently of, the particular sciences deemed successful. There is much at stake here, as captured by the following questions: Is philosophy autonomous from the sciences? Is philosophy's role to support or to criticize the sciences? Have the sciences epistemologically outgrown philosophy? Not surprisingly, the impasse that results between the two positions in this debate is often diagnosed in terms of the radically different assumptions that they make about the nature of knowledge. However, I see the problem here as being quite the opposite, namely, that the two sides have yet to fully disentangle themselves from one another. I show this to be especially true of the naturalist, who often shortsells her position by unwittingly reverting to classicist argument strategies. But after the naturalist has disentangled her position from the classicist's, she needs to

address specific classicist objections in naturalistic terms. In that sense, the naturalist needs to "incorporate" the classicist. Otherwise, a rhetorical impasse *will* result.

(D + E) *Reflexion*: Disciplinary histories of science tend to suppress the fact that knowledge is *in* the same world that it is *about*. No representation without intervention; no discovery without invention. Yet knowledge is supposed to pertain to the world prior to any "artificial" transformation it may undergo during the process of knowing. The natural sciences can suppress the transformative character of knowledge production more effectively than the social sciences. The discourses of the natural sciences are relatively autonomous from ordinary talk, and their techniques—"laboratories," in the broadest sense—for generating and analyzing phenomena are relatively insulated from the normal course of events. By contrast, because societies have placed some fairly specific practical demands on the social sciences, they have not enjoyed the same autonomy and insulation. The seams of social intervention in their attempts at representation are easily seen, especially when a social science tries to explain its own existence in its own terms. The results typically turn out to reveal the discipline's blind spots, highlighting the artifice with which disciplinary identity is maintained. For example, economics has appeared most authoritative in periods of economic turbulence, during which economists are hired to dictate policy to a market supposedly governed by an "invisible hand." However, the point of revealing such paradoxicality by historical "reflexion" (a process both *reflexive* and *reflective*) is to undermine not social science per se, but only its division into discrete disciplines. For together the social sciences have the investigative apparatus needed for showing that the natural sciences, too, are world-transformative enterprises.

(P + E) *Sublimation*: Practitioners of the Sociology of Scientific Knowledge (SSK) and artificial intelligence (AI) should be natural collaborators, bringing complementary modes of analysis to their common interest in the cognitive capacities of the computer. However, most of the exchanges to date have been hostile, often based on mutually stereotyped views that reverberate of earlier debates, especially "mechanism versus humanism" or "positivism versus holism" (Slezak 1990), often filtered through the coarse-grained representations of the mass media. As science gets a longer history and becomes more permeable to public concerns, this tendency is likely to spread. The solution explored here is for each side to export ideas that are essential to the other's project. Thus, differences are "sublimated" by showing them to be natural extensions of one another's position. The AI researcher needs to see that competence is a social attribution, in order to test empirically the cognitive capacities of a particular computer. Conversely, the SSK researcher should realize that the possible success of AI would testify to the constructed character of cognition, such that not even the possession of a human body is deemed necessary for thought. Given the tendency of debates of this sort to amplify into a Manichaean struggle, the presence of the computer as a "boundary object" of significance for both sides turns out to be crucial in facilitating the sublimation process by forcing each side to map its cosmic concerns onto the same finite piece of matter (Star and Griesemer 1989; cf. McGee 1980, on "ideographs," as pieces of language that perform much the same function).

(D + I) *Excavation*: After the initial promise of studying science historically, both in the nineteenth century and especially in the work of Thomas Kuhn, the history and philosophy of science (HPS) appears to be at a conceptual standstill, not quite prepared to make the leap beyond the disciplinary boundaries of history and philosophy to the new field of STS. I diagnose this reluctance in terms of a failure, especially on the part of historians, to explicitly discuss the assumptions they make about theory and method, which are often at odds with what the social sciences have to say about these matters. Especially suspect are the assumptions about the human cognitive condition that inform historical narratives, even narratives that avowedly draw from cognitive psychology. To "excavate" these assumptions is to articulate long suppressed differences between humanistic and social scientific approaches to inquiry. While it is perhaps too much to expect historians to become social scientists overnight, nevertheless a willingness on the part of humanists to hold their research accountable to the standards of social science would tend to break down the remaining disciplinary barriers that inhibits HPS's passage to STS. It

One: The Players and the Position

Rhetorical Aim / Trade Strategy	PERSUASION (Difference Minimizing)	DIALECTIC (Difference Amplifying)
IMPORT (Negative Analogy)	INCORPORATION	EXCAVATION
EXPORT (Positive Analogy)	SUBLIMATION	REFLEXION

Figure 2. The Modes of Interdisciplinary Interpenetration

would also enable the historian to use the social scientists' own methods to keep them scrupulous to historical detail. I end by suggesting that some of the normative issues that have made philosophers impatient with historians could be better addressed by experimental social psychology, and perhaps even the "case-study" methodology traditionally championed in law and business schools.

Here I Stand

Since the reader is about to embark upon a dialectical journey from which few have returned unconfused, let me state briefly where I position myself in each interpenetration. In the case of Incorporation, I am a staunch naturalist who believes that the letter of classical epistemology has compromised the naturalist's spirit. In the case of Reflexion, I am a staunch advocate of social science who believes that its fragmentation into disciplines has undermined the social scientist's capacity for critiquing and reconstructing the knowledge

The Position

system. In the case of Sublimation, I am a staunch supporter of the sociology of scientific knowledge who agrees that yet again philosophers have injected false consciousness into another community of unsuspecting scientists, namely, researchers in artificial intelligence, but I also believe that the sociologists are being duplicitous when they make a priori arguments against the inclusion of computers as members of our epistemic communities. Finally, in the case of Excavation, I am a staunch ally of those who want to facilitate the transition from HPS to STS, but I also believe that it is naive to think that this transition can succeed if both parties simply adopt new theories and look at new data—a new social formation is also needed.

PART TWO

INTERPENETRATION AT WORK

3 INCORPORATION, OR

EPISTEMOLOGY EMERGENT

Weighing the relative merits of classical and naturalistic approaches to the philosophy of science, or epistemology more generally, must be counted among the most interesting current issues in "metaphilosophy," the study of how philosophers ought to be spending their time: Is the philosopher engaged in an enterprise that is legitimated on grounds quite apart from science, which, once grounded, can then pass judgment on the legitimacy of science? Yes, says the *classicist* (Pappas and Swain 1978). Or rather, is the philosopher really only a scientist of science, whose own legitimacy is only as good as that of the scientists she studies? To this the *naturalist* assents (Kornblith 1985). Yet the arguments that ensue from these questions are typically among the most predictable and inconclusive in philosophy today (Maffie 1990 gives the current state of play). Substantial philosophers are often reduced to textbook-style thrust and parry. Some would argue that this goes to show the general aridity of metaphilosophic dispute, while others would lay the blame on the exhaustion of epistemology and philosophy of science as fields of inquiry that can do little to justify their continuation on any ground, be it classical or naturalistic. By contrast, the diagnosis to be explored here is that the naturalist has yet to make a clean break with the classicist's position. In arguing this point, I will take as my point of departure the naturalist's failure to abide by a simple procedural rule of argument.

Two: Interpenetration at Work

Tycho on the Run

TYCHO'S DOCTRINE: SEPARATE BUT (NOT QUITE) EQUAL

Before launching into diagnostics, let me anticipate what is to follow. I will eventually comment on a mock-up of the rut into which the classicism-naturalism debate typically falls. If one were to regard this rut as inevitable and irresoluble, then one could imagine arriving at mediating positions that effect a rapprochement *with* ruts. In order to distance my own aims from such a project, this section criticizes what I will dub *Tychonic Naturalism*. The Tychonic Naturalist holds that, in formulating the metatheory of her activity, the epistemologist can do no better than to strike a balance between the classicist and the naturalist. There is no way of transcending the standard moves made by the two positions, and so mutual accommodation is the best we can do. Thus, she operates in the spirit of the sixteenth-century astronomer Tycho Brahe, who continued Ptolemy's practice of treating the earth as the static center around which the sun moved, but then followed Copernicus in having the other planets circle the sun. The Tychonic Naturalist can strike a balance between classicism and naturalism in at least two different ways, as represented by the distinguished philosophers Alvin Goldman and Rom Harré.

The calling card of Goldman's Tychonism is the very structure of *Epistemology and Cognition* (1986), a veritable summa of naturalized epistemology. The book is divided into two parts. The first, "Ptolemaic" part is a largely a priori conceptual analysis of the defining features of the epistemic process, whereas the second, "Copernican" part is devoted to empirically isolating the cognitive mechanisms that instantiate those features. Thus, after Goldman defines knowledge as the reliable production of true beliefs, he then proceeds to look in the mind for some reliable mechanisms–all along presuming that there must be some to be found. Clearly, Goldman's commitment to naturalism is a mitigated one. In particular, he does not take seriously a naturalized version of skepticism, namely, the possibility that *nothing* in our psychological makeup conforms to the concept of a reliable-true-belief-forming-mechanism. Indeed, Goldman frequently overrules a psychologist's claim to have shown that a defining feature of knowledge is empirically unrealizable. He does this by challenging the "intelligibility" of humans acting irrationally most of the time or holding

Incorporation

mostly false beliefs–to name just two epistemologically inauspicious conclusions that psychologists have been prone to draw (especially 1986: 305-23).

Goldman (1985) believes that his naturalism binds him to a version of the "ought implies can" principle, in which individual human beings must be able to follow the norms of rationality if the norms are truly to have force. This commitment seems to motivate his attempts at discrediting experimental demonstrations of irrational judgment in individuals. But why should a naturalist tie norms to the abilities of *individuals*? After all, if a popular philosophical model of rationality, such as Bayes' theorem, consistently picks the better theory to test, but individuals are unable to follow the logic prescribed by the theorem, then maybe the theorem is suited for some other sort of being. At the very least, Bayes' theorem could govern a digital computer's selection of theories, and maybe the theorem can also be used to characterize an emergent property of a certain kind of social interaction. (By comparison, Popper thought that falsification was an emergent feature of each scientist's being her own conjecturer and her neighbor's refuter.) In any case, the scope of the theorem's governance is, as the naturalist would have it, a matter for empirical inquiry. It is often forgotten that Kant first proposed "ought implies can" as an argument for the existence of a faculty that enables us to be moral agents. Translated into naturalistic terms, this means that a norm postulates a (perhaps yet to be discovered) realm of beings that are governed by it.

It would seem, then, that in Goldman's case the Tychonic spirit is moved by an interest in keeping the disciplinary boundary between philosophy and psychology intact–itself a rather peculiar interest for a naturalist to have. Why not, instead, take the empirical unrealizability of a piece of conceptual analysis to suggest that the analysis may itself be off the mark? Although naturalists typically advertise their sensitivity to the historical character of knowledge production, they nevertheless tend to respect the disciplinary boundary separating philosophy from the empirical sciences, as if the boundary delineated a historically invariant, "real" difference in subject matter. But to be truly naturalistic, one must realize that the disciplinary boundary separating psychology and philosophy has been contingently shaped over the course of history–and, even in our own day, across different nations. A good case in point is the difference between the French Piaget, the genetic epistemologist who defines the "cognitive paradigm" (De Mey 1982), and the American Piaget, the child psychologist with eccentric methodological views (cf. Kitchener 1986). Consequently, the philoso-

Two: Interpenetration at Work

phy-psychology boundary as it stands "here and now" should not be reified as if it reflected a real difference in subject matter.

On the surface, this may appear to be a radical suggestion, but in fact it is simply the sort of consideration that has traditionally led both positivists and social constructivists to be skeptical about drawing ontological conclusions from the division of cognitive labor in science. But beyond mere philosophical argument, the story of psychology's metamorphosis into a "brass instrument" lab science has become a textbook case of the long-term institutional consequences of one man's entrepreneurship (R. Collins and Ben-David 1966). Wilhelm Wundt basically imported high-rent medical skills into the low-rent district of philosophy. My point here is *not* that the naturalist ought to distrust any hard distinction that might be drawn between the tasks of epistemology and cognitive psychology. Rather, she should simply distrust any proposed distinction that is based on "conceptual" considerations which abstract from the changing historical character of the two disciplines. Instead, the naturalist should roll up her sleeves and design some epistemologically relevant psychology experiments, argue with the psychologists about methodology, and *then* decide where (or whether) the boundary between the two disciplines should be drawn (cf. Heyes 1989).

For the sake of consistency, naturalists could take a lesson from the logical positivists, who fully recognized the completely conventional character of disciplinary boundaries, and hence transgressed them whenever it seemed necessary, as in the service of "unified science" (cf. Zolo 1989: chap. 5). But just as the naturalist cannot conceptually ground the separation of epistemology from psychology, I would also argue that she cannot, simply by argument, empirically eliminate epistemology in favor of psychology. This latter move—commonly found in such radical naturalists as Willard Quine, Donald Campbell, Richard Rorty, Paul Churchland, and Ronald Giere—neglects, no less than the former, the historical dimension of the epistemic enterprise. Please do not misunderstand me: I actually agree with these radicals that the contemporary pursuit of analytic epistemology (especially of the sort championed by Roderick Chisholm, but most of Goldman 1986 as well) is best seen as the artificial continuation of Descartes' and Locke's seventeenth-century psychological theorizing. However, it is one thing to *identify* the errors fostered by such theorizing, but quite another to *eliminate* the practice that continues to grant legitimacy to those errors. For, to be truly naturalistic, one must start with things as they already are and work from there (cf. Dewey 1958). My fellow radicals often make it seem as though the replacement of epistemology by psy-

Incorporation

chology would occur "spontaneously" once people realized that the latter was the scientific successor of the former: i.e., epistemologists would simply start doing psychology or face extinction. On the contrary, I hold that this overintellectualizes the matter, as if one "good argument" could solve what is essentially a sociological problem. In a sense, my radical friends need to naturalize their conception of argument so as to make room for *burden of proof*, the rhetorical analogue of institutional inertia, which enables epistemologists to proceed unperturbed by the findings of empirical psychology (Fuller 1988a: 99-116, 1989a: 68-69). The eliminativist essentially has the rhetorical disadvantage of trying to persuade her audience that they ought to make a career shift!

But to make psychology rhetorically more palatable to epistemologists so as to encourage them to transform their practice is, at the same, to alter psychology itself. After all, psychology has come to have its character largely by having to define itself in relation to neighboring academic disciplines, such as philosophy and sociology. This has led psychology to strategically adopt and oppose developments in those other disciplines. For example, psychology has generally adopted the methodological individualism of the moral sciences and the positivism of nineteenth-century experimental physics. It is not clear that once these neighboring disciplines are transformed, psychology will have the need to continue in its usual manner. In this regard, *reductionism* is a better model for the naturalized epistemologist than *eliminativism* because traditionally reductionism has had a prescriptive thrust, namely, a call for, say, psychology and neuroscience to develop translation manuals between their two theoretical languages (Bechtel 1988: 82-87). In the course of developing such manuals, so the idea goes, the two disciplines will realize that they are talking about the same thing to such an extent that they can come to agree on a common tongue for future joint pursuits. In short, reductionism may be seen as primarily a program to synchronize the activities of conceptually neighboring disciplines by forcing them to communicate with each other. (I realize that I am strategically exaggerating this aspect of reductionism, since, in the final analysis, the reductionist, like the eliminativist, calls for the joint pursuit to be conducted largely in the terms of the currently "more scientific" discipline, such as neuroscience in the above example.)

TYCHO GOES SOCIAL–TOO LITTLE, TOO EARLY

But what happens when Goldman's Tychonism is socialized? This project has recently been undertaken by Goldman's student, Angelo

Two: Interpenetration at Work

Corlett. Although Corlett does a creditable job of bringing experimental work in cognitive and social psychology to bear on Goldman's project, he too fails to come fully to grips with what it means to "naturalize" and, for that matter, to "socialize" epistemology. In particular, he remains wedded to the classical conception of individuals as cognitive units, which prevents him from taking to heart the very empirical work he cites.

Corlett wants to avoid having to eliminate epistemology, yet he seems to think that the naturalist can simply adjudicate–presumably by special philosophical means–when psychology is or is not relevant to epistemology. In this regard, he is deviating only slightly from the Goldman line. However, if my argument has been correct so far, then it would seem that this line of reasoning presupposes a much more rigid sense of disciplinary boundaries than the naturalist is entitled to. But Corlett (or Goldman) might then ask me: If I disavow both eliminativism and his own more moderate alternative, what is the stance that I would have the naturalized epistemologist strike toward psychology? My answer is that the results of psychology should be applied reflexively to both psychologists and epistemologists, with the result defining the line of joint inquiry that the two currently distinct groups will subsequently pursue. This interpenetration of psychology (as well as the other social sciences) and epistemology will, in turn, enable a transformation of both into a single project.

Evidence for the current lack of interpenetration of epistemology and psychology comes in two forms: (1) In matters of philosophical reasoning (e.g., the reliability of introspectively based conceptual analysis), psychology does not seem to have progressed beyond the seventeenth century, yet state-of-the-art psychology is used to identify the appropriate knowledge-producing mechanisms specified by philosophical reasoning (cf. Goldman 1986: pt. 1 vis-à-vis pt. 2). (2) Philosophers use psychological findings more often to exemplify conclusions reached by "philosophical" means than to overturn such conclusions, the latter of which would be more in the spirit of using those findings as evidence.

Notice what the call to interpenetration does *not* entail. It does not entail that either epistemology or psychology has final epistemic authority over *its own* field of inquiry. Philosophical naturalists typically accord too much local sovereignty to the disciplines on which they wish to rely. This only serves to earn them the scorn of classical epistemologists, who deem the naturalists slavish followers of scientific fashion, bereft of all philosophical scruples. This objection is avoided by going "meta" and considering the consequences of applying psychology to the psychologists, something that they themselves

Incorporation

would typically not do. Perhaps the most important consequence of the reflexive application of psychology is to cast aspersions on the idea that a sharp distinction can be drawn between *individual* psychology (and epistemology) and *social* psychology (and epistemology), a distinction to which both Goldman and Corlett still adhere. Let us now turn to this point.

In one sense, Corlett hardly draws any distinction between individual and social psychology (and epistemology), since he is ultimately concerned only with functionally equivalent individuals. This should come as no surprise, seeing that both contemporary American experimental psychology and analytic epistemology are committed to methodological individualism, even in their accounts of the social. Thus, for Corlett, a "social psychology" or a "social epistemology" is "social" only in the sense that one is studying the social knowledge of *individuals*, or, in more down-to-earth terms, what people think about each other. Moreover, Corlett assumes that social knowledge is uniform across individuals, as if no epistemologically salient *differences* in social knowledge could arise from differences in, say, the class background or role expectation of individuals (hence, Corlett's neglect of ideology as a psychological phenomenon; cf. Doise 1986). Indeed, I imagine that if Corlett were to make social epistemological policy, whatever that advice turned out to be, it would be the *same* for everyone. Once again, he would be reifying a particular way of interpreting the object of experimental psychological research that has been subject to much controversy in the course of this century: Is the goal of psychology to codify the laws governing *the* normal individual or those governing the normal variation in behavior among individuals? In America, the battle was fought between Edward Titchener and James Mark Baldwin. The former, one of Wundt's original students, treated individual differences in experimental trials as "errors." The latter, influenced by the French interest in pedagogy and psychiatry, treated such differences as necessary for the smooth functioning of the social organism (Boehme 1977).

However, I would argue that the first principle of a truly socialized epistemology is that everyone should *not* be given the same epistemic advice—or be expected to take the same advice in the same way. For, if one takes seriously the idea that knowledge is a social product—i.e., the product of a certain pattern of human interaction—then it is no longer necessary to think about individuals as having common cognitive powers and interests. Rather, what is required is that the individuals' different powers and interests function together so as to collectively produce a form of knowledge that is knowledge for the whole commu-

Two: Interpenetration at Work

nity, even though no single individual could be expected to have mastered all of its parts. Indeed, the collective identity of disciplinary knowledge is more a moral, perhaps even an emotional, commitment on the part of the scientific community to accept joint responsibility for the work of any of its members than a strictly cognitive achievement. When philosophers talk about the distinctive products of science–theories (Hempel), paradigms (Kuhn), research programs (Lakatos), research traditions (Laudan)–a moment's reflection reveals that they are referring to epistemic units that could not possibly be stored in any single individual's head or, arguably, even in a single book that an individual could be expected to use with facility. Instead, these products are distributed in parts across an entire scientific community. For example, in order for a subfield of physics to become part of the physics knowledge base, many theorists, experimenters, and technicians need to be involved in research, but no one of them would claim to understand all the inferential chains that forge the subfield. It is important to underscore that the problem here is not simply that a physicist's memory is not large enough to store all the knowledge produced by her speciality.

The problem runs deeper. For, even assuming that the physicist could chunk the knowledge of her field into a manageable size, she would still be unreliable in drawing the relevant deductive and inductive (i.e., probabilistic) inferences that together turn this information into a cognitive map of some domain of inquiry. To appreciate the significance of this point, consider that the *smallest* epistemic unit that philosophers have typically found distinctive about science–the theory–is epitomized by a formalized version of Newtonian mechanics, from which the physicist is expected to calculate indenumerably many deductive inferences from factual premises about the motions of the planets in conjunction with universal physical principles. Faust (1985) and Cherniak (1986) have discussed the impossible computational load that this places on the physicist. Thus, what physicists share are little more than bonds of mutual trust and identification as, say, "solid state physicists." My point, then, is that a truly social epistemology would take the "cognitive division of labor" idea to run much deeper in our cognitive makeup than Corlett would allow.

But how does the reflexive application of psychology encourage this turn to the social that Corlett seems to lack? Let me start with the intuitive clarity of individuals as the principal causal agents in the social order. In classical epistemology, such intuitive clarity is often enough to underwrite the legitimacy of the units that one takes as ultimate. But after experimentally studying the limits of these intuitions,

Incorporation

psychologists have come to reinterpret them as heuristics, i.e., cognitive biases that work well enough in most ordinary situations but poorly beyond them. A heuristic that is relevant in this case is the tendency to attribute causation to objects that move freely against a background in one's visual field. Perhaps the main use to which we put a concept of causation in everyday life is to coordinate our actions in relation to other things in the immediate environment. The things deemed "causal" are the ones whose movements are likely to make some difference to what we decide to do, and these are typically the things that most readily catch our eye (Kahneman 1973). In such ordinary settings, we generally have no need to speculate about whether there is anything more to the object's motion than the history of its interactions with the environment, or whether the object's motion is synchronized with the motions of other visually occluded or distant objects. However, these speculations become relevant once we start wondering whether what we see is all that there is, i.e., whether individual objects are the right units for thinking systematically about reality.

Our intuitive notions about causation are ill suited for satisfying a metaphysical impulse of this sort, yet without an appropriate theory to act as corrective, these notions function as a default theory that biases our thinking toward treating individuals that move freely in the visual field as having some kind of metaphysical ultimacy. For example, naturalized studies of science have tended to see the scientist as an agent who makes things happen in the world by exercising her intrinsic powers (Giere 1988 comes close to this view). This view directly plays to our cognitive biases by converting the palpable fact that scientists freely move about the lab into a sign that they are self-moving, or autonomous, beings who can be held personally responsible for what they do. Of course, no one assents to this view in quite so bald a form. Yet it is the one we naturally fall back on when evaluating science: if a discovery is made, a scientist is credited; if fraud is committed, a scientist is blamed. We may nod sagely that these events are "strictly speaking" the systemic effects of class, status, and power (to name Max Weber's favorite threesome) acting "through" the scientists, but this learned opinion does not make the first perspective on the scientist any more intuitively compelling. Again, the bias at work is to think that we must be able to see something move something else before the first thing is called a cause: people move apparatus, but as far as the eye can see, class struggle doesn't move much of anything in the lab.

Because our understanding of science proceeds fairly smoothly with this bias in place, it will take the wiles of a more reflexive naturalist, the social epistemologist, to throw a spanner in the works. For exam-

Two: Interpenetration at Work

people find it hard to see the need to postulate power differences in order to explain a single transaction that might be observed between two people. They ask: Why not simply invoke the intentions of the specific individuals involved, and avoid reference altogether to an occult entity like power? The plausibility of power as an explanatory principle grows with an awareness that many such transactions occur in many places and times that are systematically interconnected by counterfactually realizable situations (e.g., if one party does not conform, then the other party can impose force), the entirety of which transcends the intentions either of any of the constitutive individuals or of any given observer of a particular transaction. But all that is just to say that one has to stop using the limits of one's visual field as the intuitive measure—or metaphor, if you will—of explanatory adequacy for social action (cf. Campbell 1974). A better image would be to regard the scientist as a body whose movements are the result of a variety of forces that have been imparted in the course of its interaction with other such bodies. Although we can see no strings attached to a scientist, we can see in her behavior the marks left from her interactions with various teachers, colleagues, etc. In this context, it is tempting to say that what makes each scientist distinct is simply the uniqueness of her history of interactions.

TYCHO GETS BLINDSIDED BY THE REAR GUARD

Although I have portrayed Corlett as having only made tentatively Tychonic steps toward naturalistically socializing epistemology, it doesn't follow that he is completely protected from *rearguard* objections—in particular, that the social character of science and other epistemic enterprises is a "mere contingency" that bespeaks nothing deep about the nature of these enterprises. Admittedly, Corlett could play a naturalistic card and argue against the classically nonempirical cast of this objection, as the objector seems to suggest that nothing in the *concept* of science or knowledge requires that it be social: But aren't concepts themselves sociohistorically variant constructions? Still, Corlett would want to say that not just any old fact about science is epistemically salient. How, then, would Corlett show the salience of the social? I have argued here and elsewhere (Fuller 1988a, 1989a) that the best opening gambit is to devalue the cognitive powers of the individual. Indeed, taking a cue from Karl Popper, I would go so far as to claim that the ever restless (or "progressive") character of our epistemic pursuits would be undermotivated if, as individuals, we were not innately endowed with trenchantly *false* ideas that require long-term system-

Incorporation

atic effort to overcome. In short, while science may not require human beings for its conduct, it does require beings whose cognitive biases and limitations are comparable to those of humans, and who then see in science a way of collectively transcending their finitude as individuals.

Exhaustive as they may seem, the above considerations are unlikely to move our next Tychonist, Rom Harré whose faith in the cognitive competence of humans runs deeper than Goldman's or Corlett's. For just as Tycho was moved, in part, by a respect for commonsense intuitions about the stationary character of the earth, so too Harré is moved by a respect for the richness of ordinary usage of folk psychological concepts, a richness that is typically overlooked by the flagship discipline of naturalism, experimental psychology (Harré and Secord 1979; cf. Greenwood 1989 for a sophisticated elaboration of the Harrean position, in the face of recent defenses of experimentalism). Experimental psychology seems to be the heir apparent preferred by naturalized epistemologists because its typical unit of analysis, the interface between an individual organism and its environment, most closely resembles the setting in which the problem of knowledge of the external world was classically posed by Descartes (Quine 1985). But before we respect those wishes, it is worth mentioning that it is not clear that experimental psychology would loom so large for naturalists who focused more on modeling the problems of theory choice and conceptual change that have typified debates in the philosophy of science. In that case, as even the positivist Hans Reichenbach (1938: 3-16) realized, the sociology of knowledge would be a more suitable "naturalization."

According to Harré and Secord (1979), "aggression" is not the convergence of a couple of empirical indicators in a laboratory subject, but a deep-seated human disposition that may be elicited in a variety of ways under a variety of circumstances—the sum of which may be explicated by a conceptual analysis of ordinary language and observed in the so-called natural settings of everyday life. Since the bulk of Harré's work over the past ten years has been devoted to articulating the "ethogenic" paradigm in social psychology, it might seem as though he is more inclined to naturalism than Goldman, who still argues largely with classical epistemologists and does not seem to have ever altered any of his fundamental tenets in light of psychological evidence. This is not to deny that Goldman has modified aspects of his reliabilism in light of *conceptual* considerations (cf. Goldman 1989; Symposium 1989). In this regard, Goldman is in prestigious company, since the sequence of revisions that Noam Chomsky has made to his theory of generative grammar is better explained as responses to conceptual demands of simplicity than to the empirical demands of recalcitrant psychological

Two: Interpenetration at Work

data. Unfortunately, this stress on the conceptual at the expense of the empirical also explains "the rise (and surprisingly rapid fall) of psycholinguistics" (Reber 1987).

However, just as Goldman's naturalism is mitigated, so too is his anti-naturalism: he cites particular experiments when they serve his purpose, he discredits other experiments when they do not, but he does not call into question the appropriateness of the experimental method to the empirical study of human beings. It might simply be (so says Goldman) that not everything the psychologist does is relevant to the normative mission of epistemology. By contrast, Harré wants to recolonize psychology for the version of classicism represented by ordinary language philosophy. This means that the deliveries of conceptual analysis are the primary data of psychology, to which empirical research must conform accordingly. In fact, this is the first methodological dictum that Harré and Secord (1979) lay down. That Harré has not changed his tune in recent times may be seen in the following:

> The science of mechanics made rapid advances after careful and detailed analysis of the concept of "quantity of motion" had revealed the need for a distinction between "momentum" and "kinetic energy." These conceptual distinctions did not emerge from experimental studies. They were arrived at by analysis. Once achieved they facilitated a more sophisticated and powerful empirical science of bodies in motion. (Harré 1989: 439).

In practice, Harré abandons laboratory experiment for the sort of "onsite" ethnography commonly pursued by anthropologists, which provides the interpretive freedom needed to plumb the putative depths of human expression codified by ordinary language.

I do not mean here to cast aspersions on the contributions that ethnographic inquiry can make to the human sciences, for it may do so even in ways that would cater to the experimental proclivities of the more robust naturalist–not to mention the STS practitioner, as will become clear at the end of this chapter. In particular, I have in mind versions of ethnomethodology inspired by Harold Garfinkel's work, such as "experiments in trust," in which the inquirer tests the extent of normative constraint by disrupting the "naturalness" of the settings in which a norm ordinarily operates. As the naturalist would have it, experimental intervention is a precondition for the norm to be represented (cf. Turner 1975). By contrast, the use to which Harré puts ethnography–namely, as providing instantiations of empirically unrevisable folk psychological concepts–removes that method from the arena of

Incorporation

hypothesis testing, and hence from an inquiry that can be properly called "naturalistic." It is worth dwelling on this point briefly because I will draw on it again in the course of overcoming the canonical form of the classicist-naturalist exchange.

Although philosophers commonly talk this way, it is misleading to say, without qualification, that naturalists are devotees of the experimental method, whereas classicists prefer conceptual analysis. This way of talking makes it seem as though the two sides must be engaged in mutually exclusive activities, as epitomized by the typical locations in which these activities occur, the laboratory and the lounge chair. What happens in the laboratory is supposedly *a posteriori*, whereas what transpires in the lounge chair is *a priori*. However, to put matters in this way only revives the dogmas of empiricism so as to make classicism and naturalism seem more irreconcilable than they need be. Luckily, the history of science is a ready source of counterexamples to stereotypes that harken back to a world well lost (i.e., before Quine 1953).

On the one hand, experiments have been conducted in the name of the *a priori*. That is, experiments have been used as a means of providing concrete demonstration of truths derived by conceptual means. This attitude toward experiment was typical of those seventeenth-century thinkers whom we now call "philosophers," such as Descartes and Hobbes (Shapin and Schaffer 1985). (Boyle and Newton are usually credited with turning scientific opinion toward regarding experiment as a genuine, and even preferred, source of knowledge, rather than as an illustrative device of some incidental heuristic value; cf. Hall 1963). And it is in this spirit that Harré mobilizes the ethnographic method for his brand of social psychology.

On the other hand, as will be elaborated below, conceptual analysis has been used to arrive at eminently falsifiable empirical hypotheses, and thereby forward the cause of the *a posteriori*. The real difference between the "apriorism" of the classicist and the "aposteriorism" of the naturalist lies not in the kinds of activities each pursues, but rather in the degree to which each is inclined to revise her claims in light of unintended or unexpected outcomes of those activities. In that case, if Imre Lakatos is right, there is a classical epistemologist lurking in the metaphysical hard core of every scientific research program. What typically makes conceptual analysis the mark of the classicist is the control that the analyst has over her introspections, so that, like Descartes, a certain private illumination ultimately determines that the analysis can be revised no further. However, if Descartes had believed that he needed a second, and potentially

Two: Interpenetration at Work

overriding, opinion to evaluate his introspections, then he would have been doing conceptual analysis in a naturalistic vein. In fact, as we will now see, there is a discipline that systematically offers such second opinions, ethnosemantics (cf. Amundson 1982; Lakoff 1987).

TYCHO SANS CLASS(ICISM)

Let us now return to our naturalistic rejoinder to Harré: What would it mean to employ ethnography "naturalistically" to test a particular analysis of folk psychological concepts? To get a flavor for the nature of this enterprise, consider how an ethnosemanticist might "analyze" the folk concept of aggression. She would proceed by surveying the usage of *aggression* in a particular language–say, American English–and quickly observe the variety of contexts in which it arises. To these particular facts about the word's usage, she would then add general empirical facts about natural languages, especially facts pertaining to words used in contexts too numerous to be monitored for mutual compatibility and propriety. Given this information, the ethnosemanticist would be inclined to conclude that the deep-seated disposition that Harré sees lurking beneath the multifarious character of aggression-talk is a mirage: to wit, homonymic drift passing for synonymic stability (cf. Fuller 1988a: 117-38; in fairness to Harré, it should be said that there has been a recent turn to a more "ecologically valid" ethnosemantics, in which polysemy is taken as an indicator of some measure of conceptual depth, as in Lakoff 1987).

Notice the anti-Tychonic character of this rejoinder. Harré presumes that ordinary agents already have a reasonably reliable introspective understanding of their own minds. In interesting counterpoint, Goldman (1986: 66) restricts such self-understanding to the judgments that philosophers make in "reflective equilibrium." (L. J. Cohen 1986 provides an extended defense of this point, in the aid of establishing a distinct subject matter for analytic philosophy.) In any case, Harré seems to believe that we are entitled to his presumption because he further presumes that self-knowledge is essential to our routinely successful encounters with each other and the world. The fact that we ordinarily tend to associate polysemous words with conceptual depth is taken to be a good starting point for Harré's ethogenic inquiry. Clearly, our ethnosemanticist presumes nothing of the sort (cf. Fuller 1988a: 139-62, for a defense of uncharitable interpretive principles). Yes, speakers of a natural language provide a privileged data base for the study of word usage, but their second-order musings do not provide a privileged data base for the interpretation of those data. The second-order musings–

Incorporation

what I make of the multifariousness of my aggression-talk–is just more first-order data for the ethnosemanticist to study. Why? From a naturalistic standpoint, any given individual is a biased source of information about its own activity because of the disproportionate amount of data that the individual records about itself (usually for its own purposes) vis-à-vis the amount that it records about other relevantly similar individuals. Consequently, under ordinary circumstances, an individual will have an inadequate basis for judging the representativeness of its self-reports. This bias is manifested in people's tendency to ignore what probability theorists call the "base rates" of some phenomenon's occurrence (i.e., the likelihood that something will happen, given its track record) when making predictions (Kahneman et al. 1982). For the sake of argument, I have ignored the point that much of the bias in the data that an individual records may be attributed simply to flaws in the data recording device itself (i.e., memory).

The ethnosemanticist has the interpretive advantage of the third-person perspective, which enables her to compare that individual's utterances with those of others. Needless to say, this point also applies to the naturalized interpreter's own behavior: it too is itself best studied from the third-person perspective, which recalls the joke about two behaviorists greeting each other. One says to the other, "You're OK. How am I?" The ultimate trick, however, for any naturalized interpreter is to determine exactly what the data provided by a speaker's utterances are best taken as *evidence for* patterns of neural firing, sentences in the language of thought, socially constructed contexts, or objective states of affairs? An entire branch of experimental psychology is now devoted to interpreting "verbal reports as data." Ericsson and Simon (1984) see the interpretation of verbal reports as a matter of identifying the sort of data that is regularly registered by human speech, or in more behavioral terms, a matter of determining the factors that control verbal emissions. In this respect, the project is in the spirit of the "radical translation episode" in Quine (1960), except that Ericsson and Simon do not presume that their interpretation is constrained by the need to make most of a speaker's utterances turn out to assert truths, or even reasonable beliefs.

The Tychonic Naturalist sees only three possible outcomes to our debate: the classicist wins, the naturalist wins, or there is a mutual accommodation that enables the peaceful coexistence of both sides. As we have seen in the case of Goldman and Harré, the Tychonist favors the third option, which is achieved roughly by gauging how much naturalism a classical epistemology can absorb and still be recognizably philosophical. However, the crucial outcome that is missing here is the pos-

Two: Interpenetration at Work

sibility that the two sides may be transformed in the course of debate so that each incorporates in its own terms the issues raised by the other side. This possibility is so tricky that it requires that we briefly resurrect the ghost (or *Geist*, I should say!) of Georg Wilhelm Friedrich Hegel.

Hegel to the Rescue

A MATTER OF PRINCIPLE

A supporter of naturalized epistemology cannot help but be struck by the dialectical disadvantage in which naturalists typically find themselves. Part of this disadvantage may be explained in terms of how the burden of proof is distributed in the classicism-naturalism debate, since, after all, the naturalist is the Johnny-come-lately. However, just because the naturalist must bear the burden of proof, it does not follow that she must do so by confining herself to the types of arguments used by her classicist opponent. Yet what one all too often finds in following this debate—and as will be dramatized below—is that the naturalist succumbs to appeals to conceptual analysis, transcendental arguments, and commonsense intuitions. In these sorts of arguments, she tends to be no match for the expert classicist.

By contrast, the classicist rarely slips into naturalistic appeals for her own position. The clearest exception, when the classicist does wax naturalistic, is when she defends the classical mission of providing foundations for knowledge in terms of facts about the history of philosophy—as if the fact that people have associated epistemology with the classical version of the project for over 350 years somehow contributes to the *conceptual* well-foundedness of the enterprise. If the classicist wanted to pursue that naturalistic line of reasoning consistently, then she should also point out that even today, if a survey of professional philosophers were taken, most would agree that epistemology is most closely tied to the classical project begun by Descartes. But, I imagine, the classicist would deem a fact of that sort too "sociological" for it to count as genuine support for her own position.

The upshot of the preceding considerations is that the classicism-naturalism debate would benefit from a certain methodological consistency. Naturalists should argue naturalistically, and classicists classically. They should neither be forced to argue in ways that contradict their metaphilosophic principles, nor be allowed to tailor their oppo-

Incorporation

nents' metaphilosophic principles for their own purposes. In other words, I am proposing the following procedural rule:

> *The Principle of Nonopportunism*: When either defending her own position or attacking her opponent's, the philosopher must employ only the sort of arguments that her own position licenses. She cannot avail herself of arguments that her opponent would accept, but that she herself would not.

There is a *constitutive* and a *regulative* version of this principle. The constitutive version says that nonopportunism enters into the very construction of the position taken in debate, such that if I want to hold my opponent accountable to certain standards, then I had better be sure that I can be held accountable to them myself. By contrast, the regulative version of the principle presumes that the two positions were constructed independently of each other, prior to the debate. In that case, nonopportunism circumscribes the field of appropriate engagement between the two positions. In light of the first two chapters' elaboration of my views on the conventional character of disciplinary boundaries, the centrality of interpenetrative rhetoric, and, most of all, my "normative constructivism," I generally prefer the constitutive version of nonopportunism.

The "opportunists" whom I have in mind–they who routinely violate the principle–are stereotyped sophists, classical skeptics, and sometimes reflexive practitioners of STS (more about which in Chapter 9). All are prone to throw their interlocutor's favorite form of argument back in her face without feeling compelled to engage that form themselves. An example of opportunism would be for a philosopher to cite the empirically based disagreements between various schools of psychology as an argument against endorsing the findings of any of the schools, when in fact the philosopher herself does not believe, as a matter of principle, that the data could resolve such theoretical disputes. Metaphysically speaking, the opportunists follow in the footsteps of the Sophist Gorgias, in that they share a fundamental mistrust of communication as a process that can dissolve the incommensurable presumptions that invariably separate people in the first moment of encounter. Heirs to Gorgias are thus opportunists because they believe that if common ground is not present a priori, then it cannot be forged a posteriori. Given this Hobson's choice, it is not surprising that Gorgias' most dogged opponents–from Socrates to Habermas–have argued that common ground is present a priori, be it in a realm of universally communicable forms or in a set of transcendental conditions for pragmatics.

Two: Interpenetration at Work

The point of the interpenetrative rhetoric promoted in this book is that we need not let Gorgias dictate the terms of the debate any longer. We can grant that there is no (or very little) common ground at the start of an exchange, but, at the same time, maintain that that common ground can be built through a nonopportunistic argumentation procedure.

Returning to the debate at hand, nonopportunism places some interesting constraints on permissible moves in arguments between advocates of classicism and naturalism. Two are worthy of note here. For starters, as far as dialectical resources are concerned, nonopportunism prevents the classicist from turning to her advantage the naturalist's arsenal of historical and scientific findings and methods. Likewise, the naturalist must steer clear of the classicist's repertoire of conceptual analysis, a priori intuitions, and transcendental arguments. Admittedly, the difference between these dialectical resources often boils down to matters of presentation, as many of the same points that can be made by, say, appealing to a priori intuitions can also be made by appealing to scientific findings. This last point turns out to have more metaphilosophic significance than it may first seem, which brings us to the second, subtler point.

So far I have argued only that adherence to the principle of nonopportunism would promote a fair debate between the classicist and the naturalist. But I have also claimed that it would have the epistemologically deeper consequence of dislodging the two sides from their current dialectical impasse. To see how that might happen, let me introduce a term of art, *Hegelian Naturalism*, to describe the strategy of articulating classical epistemological concerns within the dialectical constraints available to the naturalist. To play the epistemological game by Hegelian rules is to ask which side is more effective at transcending the difference in perspective that the other side poses: Who is the better synthesist? Notice that this question presupposes that the two positions in the debate have been clearly disentangled from one another–as "thesis" and "antithesis," if you will–such that the terms of disagreement are appreciated by both sides. However, the main problem with the classicism-naturalism debate is that the two sides tend to argue at odds with their respective positions, which, in turn, suggests that the terms of disagreement between them have yet to be properly identified. In that case, it may be useful, as *propaedeutic* to debate, for each side to catch the other in self-contradictions or "immanent critiques." These preparatory practices would be nonopportunistic precisely to the extent that they are meant not to silence the opponent but to enable her to articulate her position more clearly.

Incorporation

The need to make one opponent's position dialectically tractable is especially pressing in the case analyzed below, in which the classicist (Clay) must help the naturalist (Nate) tease out his own position before the naturalist can properly incorporate the classicist's objections in an attempt to transcend the terms of their disagreement. My interest here will be in playing the naturalist's hand in this Hegelian game. But first we need a canonical formulation of the dialectical rut that gives rise to the need for the type of rapprochement I have sketched. What follows is an all too typical exchange between Nate and Clay over the metaphilosophic soundness of naturalized epistemology. It follows fairly closely two recent classicist-naturalist clashes in the journal *Studies in History and Philosophy of Science* (Siegel 1989 vs. Giere 1989; Siegel 1990 vs. L. Laudan 1990b).

> NATE: Epistemology–or at least philosophy of science–is viable only as a science of science.
> CLAY: But what's so philosophical about that?
> NATE: We need to explain how science has enabled us to learn so much about the world.
> CLAY: But that presupposes that science does give us knowledge. But how does one *justify* science's claim to knowledge? That's the philosophical question you need to address.
> NATE: I don't know: you classicists have been going at it now at least since Descartes–and to no avail.
> CLAY: But all that shows is that *you* are frustrated and, hence, want to change the subject. You haven't actually shown that an epistemic justification of science is impossible.
> NATE: Look, your whole way of talking supposes that epistemology is autonomous from science. In fact, I wouldn't be surprised if you thought that epistemology was *superior* to science!
> CLAY: My private thoughts are not at issue here. All I want to argue is that epistemology must be pursued *apart* from science, if science's epistemic legitimacy is to be judged without begging the question.
> NATE: But there are no categorical epistemic principles that establish science's legitimacy. There are only instrumental principles that tell us the most efficient course of action relative to a given end.
> CLAY: But aren't there ends of science per se? And how are they justified? Doesn't that bring us back to my original concern?

Two: Interpenetration at Work

> NATE: There has been only one end in common to the multitude of ends that have led people to pursue science throughout the ages: namely, an interest in finding out what the world is like. But in any given historical case, how the scientist proceeds to find out what the world is like will depend on the other ends that she is pursuing at the same time.
> CLAY: But at most that explains particular local successes of science, not the global success that you allege underwrites the epistemic legitimacy of science.
> NATE: Well, I never said that the science of science had to be purely descriptive. After all, the cumulative instrumental successes of science strongly suggest that we have managed over the centuries to achieve a more general understanding of how the world works. Indeed, the point of proposing *theories* in science is to capture the nature of our understanding. Moreover, once articulated, theories can be used to inform future action.

By the end of the sixth round, Nate has been once again brought to saying that epistemology is only as well grounded as the science it grounds, to which Clay will no doubt reply that that is not grounding enough. And so it would seem that we have returned to the start of the exchange, each side neither deepening his own position nor budging his opponent's. What is keeping the debate in such a rut? My suggestion is that the naturalist continues to fall into dialectical grooves largely of the classicist's making.

These grooves run deep. Take the very thing that Nate and Clay are trying to justify and/or explain. To keep the debate somewhat focused, I have had both sides characterize this thing as "science." On the surface, this would seem to bias the discussion in favor of the naturalist because the word "science" signals a sociohistorically specific form of knowledge (one begun, say, in seventeenth-century Europe) that makes a point of refusing to rest on its epistemic laurels. The scientific call for the repeated testing and revising of knowledge claims goes against the classicist's interest in establishing an intuitive or conceptual terminus to inquiry. But while this captures the epistemic stance of the scientist, it is not clear from Nate's remarks that it captures *his own* stance toward science. For Nate, science suspiciously partakes of some of the properties that Clay wants to attribute to knowledge. In particular, science does not seem to be an entity clearly bounded in space and time. A telling point here is Nate's apparent indifference to whether he is talking about science as a body of knowledge, a cognitive process, a group of people, or a single individual scientist.

Incorporation

Moreover, assuming that the precise unit of analysis can be pinned down, Nate fails to clarify whether what impresses him as worthy of justification and/or explanation is how that unit operates on a day-to-day basis, only on exemplary occasions, or cumulatively over the long haul (starting when?). This point is related to Nate's failure to see science as itself something whose epistemic legitimacy may change over the course of its own development. For example, if Nate followed Karl Popper (1970) in holding that science is epistemically impressive only during its revolutionary phases, then his attitude toward everyday science would not be too far from Clay's. Both would then bemoan the normally unreflective attitudes that scientists display toward the epistemic foundations of their enterprise. Nate and Clay would, of course, continue to diverge over whether there could be more to epistemology than relentless self-criticism, but Nate at least would begin to see that Clay's lingering doubts about the epistemic legitimacy of science is based on something more than mere philosophical one-upmanship.

Before we consider how Nate might begin to incorporate Clay's concerns into his own epistemological horizon, it is worth lingering over just how incommensurable their starting points might be in understanding the relationship between what has been historically identified as "science" and "knowledge." Given the weight that Nate gives to the instrumental success of science, we can easily imagine him telling a story of science emerging as a *by-product* of our biological need to solve problems. I stress "by-product" because, on this view (associated with both Dewey and Popper), "science" is the repository into which ideas and techniques enter once they have been crafted to solve particular life problems. "Scientists," then, are people who have the leisure to develop a discourse that interrelates these artifacts, especially so as to reveal ways in which the achievements of some of the artifacts overcome the limitations of others. This discourse—which is really all that would interest Clay in the story—is the one whose utterances are routinely evaluated as being "true" or "false." Now, by believing this story, Nate is in a position to have any of the following attitudes toward the relation of "science" and "knowledge":

> 1. Nate may think that pursuing science for its own sake is the best way to increase human problem solving ability, even though it involves an indirect route. Knowledge, still defined as problem solving, will thereby increase. In that case, the role of science in our pursuits will have changed from mere by-product to explicit aim. (This captures the spirit of Popper's

Two: Interpenetration at Work

[1972] "evolutionary epistemology.")

2. Nate may think that science is worth pursuing only insofar as it contributes to human problem solving ability, which is judged by welfare standards that are independent of those used to judge the progress of pure science. In that case, an overzealous pursuit of science could lead us to produce "useless truths" that do not deserve the title of knowledge. (This captures the "finalizationist" school of philosophers of science who follow Habermas [cf. Schaefer 1984].)

3. Nate may hold a historically informed combination of (1) and (2). At first, (1) was a good strategy. Unfortunately, since 1945, the magnitude of science has made its pure pursuit a very uneconomical way of addressing human problems; hence, the turn to (2). The exact turning point came when it became impossible to tell whether a particular scientific claim was true or false without deploying enormous amounts of resources to create an artificial environment for testing the claim. (This is in the spirit of Feyerabend's [1979] call for downsizing the scientific enterprise.)

Notice that none of these attitudes views either science or knowledge as existing in a vacuum for all times and places. There is even the suggestion that what Clay calls "knowledge," while relevant to the discursive development of science, may not be particularly relevant to what Nate calls "knowledge," especially once the pure pursuit of science is called into question, as in (2) and (3). However, in what follows I will minimize the level of potential incommensurability by having Nate adopt (1) as his attitude toward knowledge and science.

THE PRINCIPLE IN PRACTICE

Suppose now that Clay were asked what had transpired in his exchange with Nate. He would probably say that out of frustration with classical epistemology, Nate surrendered his philosophical scruples for a crypto-intuitionism, one which enabled him to intuit an omnibus conception of science as the best form of knowledge. Thus, whereas Nate would think of himself as having changed the rules of the epistemological game, Clay would argue that Nate has merely slid into the dialectically least tractable position in the classicist's game, the proffering of intuitions. Of course, the classicist has at her disposal the means

Incorporation

with which to call intuitions into question, namely, by challenging their "clarity" and "distinctness." Clay might therefore ask whether Nate's conception of science is internally consistent, and if so, whether it can be distinguished from other conceptions of knowledge. As we have seen, the ontological dimensions of Nate's "science" are somewhat uncertain, which initially make it difficult to assess the clarity and distinctness of his conception. But according to the principle of nonopportunism, we should not expect Nate to be impressed by Clay's attempt at a Cartesian diagnosis. Yet, if Nate were now to give his naturalism a Hegelian twist, he might come to see in his own terms what Clay means by treating the naturalistic conception of science as an unanalyzed intuition.

Empirically speaking, "science" is a cluster of disciplines that includes at least all of the natural sciences and probably most of the social sciences as well. It is fair to say that all of these disciplines are interested in "how the world works," and consequently each has its preferred surrogate for truth, or the ultimate end of inquiry. Newtonian mechanics gave science the truth surrogate of parsimony: that which explains the most by the least. Darwinian biology provided the truth-surrogate of long-term survival, which has enjoyed recent popularity among naturalized philosophers of mind (e.g., Dennett 1987: especially 237-322). Among the truth surrogates inspired by the social sciences, welfare economics has offered the greatest good for the greatest number, while electoral politics has contributed a variety of consensus models. In their studied avoidance of transcendental conceptions of truth, naturalists have typically alternated between these surrogates, as if they were functionally equivalent, or at least converged at the limit of inquiry. An early naturalist like Charles Sanders Peirce (1955: 361-74) seemed to believe that the simplest theory was the one with the highest survival value and the one that would command the consensus of inquirers, whose lives would, in turn, be made better off by accepting the theory than by accepting any of its alternatives. However, if Nate extends his naturalism to his understanding of the history of science, then he must admit that the disciplines responsible for these surrogates arose and have been maintained under circumstances that cast doubt on the claim that their "ends" are in lockstep with one another. For example, if a politically inspired naturalist claims that truth is consensus (or that a proposition is true because it enjoys the consensus of scientific opinion), a biologically inspired naturalist can respond that theories have been known to survive for long periods as the source of productive research, even though they never held most scientists in their sway. Indeed, biologically speaking, those theories may be taken to

Two: Interpenetration at Work

have avoided the excess of "overadaptation," whereby a species loses its dominant status once its hospitable environment changes slightly.

Clay would jump on this last point as evidence for the unclarity and indistinctness of Nate's conception of science. If only for the sake of conceptual coherence, Nate will have to be more selective in his endorsement of science. He can't have both survival and consensus as truth surrogates if they grant epistemic legitimacy to different theories. In that case, asks Clay, how does Nate decide between biology and politics as models of the knowledge production process? For Clay, this signals the need to transcend disciplinary differences and to appeal to a more global sense of epistemic legitimacy, one quite familiar to, and contestable by, the classicist. At this point, some naturalists (e.g., Bhaskar 1979) turn classicist by appealing to transcendental arguments. However, playing by Hegelian rules, Nate need only see the point that Clay raises without actually capitulating to classicism. He can do this by concluding from what Clay has said that "science" does not pick out a natural kind of knowledge, an epistemic essence common to the natural and social sciences (Rorty 1988). That would certainly explain the "incoherence" that Clay sees in Nate's conception. Moreover, the nonnaturalness of science is a problem only if Nate expected that all the epistemic virtues would line up behind one theory at the end of inquiry. That truth emanates from one source to which all inquiry then aspires is a Platonic residue in classical epistemology that has often been uncritically naturalized, as in the case of Peirce, via a "convergentist" account of the history of science. Nate can simply reject such a picture and argue that the epistemic superiority of science to other forms of knowledge rests on the character of the tradeoffs that it makes from among the cluster of virtues exemplified by the different truth surrogates. Thus, whereas, say, certain monastic religions value the long-term survival of their beliefs at the expense of all the other epistemic virtues, science trades off survival for some other truth surrogate, such as parsimony, after a certain point (cf. L. Laudan 1984). The history of science is, after all, full of theories that were overturned by the complications forced upon them by mounting anomalies.

In short, Nate can escape backsliding into classicism by portraying the world in which knowledge is produced as one that cannot afford to have all the epistemic virtues or truth surrogates jointly maximized. It is not merely that human beings are unable to jointly maximize the virtues because of, say, deep-seated incompetence, but rather, the end of knowledge has itself become so internally divided in the course of its pursuit that one can do no better than be an "epistemic satisficer" (Giere 1988: 141-78; cf. Fuller 1989a: 42-49). And while Nate has now finally

Incorporation

relinquished the classicist's ideal of one best theory on which all knowledge can be grounded, he manages to shore up what Clay feared was generally lacking from naturalized epistemologies, namely, a robust normative orientation that is potentially critical of current epistemic practices. After all, the history of science has only generated the differences among truth surrogates, differences that, left to their own devices (i.e., without the intervention of the epistemologist), are likely to grow with increasing disciplinary specialization. This is a far cry from the original exchange, in which Nate made it seem as though there were only two ways of thinking about the "ends of science," either a very generic end that is associated with science "per se" or very specific ends that are associated with the personal goals of the people who pursue science. As Clay then suggested in response, the former plays into his own understanding of science, whereas the latter amounts to simply accepting at face value the reasons why scientists do what they do.

The history of science creates the need for epistemological intervention as first-order empirical knowledge becomes the basis for disparate second-order conceptions of the ends of knowledge. This argument, which we have just fashioned for Nate, has a character that John Dewey (e.g., 1960) would have recognized as paradigmatically naturalistic. It supposes that the need to make value judgments arises from concrete exigency, with the judgments themselves evaluated by the exigencies to which they then give rise. Yet it should by now come as no surprise that a contemporary naturalist like Nate diminishes the force of this argument by unwittingly presuming part of the classicist's position. The telling point here is Nate's cherubic attitude toward the instrumental success of science.

In his final round with Clay, Nate seems to claim that if applying a certain theory has the consequence of increasing our control over nature, then that theory can be automatically credited with success, which, in turn, earns the theory a place in the storehouse of human knowledge. The *post hoc, propter hoc* fallacy in that line of reasoning is easily spotted. But still worse, from the standpoint of a naturalist, is Nate's failure to evaluate the consequences of the theory's application in terms of the exigencies to which it gives rise, that is, the potentially problematic by-products—process, opportunity, and transaction costs—that are brought about alongside the theory's "instrumental success." (For economists, *process costs* and *opportunity costs* refer to, respectively, the effects of doing something now on the ability to do something else later and the effects that probably cannot be brought about because of what one has already decided to do. *Transaction costs* are

Two: Interpenetration at Work

the effects of doing something borne by parties not involved in the original decision.) Indeed, there is no indication that these correlative but unintended consequences of the theory's application entered into Nate's overall evaluation of the theory. His negligence here is an instance of what Dewey and other pragmatists derided as "intellectualism." But notice that anti-intellectualism does not imply antirationalism. As Nate's responses indicate, the naturalist simply denies that there is a species of rationality associated with knowledge production that cannot be analyzed as a form of instrumental rationality. For it seems that Nate is interested only in the extent to which the world confirms or resists his prior theoretical notions, not in the overall consequences that his acting on those notions has in the world. Such intellectualism remains very much in evidence among so-called historicist philosophers of science (e.g., Lakatos 1979, L. Laudan 1977, Shapere 1984), all of whom avow at least a mitigated naturalism, yet who continue to evaluate scientific theories as if they normally had consequences *only* for the conceptual development of science.

Building the Better Naturalist

Behind Nate's latent intellectualism is a view of language—at least of theoretical language—that is shared by Clay. It is the seemingly innocent view that theorizing does not transform the world in the manner of other productive activities; rather, it merely produces causally inert "mirrors of nature." Thus, the only consequences of theorizing that concern Nate are the ones that determine the extent of the world's conformity to his theoretical expectations. But because they too have held such a view, classical epistemologists have made a point of introducing a distinctly normative dimension of "justification" alongside the empirical one of "explanation." In this way, the classicist may intervene in the knowledge production process for purposes of criticizing and perhaps even revising the foundations of that process. Since the naturalist typically fails to make such a distinction, she is easily read as having no normative interests aside from the clinical ones of assessing the extent to which theories are confirmed or the extent to which means achieve their ends. Certainly, Clay never seems to catch Nate in a commitment to anything more robustly normative. However, this is only because Nate refuses to take to heart the naturalist dictum that knowledge is part of the same world as the objects of knowledge,

Incorporation

and that, consequently, every theoretical representation is ipso facto a causal intervention.

To unpack this last claim, let us consider that theorizing–especially the sort of metascientific theorizing that a naturalized epistemologist is likely to do–can be either a *passive* or an *active* form of causal intervention. Good examples of passive intervention may be found in the ethnographic accounts of "laboratory life" (e.g., Latour and Woolgar 1986) that constitute much of the empirical base for STS. The accounts profess to offer descriptions of ordinary scientific practice, shorn of all normative epistemological baggage. And, indeed, the ethnographers tend to tell fairly prosaic tales of the labs. People and other medium-sized dry goods are shunted back and forth in a setting only slightly less structured than the average industrial plant. However, because this "neutral" description of science clashes with the expectations of readers, most of whose images of science are already very norm-laden, the net effect of these ethnographies has been to inspire a wide-ranging reevaluation of the epistemic legitimacy of science. Yet, one would be hard-pressed to find an ethnographer of science willing to admit that she is doing anything more than describing what she observed. As a form of theorizing, the ethnographies passively intervene in the scientific enterprise simply by offering a perspective that differs substantially from standing expectations and that thereby, perhaps even unintentionally, calls into question the groundedness of that enterprise (Button 1991: chap. 7 provides ethnomethodological groundings for this perspective). Of course, I have said nothing that denies the possibility or value of determining the degree of fit between actual scientific practice and the ethnographic accounts. However, such a determination would still fail to explain the *impact* that these ethnographies have had in a world where science is valued and practiced, an impact that seems to have had little to do with any formal testing of the ethnographers' descriptions.

How does this appeal to science "as it actually is," also known as *descriptivism*, turn out to be so rhetorically effective? From a rhetorical standpoint, a description is a verbal representation of some object to some audience, such that the speaker is able to change the audience's attitude toward the object without changing the object itself. Thus, the trick for any would-be describer is to contain the effects of her discourse so that the object remains intact once the discourse is done. In descriptions of human behavior, this is often very difficult to manage, as the people being described, once informed of the description, may become upset and proceed to subvert the describer's authority. A major finding of STS research is that this predicament extends even to the natural

Two: Interpenetration at Work

sciences, in spite of the fact that their objects do not seem capable of either eavesdropping or talking back. Nevertheless, natural objects typically have their own spokespeople (experts) who are capable of, so to speak, personifying the challenge that a description may pose to the disposition of the objects described. Thus, if the STSer claims that a given theory works only because the relevant people agreed, the spokesperson for nature could always rejoin that the STSer hasn't examined the depth or detail of the natural process in question. In that case, the spokesperson's plea for comprehensiveness disguises an attempt to keep the burden of proof squarely on the STSer's shoulders.

It would seem, then, that if a describer, such as the STSer, wants to secure for her descriptions the aura of detachment that comes from representing things as they are, then she should construct her descriptions in a language that only the describer's intended audience will understand; hence, there is an elective affinity between capturing the world "as it actually is" and operating from an autonomous disciplinary standpoint. This, I believe, explains the sense of objectivity that often accompanies the introduction of technical terminology. Not surprisingly, the call to descriptivism in STS has invited the cultivation of arcane "observation languages" that only fellow STS researchers–and not the scientists under study–can understand (Segerstrale 1992). Although this development can be portrayed as a step toward the disciplinization of STS, it ultimately goes against the democratizing mission of the field. Ironically, the recent "reflexive turn" toward integrating the STS practitioner into her description of science, an attempt which prima facie appears to aim at a more comprehensive picture of science, actually exacerbates descriptivism's tendency to provincialize audiences, as "comprehensiveness" becomes relative to whether the author's presence is integrated into *her own* text (cf. Ashmore 1989).

By contrast, a theorist actively intervenes in the knowledge production process when she tries to remake the process in the image of her theory. This rather straightforward idea, whose complex implications are highlighted in this book, is typically neglected by intellectualist accounts that portray theories as things that predict and explain, but not construct, phenomena. However, construction is arguably the most important role that theorizing plays in the social sciences (cf. Hacking 1984). As my remarks on descriptivism suggest, I disagree with those (e.g., Hacking) who believe that theoretical intervention distinguishes the social from the natural sciences (though I may be willing to grant a certain lack of awareness of this process on the part of natural scientists). Nevertheless, I will focus my remarks in what follows on the styles of active theoretical intervention one finds in *anthropology, economics*, and *psychology*.

Anthropologists commonly draw a distinction between an insider's everyday, *emic*, knowledge of social life and an outsider's scientific, *etic*, knowledge (cf. M. Harris 1968). Often this is described as the difference between the "first-person" perspective of the native and the "third-person" of the analyst (e.g., Fuller 1984). And often these two perspectives are made to look mutually exclusive, complementary, and exhaustive. Accordingly, if the anthropologist is to abide by an agent's normative categories, then she must also abide by the judgment calls that the agent makes on the basis of those categories (i.e., go emic); otherwise, the anthropologist is simply importing her own alien categories into the agent's situation and thus at least implicitly questioning the validity of the agent's categories (i.e., going etic). However, the journey from emic to etic affords a rhetorical way station, a *second-person* perspective, as it were. It involves appending to the agent's own categories a tighter procedure for accounting for the agent's behavior. As a check on the agent's self-explanations, trained external observers (and, in more recent years, cameras and other more reliable recording devices) can be introduced into the situation. Not surprisingly, if one examines Francis Bacon's and other early justifications for the experimental method as a privileged source of knowledge, they spring from an awareness that if we were to scrutinize each other's behavior a little more closely than we normally do, the surface rationality of everyday life would yield to an assortment of biases and liabilities that "succeed" largely because they remain unchecked.

Bringing this anthropological insight back to the history of our own culture, Shapin and Schaffer (1985) have masterfully analyzed the alignment of interests in seventeenth-century Britain that ultimately authorized experimenters to speak for a deeper analysis of ordinary experience, thereby overriding the accounts of both naive observers (the emicists of the day) and learned scholastics (the eticists). The modern scientific mentality emerged once people started to regard the tighter accounting procedures as a *de*contaminant, rather than as a contaminant, of everyday life–that is, not as artificially restricting our intercourse with nature, but as removing the obstacles that normally inhibit such intercourse. Once this long-fought battle was won, proposals of varying degrees of merit were made to reconstitute ordinary language in scientific terms. Thus was born positivism's "popular front," that gallery of linguistic reformers extending from Jeremy Bentham ("science of legislation") to Count Korzybski ("general semantics"). But more lasting perhaps have been the corresponding efforts to reconstitute the natural world as a place in which experimentally relevant signs can be

Two: Interpenetration at Work

detected and interpreted more easily. The interesting rhetorical point here is that the difference between a speaker's ability to establish a mnemonic association between a place and a part of her speech and an engineer's ability to manufacture collective memory by designing public spaces that remind the citizenry of normatively acceptable and unacceptable forms of intercourse is one of degree not kind (cf. D. Gross 1989; Lowenthal 1987).

Economics has recently become a favorite naturalistic model for reconceptualizing normative epistemology (Giere 1989b; Goldman 1992: chap. 12). Often what seems to attract philosophers to this field are the qualities of economic modelling that most resemble analytic philosophical reasoning: specifically, its abstract, reductive, rigorous, a priori character. However, to focus on these qualities is to obscure how economic models function in policymaking (cf. Lowe 1965). For example, a theoretical model of the market sets the standard that defines normal and abnormal economic behavior, as well as the obstacles that need to be overcome to approximate the market ideal more closely. Increasingly, economists interested in socially embedding the policy process have challenged this way of deploying models, as it presumes the normative standard–the ideal market–to be a fixed equilibrium toward which economic activity eventually gravitates, with or without help from the government (e.g., Block 1990: especially chap. 3). Such an orientation neglects irreversible moves away from the original state of equilibrium–many of them beneficial–that are produced by innovative entrepreneurship, the institutional absorption of transaction costs, and simply a change of scale in the economy (cf. Georgescu-Roegen 1970). In this, the real economic world, it no longer makes sense to think of abstract models as "rigid rods" in terms of which actual economies are gauged and corrected. Indeed, naturalists attracted to economics would do well to study the recent attempts to formalize a more "relativistic" (in the Einsteinian sense), even stochastic, conception of market norms for real economies (cf. Mirowski 1986, 1991).

However, it is to ongoing debates in psychology over the "external validity" of experiments that any naturalist of a Hegelian frame of mind should turn to gain the most immediate insight into the problem of reconstituting in empirical terms the classicist's conceptually derived epistemic norms (cf. Berkowitz and Donnerstein 1982; Fuller 1989a: 131-35). Critics of the experimental study of human beings typically argue that subjects' performances in the laboratory are too artificial to form the basis of generalizations about normal human behavior. Defenders then respond that experiments are designed to determine the contribution that an isolated variable (or set) makes to an overall ef-

Incorporation

fect. If the effect is a positive one, then the point would be to restructure the environments outside the lab so as to make them more like the conditions that enabled the variable to contribute to the effect observed inside the lab. Thus, if the variable in question is a heuristic that enabled subjects to solve artificial problems more effectively, then the task ahead would be to transform normal problem-solving settings into ones in which the heuristic would also work. Since these heuristics tend, as a matter of fact, to be drawn more or less explicitly from epistemic norms that epistemologists and philosophers of science have proposed, it would seem that a Hegelian Naturalist could easily reinterpret classical talk of "ideal epistemic agents" and "rationally reconstructed histories" as first passes at specifying the laboratory conditions in which certain norms could be demonstrated to have epistemically efficacious consequences (cf. Gorman and Carlson 1989, Fuller 1991).

Perhaps it is not so surprising that naturalists have failed to take to heart their own dictum that every theoretical representation is a causal intervention. After all, naturalists are often portrayed as either hostile or indifferent to metaphysics. Quine (1953) is a good example of someone who manages to project both images at once. In any case, the naturalist is typically not seen as engaged with the problems of universals. However, as we have been seeing in the course of this chapter, the naturalist loses ground to the classicist precisely when she ignores ontological considerations. Because Nate is concerned with locating the consequences of theorizing in conceptual space, but not in physical space, he unwittingly adopts the classicist's transcendental conception of language and, as a result, short-circuits the interventionism that gave John Dewey's naturalism its distinctive normative slant. That ontological considerations should loom so large for the naturalist should, upon reflection, come as no surprise. After all, it is the naturalist—not the classicist—who believes that there is something epistemically salient about knowledge being both *about* and *in* the world. And what it is to be in the world, specifically, how and where knowledge is embodied in the world, is a matter for ontology to decide.

When the classicist defines knowledge as "justified true *belief*," she typically assumes that the same belief can be embodied in many different ways. These "multiple instantiations," to use the expression of choice in these matters, include states of consciousness, unconscious states of the brain, the linguistic and non-linguistic behavior of human beings, and perhaps even the behavior of non-human beings. Admittedly, recent classical epistemologists have had little to say about how it is possible that all these different sorts of things embody

Two: Interpenetration at Work

the same belief. But, in large measure, this silence simply reflects the post-Kantian tendency to treat epistemological questions as separate from questions of metaphysics and even the philosophy of mind. By contrast, for example, in the Middle Ages, the problem of knowledge was not thought to be adequately addressed unless the philosopher could account for the multiple instantiation of a belief, or what was then called the "communicability of the form of the belief." However, nowadays the classicist is content to presume such communicability and move on to the business of characterizing a special epistemic relation in which the communicated form stands to some external reality. In short, she asks: What makes my (true) belief that S is P a belief *about* S's P-ness? The fact that a particular instance of my belief that S is P inhabits the same world as—and hence stands in some causal relation to—a particular instance of S being P is immaterial to the classicist's epistemological concerns. But, as we have seen, it *is* material to the naturalist's. For when the naturalist asks what should one believe, she is at once making implicit reference to a vehicle for instantiating a belief (whether it be a neural network, a piece of electronic circuitry, or a pattern of social interaction), the likely causal trajectory of that vehicle in relation to other things in the world, and the relative desirability of the possible outcomes of its interaction with those things. This produces an epistemology that looks more like science policy than literary criticism. It is quite different from the theories of knowledge to which classicists are accustomed, but at least it is one worth arguing about.

Conclusion: Naturalism's Trial by Fire

Just because arguments remain unresolved, it doesn't follow that they die or go away. David Hume understood this point very well and tried to do something about it. He prescribed that books whose claims were grounded in neither logic nor experience should be cast into the flames. Hume, one of the acknowledged progenitors of naturalism, thought that books of metaphysics and theology should take their rightful place amidst the timber in his fireplace. However, one may wonder, now over two centuries later, whether epistemological texts—including naturalistic ones—deserve a similar fate. *Specifically, if the epistemologist is neither palpably improving the production and distribution of knowledge in society nor accurately describing current prac-*

Incorporation

tices, then what exactly does she think she is doing? From canvasing the works of Goldman (1986, 1992) and his students, and threading through Nate's trials in the hands of Clay, I worry that an unwholesome third way has been paved: epistemologists are devoted to describing what an improved state of the knowledge system "would look like"–with the subjunctive left dangling in midair. In practice, the accuracy of such a description is relative to the ideal that the particular epistemologist has in mind, her "intuitions," as it is sometimes called (cf. L. J. Cohen 1986). These intuitions may be conceived and refined before an audience of fellow epistemologists proffering alternative intuitions, all of whom are generally far removed from the people who would need to be persuaded in order for any of these intuitions to be realized. This is a particularly ironic situation for a "naturalized" epistemologist to face, as it suggests that she is causally insulated from the workings of the very enterprises whose norms she would legislate!

This irony bespeaks the ultimate violation of the Principle of Nonopportunism that today's naturalists tend to commit, namely, to accept without question the classicist's conception of *what it is to be a norm*. For the classicist, a norm commands our attention if it makes sense "on paper" or in a discussion with our similarly trained friends. The relevant criteria for evaluating norms, then, include aesthetic satisfaction, logical coherence, and overall intellectual and pragmatic suggestiveness. Missing from this list are criteria specifically associated with *governance*, such as the propensity for gaining the consent of the governed so as to enable maximum improvement of their lot at minimum cost to their current way of life. An interesting source of inspiration in this regard is the positivist theory of law, as advanced by Hans Kelsen, a legal theorist who participated in the Vienna Circle (cf. Moore 1978). Not that as rhetorically informed theorists we could simply apply Kelsen's views off the shelf. His tendency to reduce governance to the threat of sanctions merely highlights, in an unproductive and self-fulfilling way, the extent to which norms are introduced to compensate for already existing behaviors. However, he got the basic point, which is lacking in much of what passes today as naturalized epistemology, namely, that a statement is not a norm, regardless of its content, unless it has the power to bind action.

4 REFLEXION, OR THE MISSING MIRROR

OF THE SOCIAL SCIENCES

Here is a possible story about how science developed. Science originally arose in the area where humans displayed the most knowledge and interest, namely, themselves. Gradually, the human cognitive grasp moved outward, first toward the nonhuman things with which they had the most in common, and ultimately to the more remotely nonhuman, until humans were able to make sense not only of nonearthly things (e.g., the heavens), but more important, of a perspective that was literally a "view from nowhere." Such was pure objectivity (cf. Nagel 1987). The general strategy behind this outward reach would be for humans to model the nonhuman as much as possible on facets of themselves, and then use the points of disanalogy as the basis for an autonomous body of research that eventually issues in a full-fledged science. Thus, biology would be expected to have spun off from sociology, after a critical number of nonhuman properties had been recognized in animals. In light of this story, the newest science should be the most nonhuman study of them all, cosmology, which conceives of reality as a "universe" of which humans are an infinitesimal part.

Now, of course, this story is not only fictional but the exact opposite of the true story. What went wrong? As a first approximation, it might be said that the story took a little too seriously the idea of explaining the unknown in terms of the known. On second thought: perhaps a little too uncritically. For one of the most instructive themes in the history of science is that more knowledge is better than less *only after science is already in place*, and quality controls have thus been instituted for the production of knowledge. Otherwise, less knowledge is generally better, especially if it is of things–such as observable physical objects– whose remoteness from the human condition makes it easier to provide an objective evaluation. Now, in what exactly does this "remoteness" consist? I would say it consists in a *rhetorical impoverishment*, a lack of

Reflexion

alternative discourses for characterizing the phenomena in question. Rocks, streams, and stars–the stuff of which both early Greek cosmology and Renaissance physics were made–have rarely been elaborated with the richness or complexity of human creations. Consequently, these things provide a more natural basis for standardized observation languages, which in turn enable both smoother communication and easier conceptualization, which in turn offer the possibility for greater manipulation and control of the motions of the things themselves (cf. Dear 1987 on the influence of this line of thinking during the establishment of the first scientific societies). While ordinary language affords many ways of imagining the agents and resultants of change in human behavior, the number is considerably smaller in the case of, say, rock behavior. Even today, textbooks in the more "scientific" of the social sciences–psychology and economics–will routinely introduce their subject by talking about the need to regiment our common talk about human beings in order to sort out the wheat from the chaff in what we say about ourselves. In part to preempt student worries that such textbooks merely clothe the obvious in jargon, they will typically point out that too much information–especially when its relevance and reliability remain unanalyzed–is often worse than too little (e.g., MacKenzie and Tullock 1981: chap. 1).

The lesson to learn from our false story is that the development of science involves something other than the spontaneous accumulation of knowledge–or, for that matter, the spontaneous generation of ideas. On the contrary, it requires *discipline*, the cultivation of a consistent perspective by adopting a language and techniques that focus the inquirer's attention, generally to the exclusion of other potentially observable matters. Perhaps this is the point most frequently taken from Thomas Kuhn's seminal work, *The Structure of Scientific Revolutions* (1970). However, it is a point worth belaboring because, with the exception of Toulmin (1972), only in very recent years have disciplines been studied in systematic detail. The variety of approaches to them may be found in Graham, Lepenies, and Weingart (1983), Willard (1983), Shapere (1984), Whitley (1985), and Bechtel (1986). In terms of the things one might study about science–including theories, concepts, research programs, and the like–disciplines are probably the only units that require the cooperation of the rival historiographical approaches in science studies: the *internal* approach, devoted to charting the growth of knowledge in terms of the extension of rational methods to an ever larger domain of objects, and the *external* approach, devoted to charting the adaptability of knowledge to science's ever changing social arrangements (cf. Fuller 1989a: pt. 1). In a nutshell, disciplines mark the

Two: Interpenetration at Work

point where methods are institutionalized, where representation is a form of intervention, where, so to speak, *the word is made flesh* (cf. Fuller 1988a: chap. 8).

My own contribution to this discussion has been twofold. On the one hand, I have observed how the referential character of a discipline's discourse draws attention away from the discipline's source of power, as the audience is beckoned to focus on certain prescribed objects whose identities are detached from the speaker's. At its most extreme–the sign of a successful science–all the objects become so externalized from the speaker's identity that conceptual space is no longer available to hold the speaker accountable by the standards imposed on the prescribed objects. Thus, a piece of technology is routinely regarded as the "application" of physics, yet the physics is portrayed as being already embodied in the technology rather than as something which the technologist physically adds (Fuller 1988a: especially 188). On the other hand, I have also noted that once objects have been externalized in a disciplined manner, they serve as standards against which to evaluate and calibrate human performance, such that the history of science reveals the "human" to be a "floating signifier," the shifting residue of our behavior that resists standardization. I have examined this issue most closely in terms of the use of computers to model thought (Fuller 1989a: pt. 2).

In what follows, I argue that the rhetorical character of disciplinary boundaries in the social sciences provides an especially good context for examining the embodiment of knowledge as a source of worldly power, a topic typically neglected by epistemologists and philosophers of science, who still tend to think of knowledge as a politically indifferent, or "disembodied," phenomenon. I start with what seems to be a technical problem in the philosophy of science, namely, whether it is possible to demarcate criteria for demarcating science from nonscience. Recent philosophers have despaired of finding such "metaboundaries," and as a result have begun to call into question the very identity of the philosophy of science. Against this line of reasoning, I argue that the failure of the demarcation project only shows that attempts to study science scientifically, as the philosophers have wanted to do, tend to result in science deconstructing *its* identity. But, clearly, the epistemic authority of science has worked to block such self-deconstructive moves in the normal course of inquiry. How? I propose a strategy for addressing this question, namely, to examine how science exercises worldly power by rhetorically drawing our attention to the fact that scientific knowledge *represents* the world and away from the fact that it also *intervenes* in the world (cf. Hacking 1983). (In

Reflexion

the text I speak of this duality of knowledge in terms of its being both *about* and *in* the world.) Because the social sciences have throughout their histories been fighting an uphill battle to secure epistemic legitimacy, it is easier to see the rhetorical seams of their attempts to represent without appearing to intervene in the world. After surveying the canonical historiographies of five social sciences for this theme, I focus on the battle fought between economics and political science over the contested field of "politics." Finally, I return to the subversive, and hence relatively unexplored, possibilities for studying science scientifically. These include various social sciences of science, as well as deconstructions of the natural sciences' historical ascent to worldly power.

Studying disciplinary *boundaries*–their construction, maintenance, and deconstruction–adds a new dimension to the "interface" role already played by disciplines. And rather than merely continuing a recent line of inquiry, this one goes to the heart of what has made philosophy of science a distinct specialty throughout most of this century. Philosophers of science are most familiar with disciplinary boundaries from Carnap's and Popper's quests for *demarcation criteria* that systematically discriminate the sciences from nonscientific (and especially pseudoscientific) forms of knowledge. Disciplinary boundaries provide the structure needed for a variety of functions ranging from the allocation of cognitive authority and material resources to the establishment of reliable access to some extrasocial reality. Historically, however, philosophers have not agreed on the specifics of the demarcation criteria, since they have drawn the science-nonscience boundary largely in order to cast aspersions on the legitimacy of particular pretenders to the title of science (L. Laudan 1983). Popper's (1957) attempt to use the criterion of falsifiability to undermine the scientific credibility of psychoanalysis and Marxism is a clear case in point. An implication of this point is that there may be no properties common to all disciplines deemed scientific–except perhaps the approval of the person doing the deeming. If that is the case, then there is no ahistorical essence to science. And insofar as the philosophy of science itself has been devoted to the divination of such an essence, it may be that this philosophical subdiscipline, as traditionally conceived, may be an enterprise doomed to failure.

Notice what has happened here. In trying to bound the boundary of scientific from non-scientific disciplines, philosophers have come to discover only that no such metaboundaries are to be found. This suggests, somewhat unwittingly, a strategy for subverting existing disciplinary structures, namely, to show their long-term *lack* of discipline.

Two: Interpenetration at Work

More generally, the philosophical project shows that there is no epistemically privileged way of conferring epistemic privilege. Although one might think that this insight would enable philosophers of science to radicalize their understanding of knowledge production, the contrary has in fact been the case. If anything, philosophers have used the insight more against *themselves* than against science. In other words, they have debunked the idea of a demarcation criterion without debunking the things (i.e., the "sciences") that are supposed to be demarcated by such a criterion. Instead of taking seriously the possibility that the demarcation criteria proposed by philosophers might be getting at some privileged, albeit nonepistemic, means of conferring epistemic privilege, Laudan (1983), for one, has advised that the whole project be scrapped: i.e., philosophy of science should be reabsorbed into epistemology. In effect, this would be to turn away from such relatively social units of epistemic analysis as "sciences," "paradigms," and "research programs" to more subjective and purely formal units, such as "beliefs" and "theories." Similar regressive moves have been made by Arthur Fine (1986), who would have the philosophy of science wither away, leaving the history and sociology of science in its wake, and Dudley Shapere (1984), who would have philosophers be content with raising successful scientific practice to methodological self-consciousness.

While there is a certain perverse nobility in sacrificing the identity of one's own discipline for the sake of another, there is a sense of protectiveness as well. For if philosophers were to follow the demarcation problem to its full logical consequences, then they might be forced to reconceive the nature of science itself.

Why the Scientific Study of Science Might Just Show That There Is No Science to Study

There are two senses in which trying to demarcate demarcation criteria is a reflexive enterprise. I have so far focused on one sense, whether the concept of a demarcation criterion is itself demarcatable: i.e., Have there been any significant properties common to the criteria proposed through history? The fact that these criteria have shared few, if any, significant properties seems to provide indirect evidence that no principled science-nonscience distinction can be drawn. Notice that *no more* than an indirect link between the premise and the

Reflexion

conclusion can be asserted here, since it is entirely possible that science does have an essence (in the classical positivist sense of there being an optimal methodological route to knowledge) but that most philosophers–preoccupied as they are by the epistemic squabbles of their times–have failed to grasp it. Indeed, maybe one philosopher–Popper, say–got the criterion right, but his immersion in local squabbles has tended, after the fact, to cast aspersions on the universality of his claims. In our post-Enlightenment age, we are prone to overlook this point: that some periods in history may be better than others for uncovering truths that apply to all periods. It suggests a deeper, second sense in which demarcating the demarcation criteria is a reflexive enterprise. For example, Laudan's (1983) negative conclusion is persuasive, in large measure, because he claims to have used scientific means for determining whether a science-nonscience distinction can be drawn. Specifically, Laudan has taken a representative sample of opinions from the history of philosophy and has found little mutual agreement. Indeed, he claims that one could predict the criterion that a particular philosopher proposed simply on the basis of what she took to be the disciplines that were granted undue cognitive authority in her day. Thus, Laudan presents the lack of historical continuity in the demarcation proposals as inductive disconfirmation of the claim that science has an enduring nature. In this sense, then, scientifically studying the nature of science may reveal that science has no nature to study.

Admittedly, at this stage of the game, I have offered little more than an engaging abstraction by way of radicalizing our understanding of knowledge production. Let me now turn to some relatively concrete research proposals.

Although philosophers may be right that there is no epistemically privileged way of conferring epistemic privilege, it does not follow that there is no *nonepistemically* privileged way. What I have in mind is that behind the variety of demarcation criteria may lie a function that must be regularly performed in maintaining the social order. The relative constancy of the motives that philosophers have had for proposing such criteria already hints at what this function might be. I have elsewhere defined the function in terms of the Baconian Virtues (Fuller 1988a: chap. 7). A discipline is deemed to possess the Baconian Virtues once it is credited with producing the sort of knowledge that is necessary for maintaining the social order. That is the function of science which philosophers of science have perennially wanted to have a hand in determining. The natural sign that a discipline possesses these Virtues is that the esoteric character of the knowledge it produces serves only to enhance the discipline's perceived centrality to the

Two: Interpenetration at Work

society; hence, a "cult of expertise" develops, which, in turn, enables the discipline to have access to vast political and material resources, including the seats of power themselves (Abbott 1988: chaps. 3-6). For a vivid sense of the epistemic variation in this social function, consider the shift in qualifications (more pronounced in Europe than in the United States) for the diplomatic corps between the nineteenth and twentieth centuries. Earlier it was believed that an education in the classical liberal arts would deepen the diplomat's appreciation of the values and ideals shared by the representatives (mostly European) that he would most likely face in negotiation, values to which he could then strategically appeal as rallying points for agreement (Grafton and Jardine 1987). Over the past hundred years, however, the humanities have been eclipsed by engineering and economics as the preferred educational background for diplomats, reflecting a less elitist, as well as less personalized, sense of worldly power (cf. Ben-David 1984: chap. 6 ff.).

Two points stand out in trying to develop a nonepistemic way of conferring epistemic privilege: the first pertains to implications for locating the center of the knowledge production process and the second to implications for "de-centering" it.

In one sense, the nonepistemic route is a search for an implicit principle of social organization. The sociologist Niklas Luhmann (1979) has introduced the concept of *self-thematization* to capture the fact that any well-bounded social system is marked by the presence of Kuhnian "exemplars." In a self-thematized system, one can point to a component activity that synecdochically represents the working of the entire system. The performance of the system's other components can then be evaluated in terms of that exemplary component. Thus, if the scientific exemplar is Faraday's experimental work on electromagnetic induction (as was becoming the case in the second half of the nineteenth century; cf. Gieryn 1983), then one should expect some disciplines to start emulating the exemplar in their own work, perhaps even in dubious or superficial ways, so as to draw on the exemplar's epistemic authority. Other disciplines, however, may be so removed from the exemplar that they would first have to erase their current identities before being taken seriously as sciences. The humanities have increasingly become disciplines of this sort, and it is interesting to see how they have dealt with this social relocation from the center to the periphery of the knowledge production process. For well into the Scientific Revolution, knowledge of rhetoric and the classical liberal arts was held to be the key to worldly power. Indeed, the experimental tradition responsible for the ascendancy of the natural sciences first laid claim to the scien-

Reflexion

tific exemplar by appearing to be more powerful in the terms set by rhetoric (Shapin and Schaffer 1985). I shall return to this point at the end of the chapter.

Nevertheless, by the late nineteenth century, the sense of "worldly power" had sufficiently changed so that few took seriously the idea that rhetoric and the other humanistic disciplines were among its sources. Stephen Toulmin (1990) has recently documented the transition as a shift in the seat of knowledge-power from (personal) "influence" to (impersonal) "force." However, a more illuminating contrast may be in terms of the stage setting needed for the display of power. Take the idea of "law" before and after the emergence of the experimental paradigm in natural science. Before the rise of experiment, a law was a norm whose validity resided in its usefulness as a standard against which to evaluate and shape behavior. To explain something by such a law was to understand it in terms of its normative status, that is, as either conforming to or deviating from the norm. Thus, the expression "natural law" was indifferent to epistemic and juridical usage (cf. Zilsel 1942). A "monstrous" birth was more than an accident to be explained; it was also a misfortune to be justified. However, the prediction of behavior had little to do with these two tasks. Knowing the laws of nature, or "nomothetic knowledge," did not entail that all the relevant phenomena were already disciplined by those laws; rather, possession of such knowledge entitled the possessor to use whatever influence or force it took to make the phenomena conform to the laws—including the elimination of certain persistently deviant creatures. For, even the act of elimination could be represented, in Aristotelian fashion, as the human completion of nature by bringing the normative to full "self-realization."

However, the rise of experiment changed the character of nomothetic knowledge, so that it no longer referred to the *original authorization* of force in the name of normative order, but rather to the *subsequent achievement* of normative order, once such force, or discipline, had been imposed. This point is most apparent in what positivists (e.g., Hempel 1965) call the "symmetry" between explanation and prediction, according to which a scientific law does not explain a range of phenomena unless the law can also be used to predict those phenomena. (One might think of this as the principle of the unity of theory and practice in science: if I truly understand something, I can then control it.) Of course, given their abstract character, scientific laws are fairly useless in predicting phenomena unless the phenomena have been themselves disciplined in advance, say, by minimizing the number of interacting variables in a laboratory environment. That highly disciplined

Two: Interpenetration at Work

setting, endemic to all experimental inquiry, is typically articulated in a ceteris paribus clause attached to the law, which implicates the various background conditions that must be maintained in order for the desired lawlike regularity to be displayed in the laboratory. A major finding of on-site studies of science in action is the vast human and material resources–often quite specific to lab locales–that are required for maintaining such background conditions on a regular basis (cf. H. Collins 1985). However, all of this now happens behind the scenes of science, to be ferreted out by diligent sociologists, whereas before the age of experiment these events transpired in the open, indeed, in public trials.

In sum, the transition from humanistic to experimental epistemic cultures may be captured as follows: the former issues laws to license overt politicking in the name of science, whereas the latter issues laws only to reward such politicking that has succeeded by conferring on it the name of science. In humanistic culture, as in the work of Aquinas and the Thomistic tradition, the "natural law" of the judges and the philosophers are the same, whereas in experimental culture the "positive law" of judges enters just when the "positive laws" propounded by social scientists fail to significantly constrain people's behavior. As a methodological heuristic, one might thus see the project of deconstructing sources of worldly power in terms of reversing the transition from humanistic to experimental cultures by revealing the moments of politicking when the voices of various parties are amplified, silenced, or in some other way strategically represented by having their ends translated into the means to someone else's ends (cf. Callon and Latour 1981).

Not surprisingly, in the last hundred years, as the humanities have decisively lost their claim to worldly power, their epistemic aspirations have also changed. Instead of touting their role in the creation and maintenance of cultural values, the humanities have turned to the academic pursuit of knowledge "for its own sake," that is, without regard to the social consequences that it might have beyond its own institutional boundaries (Ohmann 1987: pt. 1). In *Social Epistemology*, I tagged this move "the retreat to purity" (Fuller 1988a: chap. 8). One way of understanding what happened here is by appealing to the social psychological concept of *adaptive preference formation*, a class of strategies for rationalizing failure whereby one adapts aspirations to match expectations (cf. Elster 1983). Indeed, a sign that a discipline is receding from the scientific exemplar may be that it refuses to have its significance evaluated in terms of concrete outcomes, especially the production of real world effects.

Reflexion

However, in citing the recent history of the humanities as an adaptive preference formation, two points need to be kept in mind: (1) that the humanities have not always been so coy about being judged by the consequences of their practices (e.g., the claims to worldly power made for humanism during the sixteenth-century Renaissance and the early-nineteenth-century German university movement); and (2) the success of a discipline's claims to worldly power is based largely on *folk* perceptions about the discipline's ability to transform the world, which, in turn, serve to define the exemplar of worldly power itself. Indeed, these folk perceptions typically reach hegemonic proportions, affecting central and peripheral disciplines alike. Thus, humanists have only now, under the guise of "postmodernism," begun to question the epistemic authority of the natural sciences, not so much to shore up their own authority—of which they remain skeptical—but to erase the very possibility of epistemic authority (Lyotard 1983).

But postmodernist sophistication continues to be no match for commonsense appeals to the epistemic superiority of the natural sciences that point to astronauts going up into space or nuclear bombs being exploded (Fuller 1990). Of course, it would be misleading in the extreme to suppose that at some point someone actually demonstrated that the natural sciences were more efficacious than the humanities, since both the historical record and current divisions of labor reveal that technologies have been developed and maintained in spite of users' ignorance of the relevant physical principles. Indeed, on philosophical grounds alone, one of the main reasons for always trying to refute a standing explanation of some fact—such as why a piece of technology works—is that a plethora of hypotheses will always be able to save the phenomena that are taken as evidence for this fact (cf. L. Laudan 1984: chap. 5). Yet, none of this has prevented increasing numbers of significant real world effects from being persuasively presented as embodying natural scientific knowledge. That is, people have been persuaded to *presume* that the efficacy of natural scientific knowledge is behind effective technology, which then serves to *preclude* a demonstration that natural science has indeed generated the relevant effects (cf. Mulkay 1979; Fuller 1988a: chap. 4).

Well, in a world where presumption does not make for strict demonstration but "merely" worldly success, how do the natural sciences manage to succeed? Clearly, some artful rhetoric must be involved to enable most people to routinely ignore the empirical and conceptual doubts raised here. In other words, what might otherwise be taken as grounds for skepticism (of the efficacy of natural scientific knowledge) is instead just taken for granted. (I have elsewhere called

Two: Interpenetration at Work

this phenomenon, whereby lack of explicit refutation is taken as implicit confirmation, *the inscrutability of silence*; Fuller 1988a: chap. 6.) It should be clear by now that simply referring to popular ignorance or impressionableness will not do, since it is precisely the appearance of these factors that an artful rhetoric brings about, and hence can explain. My own speculations in this direction turn on the difference between the contexts in which the social and natural sciences are held accountable for their knowledge claims.

To impute "success" to an enterprise is to imply both a standard of evaluation and a procedure for evaluating cases. One of Arthur Fine's (1984) arguments against the need for a distinct specialty called "philosophy of science" is that the natural sciences' alleged track record of theoretical and empirical successes does not withstand close historical scrutiny. However, it is usually difficult to scrutinize the record because, until quite recently, historians of the natural sciences have had a remarkable capacity for recalling the same cases as successes and forgetting virtually all the failures (cf. Fuller 1988a: chaps. 3, 9). In contrast, historians of the social sciences tend to disagree about what should count as successes and failures, and so less of the actual history is consigned to silence. Consequently, these histories often center on disputes that seem to be of greater import than their inconclusive and temporizing outcomes. And so, as less of the actual history is typically suppressed by historians of the social sciences, the success rate of those disciplines seems weaker.

Interestingly, this contrast is reflected in the microhistories that are constructed in the literature reviews which preface articles in the natural and social sciences. Hedges (1987) has found that physics articles routinely report greater cumulativeness in their lines of research than psychology articles, and not because physics research is so much more replicated and extended than psychology research. Rather, the reason seems to lie in the statistical techniques that the two disciplines use to analyze and synthesize data. In brief, whereas physicists are inclined to intuitively throw out studies that would make for extreme data points and to use fairly elementary statistical techniques to elicit clear empirical regularities, psychologists tend to integrate every available study into a complex statistical formula that, not surprisingly, lowers the level of certainty and reliability of their findings. This difference, in turn, represents the legacy of statistics to scientific reasoning, which took root first in the study of social phenomena, where it was presumed that special methods would be needed to tease order out of chaos (Hacking 1990). For nearly two centuries after

Reflexion

Newton's *Principia Mathematica*, the natural sciences were not felt to have any such need (Porter 1986).

The reader should begin to see that the "success" of the natural sciences may be an artifact of their relatively loose accounting procedures. As Porter (in press) has observed about the increasing political relevance of cost-benefit analysis, people are easily fooled into thinking that the rigor in the calculations that can be done once numbers are assigned to qualities implies a similar rigor in the method by which the numbers were assigned to the qualities in the first place. This fallacy is partly responsible for the success of the natural sciences–but hardly the whole story. For to measure success is not only to evaluate cases in a certain (perhaps pseudorigorous) way, but also to set the standards of evaluation so that the successes can be easily seen against a backdrop of insignificant failures. Here, too, a double standard exists for the natural and social sciences. We are typically satisfied with the effects and entities that natural scientists, especially physicists, reliably generate in their labs, however unrelated to our ordinary experience of the natural world these things may be. In some vague way, we generally accept the idea that such artifices unlock the secrets of nature, but ultimately our marvel at the artifices themselves compensates for the vagaries involved in translating the "internal" validity of the laboratory into the "external" validity of the real world (cf. Fuller 1989a: pt. 3).

Yet awe is not the order of the day in the public reception of social science. If it were, then B. F. Skinner's intricate schedules for shaping pigeon behavior should have inspired as much intrinsic fascination as the discovery of a new microphysical particle. But not only does the external validity of experimental research need to be explicitly demonstrated (not presumed) in the social sciences, but it typically needs to be demonstrated as contributing toward solving an important social problem (cf. Campbell and Stanley 1963). If such high expectations were routinely imposed on the natural sciences, they too would seem singularly unimpressive. Luckily for natural scientists, when the most is expected from their work, such as in the launching of a spacecraft or the deployment of a new drug, it is subject to the least public scrutiny, whereas when scrutiny is highest–that is, when an experiment is closely observed by a host of interested lay parties–the spectacle of the laboratory is taken to be largely its own reward. And herein lies the full secret of the success of the natural sciences: they *inversely* vary the levels of expectation and scrutiny.

From these last remarks follows the second point about the nonepistemic route to epistemic privilege. It is one that recalls the ongoing debate between structural functionalists and ethnomethodologists in

Two: Interpenetration at Work

American sociology (e.g., Knorr-Cetina and Cicourel 1981). Functionalists start with the idea that society is a natural kind tractable to scientific analysis. Consequently, they pitch their descriptions at the level of abstraction needed for characterizing "functions" that all societies need to perform in order to maintain themselves. Since societies differ in so many details, the level of abstraction will have to be quite high to pick out universally shared functions: e.g., "education," "economy," "religion," "law." But as ethnomethodologists have then pointed out, once we turn to a microanalysis of actual societies, we find few, if any, constraints on the range of activities that may perform a given function in a given society. This fact alone is grounds for thinking that the picture of societies having to perform certain functions is little more than a theoretician's fantasy.

Likewise, it may be argued, it is telling that in order for both the humanistic and the experimental disciplines to be seen as having performed the same "social function of science," one must quickly add that the sense of "worldly power" attached to these two forms of knowledge is rather different. Why not, then, simply deny that there is any interesting sense in which medieval *scientia* and twentieth-century "science" have played the same role in society? They are simply two institutions that have etymologically continuous names, but nevertheless perform rather different social functions in rather different societies. In that case, might the very idea of a social history of knowledge production that extends from ancient Greece to the present be a figment of the sociological imagination fostered by an illusion of what Wittgenstein would call "surface grammar"?

The Elusive Search for Science in the Social Sciences: Deconstructing the Five Canonical Historiographies

My argument has pursued some of the ways in which the scientific study of science may reveal that there is no science to study. After examining the recent deconstruction of the demarcation criteria problem, I considered the Baconian Virtues as a basis on which the attribution of "science" to a social practice could be made. But it would seem that this approach is, in a sense, a little too promising, for it throws into doubt not only the existence of any univocal conception of science, but any univocal conception of power as well–at least if power is defined in terms of the production of real world consequences. The nature of those consequences and the social function that they are taken to serve have changed considerably over time. Indeed, these changes left us wondering at the end of the last section whether a continuous history of science is to be had at any level of analysis whatsoever. I will now argue for an affirmative answer to this question, using as my linchpin the idea that the ultimate ground for the "Knowledge Is Power" equation is *rhetorical*. For the thread that connects the history of science from the Greeks to the present day is that people come to be convinced that particular forms of knowledge are embodied in the world—in skillful people and crafted goods—and are, in that sense, the hidden sources of power over the world.

Since rhetoric works only on "receptive audiences" and experiment works only given the proper "initial conditions," the only sort of power that we can be sure that a form of knowledge generates is an interest in the production of particular effects. People come to be convinced that certain deliberately staged events are exemplars of the knowledge-power nexus, that more "natural" events are to be interpreted charitably in terms of these exemplars, and that potentially troublesome events are to be ignored altogether. Rhetoric declined as a discipline during the Scientific Revolution, as people came to scrutinize—as they had not done before—the relation between rhetoric's claim to knowledge and the real world consequences of having that knowledge. However, rhetoric is not alone in its vulnerability to this sort of critique. Social constructivists in science studies today are making a similar move on experimentally derived knowledge in the natural sciences, arguing that these sciences seem to be epistemologically sound only because we have learned to turn a blind eye to the many times when avowed methodology and actual practice diverge (cf. especially H. Collins 1985; Woolgar 1988b).

If my line of inquiry is on the mark, then it would seem to have serious reflexive implications for how the history of the social sciences is conceptualized. On the one hand, the perennial need to persuade, or move people generally, is a robust social fact worthy of scientific treatment; on the other hand, if a scientific treatment of this phenomenon is possible, then that is only because we have been persuaded to see knowledge as having been embodied in particular ways on particular occasions. One clear consequence of this reflexive tension is that histories of the social sciences tend to self-deconstruct. The closer they get to detailing the knowledge production process, the more the authoritativeness of their accounts is jeopardized, for the reader is implicitly provided with the analytic tools to subvert the distinction between

Two: Interpenetration at Work

knowledge as being *about* the world and knowledge as being *in* the world. However, this serves only to cast aspersions on the accounts themselves, including (ironically) their generalizability to the natural sciences. Yet if the history is written so that knowledge is exempted from explicit consideration as a social fact, then the reader is sustained in conceptualizing knowledge in disembodied terms, in which case the social sciences can, at most, draw on the epistemic authority of the natural sciences. The task of managing this tension is given to the social sciences' *canonical histories*.

A history is "canonical" if it is written by a practitioner of a discipline with the express purpose of painting a panglossian picture of the discipline's development. Let us consider what this does *not* mean. First, true to Dr. Pangloss, it does not mean that the events recounted in the history are, when taken by themselves, exemplary. For example, the history of political science is canonically emplotted as successive attempts to recover the original (largely Greek) unity of theory and practice from successive practitioners of the science. Indeed, as will become clear below, political science is as close as one could get to a discipline whose history has been written by the *losers*. As for the second disclaimer, just because a history is canonical, it does not follow that it must tell a story of ever greater scientificity. On the contrary, the author may be forced to conclude that the discipline has not, on the whole, enhanced its scientific credentials–or perhaps even its disciplinary autonomy. Yet the author would try to show, in proper panglossian fashion, that "it has all been for the better:" for example, the discipline comes to recognize, more than it did originally, the inherent richness of the phenomena which it tries to understand, which, in turn, encourages the historian to fuzzify her discipline's boundaries. Because the authors of canonical histories are disciplinary partisans entrusted with structuring a vast amount of information for use by fellow partisans and sympathetic onlookers, one would expect to find the authors fusing their own perspectives with those of their subjects. Thus, they have a tendency to impute to a historical personage foreknowledge of the role that her actions ultimately played in the formation of the discipline.

Several sorts of questions may be posed of canonical histories. I elaborate three below. They are followed by brief accounts of the canonical historiographies of five social sciences: anthropology, sociology, political science, economics, psychology (cf. Hoselitz 1970). These accounts are intended to offer guidance to the reader interested in answering the three sets of questions, but I do not pretend to have provided conclusive answers here. At the end of the accounts, however, I

Reflexion

sketch a case in which impressions made on readers by the histories inscribed in two works of roughly the same vintage and scope helped determine which discipline gained control over a contested domain.

1. To what extent did the practitioners of these disciplines think that the "well-boundedness" of their discipline rested on its resembling a scientific exemplar? Could one have a well-defined yet nonscientific discipline of, say, anthropology? Or would a nonscientific anthropology be essentially indistinguishable from other humanistic studies? In short, is every disciplinary boundary question also an implicit question of demarcation in the sense that interested the positivists and Popper? My own view suggests that this is largely the case.

2. To what extent do these debates occur during a time when it is unclear which disciplines are exemplars of science? As mentioned above, a significant shift occurs in the second half of the nineteenth century, when most of these debates begin. Did each of the defenders of "science in the social sciences" propose to construct their discipline on the model of experimental physics? This is certainly the received wisdom, which has been recently given an interesting systematic treatment by the philosopher Peter Manicas (1986). Manicas argues that the foundational debates in the social sciences presupposed the faulty epistemological understanding that natural scientists had of their own activities, most of which were justified in terms of misunderstandings of what Newton had achieved.

3. To what extent do the differences between earlier and later historical accounts reflect a difference between the relative openness and closure of the disciplinary boundaries? I would expect an affirmative answer to the following three elaborations of this question: Do the later accounts present a continuous narrative where a fragmentation of perspectives was presented in the earlier accounts? Do the later accounts present the disciplines as more "internally driven" (i.e., fewer references to events in other disciplines or society in general as influential) than the earlier ones? Are contemporary concepts and theories attributed to disciplinary founders in the later accounts, even though the names of these entities had not yet been introduced when the founders originally wrote?

Two: Interpenetration at Work

Let me digress for a moment on the second set of questions, as that will enable us to see, in short compass, the rhetorical rootedness of historical attempts at demarcating the sciences on methodological grounds. I suspect that Manicas overstates the extent to which experimental physics functioned as an exemplar of the nascent social sciences, especially at the start of the foundational debates. On the one hand, Manicas (1986: chap. 1) rightly notes that the initial impetus for "science in the social sciences" came from nineteenth-century positivism, which pointed to "prediction and control" as the natural signs of epistemic power in the physical sciences. On the other hand, it is not clear that prediction and control became the twin virtues of physics *until the social sciences*–or, more precisely, positivism on behalf of the social sciences–*started aspiring to those virtues themselves* (Scharff 1989).

Of course, precedents can be found in Francis Bacon and other Scientific Revolutionaries for modest claims to predicting and controlling nature by experimental science. Yet those twin virtues were usually presented as a means to the ultimate end of predicting and controlling *human beings* (cf. Sorell 1991). Indeed, as I observed in the previous section, the levels of expectation and scrutiny for the social sciences are typically higher than for the natural sciences, reflecting perhaps a cultural sensibility in which the natural sciences figure as methodological rehearsals for the truly efficacious knowledge to be gained by the social sciences. In any case, until the rise of positivism, the success of, say, Newtonian mechanics was traced not to the twin virtues of prediction and control, but rather to Newton's ability to provide a unified explanation of the disparate motions of the earth and the heavens under an economical set of principles. Indeed, it was only with Laplace's *Celestial Mechanics* in 1799 that Newtonian mechanics became able to predict the positions of the planets at a level of precision and accuracy that rivaled Ptolemy's (Toulmin 1972: 378). Throughout the seventeenth and eighteenth centuries, "moral certainty" was attributed to experimental knowledge, which is to say, it had the epistemic status traditionally reserved for persuasive, but logically inconclusive, arguments (Hacking 1975). Indeed, the first manual to present a "method" for the direction of the mind had been written by rhetorician Peter Ramus, who antedated Descartes by a century (Ong 1958). And the first manual of scientific method to show the hand of Descartes, the Port-Royal Logic, draws the discovery-justification distinction–so crucial for modern discussions of scientific epistemology (Reichenbach 1938: chap. 1)–in terms of the old rhetorician's distinction between contexts of invention and instruction (Arnauld 1964: pt. 4). In this light, Kant's

Critique of Pure Reason may be read as trying to show that the mind's receptiveness to Newtonian mechanics is indicative of something deeper than the persuasiveness of experimental reasoning to a given community at a given time; rather, Kant believed that Newton raised to self-consciousness the very conditions necessary for rendering the world intelligible to anyone at any time and any place (Beiser 1987). Thus, the categories that Kant took to underlie the intelligibility of modern science, especially causation, are no mere scholastic topoi or *aides-mémoire*, but the structure of human understanding itself.

The three sets of questions just enumerated can be answered in terms of the canonical historiographies surveyed below. Each of the five disciplines presupposes a model of the human subject, all of which together come close to providing a de facto criterion for demarcating these disciplines from one another. The criterion is encapsulated in two questions.

> 1. The *ontological* question: Is the subject's behavior determined principally by things happening inside her or by things happening outside her?
>
> 2. The *epistemological* question: Is the subject typically aware of the things that determine her behavior?

The following paragraph provides an idealized survey of the answers that brings out the implicit terms of peaceful coexistence among the social science disciplines. The complexity of the actual disciplinary histories is then elicited in the rest of this section.

Since the subject matter of *anthropology* is usually defined as incorporating the inquirer in some capacity, all four logically possible answers to the two questions are conceptually permissible. However, the other four disciplines have characteristic biases. With its emphasis on communal bonds, traditions, and norms, *sociology* tends to conceive of the subject as internally but subconsciously determined. Diametrically opposed to this conception is *political science*, whose concepts of power and forces imply a subject who is externally determined, yet sufficiently aware of these entities to turn them to her advantage. *Economics* also presents a subject who is aware of the determinants of her behavior, but these are now such internally defined entities as utilities and expectations. Finally, from its inception as an experimental science forty years before behaviorism, *psychology* has been fixated on the image of the atomized organism as a function of its environment. Consequently, it presumes that the subject is determined by forces

Two: Interpenetration at Work

outside her, forces of which she need not be aware. While I have undoubtedly stereotyped matters somewhat, a good test would be to see how different social sciences handle cognate areas (e.g., "social psychology" is handled by sociology and psychology), especially the extent to which the difference in treatment is attributable to a difference in how the human subject is conceptualized.

ANTHROPOLOGY

Although anthropology is now called the science of "culture" in order to highlight the discipline's commitment to studying the humanly mediated, "artificial" (as opposed to "natural") features of reality, Edward Tylor's original focus on "culture" was meant to stress the habitual character of the human condition, which signaled our evolutionary continuity with the rest of the animal kingdom (cf. Lévi-Strauss 1964: chap. 17). From these two understandings of culture have come the two main research sensibilities, which Marvin Harris (1968) has christened "emic" (symbolic-idealist) and "etic" (ecological-materialist). They are the sources of the two main traditions of canonical histories of the field. The emic tradition, illustrated by Malefijt (1974), portrays anthropology as the general study of humanity with quite open disciplinary boundaries and a commitment to methodological eclecticism. The etic, illustrated by Harris, portrays anthropology as increasingly turning to natural scientific models (borrowed especially from systems ecology and evolutionary biology) in order to overcome the folk beliefs of our own and other cultures. Each sort of history tends to blame the excesses of the other for anthropology's notable embarrassments: e.g. emic historians locate the roots of racism in etic attempts to reduce social to biological phenomena (e.g., Social Darwinism, ethology), whereas etic historians trace it to the emic tendency to reify perceived cultural differences (e.g., Romanticism, Nazism). And the excesses do seem to go both ways. If it is common nowadays to fault etic anthropologists for importing Western standards of efficiency to evaluate the rationality of native practices, that is not to say that a more ethnographic approach is guaranteed to remedy matters. After all, in the nineteenth century, racism was more virulent among those who actually observed the native cultures firsthand than among the armchair systematists, like Tylor, who tended to make a priori pronouncements about the unity of humankind, admittedly to justify bringing the natives up to Western standards (Stocking 1968). Until very recently (e.g., Clifford and Marcus 1986), both camps of anthropologists have been unreflexive in a crucial respect, namely,

Reflexion

they have refused to think of their discipline as a culture in its own right. Such refusal has been a by-product of their studiously trying to avoid charges of ethnocentricism. In an interesting sense, the objectively oriented etic anthropologist has been less guilty of this charge than her relativistically oriented emic colleague. For whereas the former typically admits the presence of her own disciplinary culture (and its superiority to the natives'), the latter presents herself as ideally a (universal) mirror of whichever native culture happens to be under study, which usually presupposes that the natives see just as much of a difference between their culture and the anthropologist's as the anthropologist sees. Ironically, anthropology's recognition of its own presence in the study of culture has coincided with the global interpenetration of cultural boundaries through mass communications (i.e., "globalization"), which, in turn, has led many anthropologists to wonder whether the very idea of their discipline as a culture might itself be an anachronism (Marcus and Fischer 1986).

SOCIOLOGY

Lévi-Strauss (1964: 354) distinguishes the methodological orientations of Emile Durkheim and Bronislaw Malinowski in a way that is emblematic of the ontological difference between sociology and anthropology: the former studies society as a thing, while the latter studies things as social. Despite Talcott Parsons' (e.g., 1951) efforts, sociology has forsaken any pretension of providing a unified theory of the social sciences in favor of pursuing an autonomous disciplinary course, one with a conspicuously provincial sense of the scope of "social theory," namely, whatever issues from the pens of "theoretical sociologists," who have themselves become largely autonomous from empirical sociological research (cf. Giddens and Turner 1987). Moreover, canonical histories of the field (e.g., Martindale 1960; Bottomore and Nisbet 1977) have consistently presented an endless proliferation of "schools of thought." Systematic works, starting with Parsons (1937), resemble medieval encyclopedias, even to the point of being typically criticized for not having included all the major schools. Contemporary systematists, such as Collins (1988), are preoccupied, less with sorting out the wheat from the chaff in these schools (and perhaps denying the validity of some altogether) than with delimiting the relevance of each school within an all-encompassing picture of society. Yet the canonical histories portray these schools as having some rather peculiar qualities: e.g. they are not necessarily linked to particular academic lineages, nor are they populated by unique sets of members, as in the case of "structural func-

Two: Interpenetration at Work

tionalism," whose founders supposedly include Weber and Durkheim, and sometimes even Marx. In addition, sociological schools are presented as laying claim to the entire subject matter of the discipline (much like paradigms), although it would seem that each school emerged from, and is most plausibly situated in, only part of that subject matter. This serves to confer on sociology a spurious sense of internal division, one due more to lack of communication than to genuine disagreement. For example, the microperspective of Simmel's formalism and the macroperspective of Durkheim's functionalism appear in conflict only if one neglects the fact that Simmel is usually talking about specific sorts of face-to-face interaction, while Durkheim is talking about systemic features that are necessary for any social interaction to take place. All this points to a major reflexive weakness of sociology: namely, that the discipline is not well constituted as a social unit. To classify schools of thought by common themes rather than by academic lineages is to suggest that the same ideas can arise independently in different social settings, which, in turn, calls into question the extent to which knowledge is socially determined. Indeed, histories of the field typically begin by invoking "the experience of modernity" shared by such otherwise disparate figures as Marx, Weber, and Durkheim. Until an account is explicitly given of how an abstract experience like "modernity" can emerge from the quite different social settings in which these thinkers lived, the locus of sociological inquiry will remain in doubt.

POLITICAL SCIENCE

Recent canonical histories (e.g., Wasby 1970) have made the discipline seem like the social scientific equivalent of comparative morphology: i.e., devoted to producing "middle-range theories" consisting largely of correlations between types of political structures and functions. This is very much in the spirit of political science's inauspicious origins as the loser of the *Methodenstreit* over the scientific status of political economy that took place in Germany in the 1890s (Manicas 1986: chap. 7). Today's political scientists have descended from proponents of the "historical method" in economics, who were interested in assembling the social record of various nations for purposes of discovering statistical tendencies that would simultaneously contribute to social science and social policy. In addition, histories of the field vacillate over whether the middle range is to be subsumed under more general theories of human nature or to be further deployed as tactics in political practice (cf. Ricci 1984). It is quite striking that political scientists who tried to

Reflexion

transcend the middle range by using interdisciplinary research as a vehicle for modeling political complexity (e.g., Harold Lasswell, Karl Deutsch, Murray Edelman) have been repeatedly treated with respect during their professional careers, but promptly forgotten thereafter (J. Nelson 1987). Indeed, the most persistent pursuit of "grand theory" in political science has taken the form of a subterranean "classical tradition" that follows Machiavelli, Hobbes, and Locke in deriving political counsel directly from a general account of human nature, bypassing the middle range entirely (cf. Skinner 1987). This tradition regards twentieth-century political science as little more than a temporary aberration. At the start of the century, this movement was associated with such cultural conservatives as Michael Oakeshott and Leo Strauss, but now it has taken on a more progressive cast, as the study of politics seeks to reestablish its ties to civic responsibility (e.g., J. Nelson 1986). The main reflexive difficulty here is that political science's need to have an independent object of inquiry politically incapacitates its practitioners (Karp 1988). In the case of the dominant tradition, "political realism" portrays the politician as someone who mediates conflicting "forces," the macrohistorical nature of which resists the intentions of the individuals who serve as its vehicles (cf. Keohane 1986). These forces, often called ideologies, are discussed as if they were reified states of mind (e.g., "terrorism" as a contemporary political force). However, more like physical forces, political ones may be constrained by particular human interventions, but never fully eliminated. Thus, not only the twentieth-century political climate, but also the ontological demands of political science, serves to stereotype politics as a Bismarckian *Realpolitik* where the goal is always the containment, but never the resolution, of political differences. As for the subterranean tradition, it typically falls back on an equally incapacitating eschatological model of political history, whereby the scholar keeps alive the classics until "the time has come" to reunite political theory and practice (Gunnell 1986). Of all the social sciences, this is the one with the severest cross-cultural identity problem. Should political science be housed in the law school (as in Germany), the school of public administration (as in France), or the college of liberal arts (as in the United Kingdom and the United States)?

ECONOMICS

Histories here seem to be most preoccupied with the scientific status of the discipline. Earlier histories (e.g., Robbins 1937) portray progress in the field in terms of the gradual conceptual, if not practical, separation

Two: Interpenetration at Work

of the economist's value judgments from those of the economic agents. Later histories (e.g., Blaug 1978) shift the motor of progress from this sort of conceptual analysis to the increasing willingness of economists to submit their theories, however value-laden they may be, to quantitative tests. This shift matches a move in the preferred philosophy of science from positivism (via Max Weber) to Popperianism, as well as a trend toward portraying Neoclassicism as having definitively triumphed over Marxist, Keynesian, and other institutionally oriented economic paradigms. The reflexive worry that dogs the history of economics turns on the tension between the rationality of the economist and that of the economic agent. Classically inspired approaches presume that the object of economic inquiry is the natural order that emerges from the sum of individually rational and self-interested agents. But given the instability of Western economies during the period that has coincided with the history of economics, it would seem that, as a matter of fact, there is no particular sense of order, or "equilibrium," toward which the economy naturally gravitates (Deane 1989). Is this because economic order can be maintained only through intervention (à la Keynesianism) or because the natural order has been subject to external interference (à la Neoclassicism)? In either case, to preserve her object of inquiry, the economist must apply her own scientifically defined sense of rationality to compensate for the inadequate rationality displayed by the agents themselves (Lowe 1965). The need for two senses of economic rationality is usually attributed to differences in economic scale (Georgescu-Roegen 1970): that is, individuals are good utility maximizers relative to their own cognitive horizons, subject to periodic revision, which unfortunately are insufficient for understanding the entire economic system (Keynes 1936). The economist's special sense of rationality occurs both retrospectively and prospectively. When regarding the past, economists have become particularly adept at interpreting seemingly noneconomic practices, such as in the emergence of an elaborate system of patent law to provide both stimulus and protection to innovation, as latently economic (cf. Lepage 1978: chap. 3). When regarding the future, economists exhibit their rationality in the standard uses to which statistics are put as "indicators" of long-term trends that normally escape the deliberations of the economic agents (McCloskey 1985: chap. 9). This, in turn, has enabled economists to enter the political arena more readily than other social scientists, including political scientists. At the same time, however, the public demand for economists reflects the sense in which the concept of economic rationality is an artifice of economic science. Indeed, as they continue to mask the manufactured character of economic rationality behind spurious analogies with

Reflexion

idealized closed systems in the physical sciences, economists run the risk of communication breakdown among themselves, as well as with the public (cf. Klamer et al. 1988).

PSYCHOLOGY

Canonical histories of psychology conform best to the canons of ordinary historical scholarship, in terms of pinning down who did what when and where (especially Boring 1957). This may have to do with the field's strong sense of academic lineage, owing largely to its origins in the German research university. Thus, a succession of students pass through a laboratory which is under the directorship of a major professor. In this regard, psychology is (ironically) the social science that is most self-consciously a social construction, even to the point of conducting research under a distinctive identity (i.e., "experimenter" and "subject") in a distinctive setting (i.e., the lab) (cf. Morawski 1988). As opposed to the situation in sociology, schools of psychology are clearly identified with nonoverlapping sets of people. The cross-fertilization of schools is nonmysterious, typically involving the member of one school spending time in another school's lab and then revealing the effects of that contact in her subsequent work. However, recent histories (e.g., Schultz 1981) have had difficulty accommodating post-World War II developments to the canonical lineages, which makes these histories the least reliable in the social sciences as guides to current research. It may be that psychology has become too technical for its own historians to fully grasp, or, more likely, that the field has implicitly relinquished the narrative principle of earlier histories. That is, psychology aims to provide no longer a general theory for the entire human psyche, but merely special theories of perception, cognition, motivation, and other processes that can be read off the behavior of individuals. Moreover, these processes are increasingly being treated not as uniquely human, but as instances of still more general processes that are studied by other disciplines: e.g., the modeling of cognition on computers, the subsumption of motivation under animal ethology. Indeed, the recent fragmentation of psychology may be one vindication of Foucault's (1970) prognosis for "the death of man." As such, it calls into question the extent to which the research programs that today travel under the name of psychology are unified by anything more than a common physical object (the individual human) that is studied in a controlled physical environment (the laboratory). This conclusion would be especially ironic for a discipline that has repeatedly taken great pains to distinguish its subject matter from that of the physical sciences, starting with

Two: Interpenetration at Work

Wundt's distinction between immediate and mediate experience and culminating with B.F. Skinner's strategy of treating the behaving organism as a physiological "black box."

How Economists Defeated Political Scientists at Their Own Game

It may be useful, at this point, to consider a case in which a difference in canonical histories seems to have contributed decisively to one discipline's overtaking another in some contested domain (cf. Fuller 1989b). The domain in question was "politics," that amorphous mix of classical philosophy, legal history, and folk sociology typically studied by aspiring civil servants at British universities in the nineteenth century (Collini et al. 1983). The contesting disciplines were economics and political science, and the contest has had far-reaching consequences, for economists in the twentieth century have come to dominate the higher echelons of "policy," be it as art or science, even though political science would seem to provide a richer and more relevant background for the aspiring policymaker or analyst. Indeed, some political scientists study economists for their policymaking skills (e.g., Galbraith 1988), a compliment that an economist would never think of repaying. Indeed, America's leading political scientist of this century, Harold Lasswell (1948: 133), bemoaned his discipline's lack of "public image," so that even Harold Laski, who succeeded Graham Wallas in the Chair in Politics at the London School of Economics, felt compelled to call himself an economist in order to enhance his credibility. Economics and political science epitomize stories of success and failure, respectively, in the art of discipline building, artful and artless ways of reading one's past into the future.

By 1910, two books had become famous in Britain for presenting a "scientific" study of politics, Alfred Marshall's *Principles of Economics* and Graham Wallas' *Human Nature in Politics.* Given the diverse constituency for such a science–academics and bureaucrats, as well as journalists and practicing politicians–it might be thought that Wallas had the natural advantage: his book was shorter and easier to read, and showed the signs of being written by someone who had spent time away from the academy and in the halls of Parliament. Wallas had the advantage of networking from London, while Marshall remained

Reflexion

ensconced in Cambridge. However, the self-images that we have found to be characteristic of political science and economics can be equally seen in the historical perspectives that Wallas and Marshall, respectively, adopt toward the domain of politics. I will first consider the stance that each takes toward his discipline's past, and then look at how each handles a contemporary rival for epistemic authority, psychology.

There are two ways of measuring the control that Marshall and Wallas exert over the histories of their disciplines. The first is the ease with which they transfer analogies from the past to the present, thereby establishing a continuous line of inquiry. On the one hand, we have Wallas' (1910: chap. 4) Sisyphean struggle to transfer the concept of democracy from the rationally tempered, homogeneous Greeks to the volatile, heterogeneous nations he finds in early-twentieth-century Europe. On the other hand, we have Marshall's (1920: chap. 1) nimbler efforts at showing that modern capitalist economies preserve what was worthwhile in precapitalist conceptions of trust and cooperation. The increase in dishonesty that critics of capitalism regarded as having decisively corrupted the economic scene, Marshall explains simply in terms of the appearance of more opportunities to express dishonest sentiments the extent of which has remained unchanged in the course of history. In fact, Marshall managed to strike a blow for both science and morality by refusing to accept at face value the eighteenth-century dogma that "private vices make for public virtue." Rather, he took it as an empirically open question for economists to determine who the true "captains of industry" really were, that is, the people whose enterprise made the decisive difference to economic growth. They did not necessarily correspond to the Dickensian stereotype of the rapacious industrialist (Collini et al. 1983: 335).

The second way of measuring the two authors' control over the historical record is by the cooperation or resistance that their predecessors are seen as providing. Quite telling in this respect is the contrast in metaphors that are elicited from Marshall and Wallas when they speak of "founding" a science: Marshall tends to see the process as a matter of "building an edifice," whereas Wallas regards it in terms of "clearing the ground." And while Wallas' image could be reasonably read as a precursor to Marshall's in some unified project, it is significant that they are writing at the same time, with presumably the same information on hand, which suggests a difference in the standards and expectations that the two authors bring to the task of founding a science. Clearly, Marshall's epistemic criteria are better adapted to the knowledge of his day than Wallas'. For his part, Marshall (1920: App.

Two: Interpenetration at Work

B) offers a brief but striking portrait of economics in the second half of the nineteenth century as an international, cooperative enterprise devoted to completing the consumption side of the economic equation after the classical political economists had mastered the production side in the first half of the century. If anything, Marshall overplayed the cooperativeness of his contemporaries, even with regard to their own understanding of what they were doing. Wallas (1910: Intro.) was, once again, less effective in his quite opposite attempt to condemn the two major schools of political thought in late Victorian Britain, the bureaucracy-oriented utilitarians and the university-oriented idealists. He presented them as exemplifying the practice-theory split that perennially stymies the possibility of a science of politics. Regardless of the justness of Wallas' charge in his own day, it served only to aggravate his potential allies within the political establishment and to discourage his potential allies outside of it (Collini et al. 1983: chap. 12).

With regard to the psychology of his day, Marshall believed that he had to challenge directly that discipline's authority, given the wide scope that economics needed in order to cover politics completely: namely, "men [sic] as they live and move and think in the ordinary business of life" (Marshall 1920: 14). His strategy here was to argue that economics presupposed a psychological theory and method more fundamental than the ones normally treated under the rubric of psychology. For if economics accepted without question the psychic diversity of motivation, then the quantitative basis of the discipline would be undermined, given that values cannot be calculated unless they are reducible to a common currency of utilities. Thus, the psychologist would reinforce the very difference between, say, smoking a cigar and drinking a cup of tea that Marshall needed to eliminate in order to render the two activities amenable to economic analysis. Marshall's solution was to argue that smoking (a certain amount) and drinking (a certain amount) could be seen as functionally equivalent means to some common end. In other words, as long as psychological states were regarded as always transitional to some other state, then the possibility of exchanging the states at some rate was left open to the economist as a distinct and important field of study—one which highlighted the dynamic character of the agent, in contrast to the more contemplative, and hence less realistic, image depicted by the introspective psychology of Marshall's day.

By contrast with Marshall, Wallas heavily relied on the instinct psychologies of William MacDougall and William James that were popular at the start of this century. Wallas tried to argue that quasi-biological political impulses gave content and purpose to the rational

Reflexion

calculations that economically oriented thinkers emphasized, but that at the same time these impulses occasionally had effects on political behavior (e.g., riots and upheavals) that escaped calculation. However, Wallas ran into reflexive difficulty in trying to define the political in this way, for, on the one hand, if the quasi-biological substance of politics interrupts calculation too frequently, then the psychology of anyone attempting a rational science of politics (such as Wallas himself) should be held suspect; but, on the other hand, if the impulses rarely bubble to the mind's rational surface, then they can safely be ignored, and we would then seem to be back to Marshall's vision of economics. As Marshall himself realized, Wallas' dilemma fed nicely into the commonsense notion of "politics" as "mere expedients," tactics that temporarily worked but failed to probe the deeper, long-term tendencies that only a deductive science like economics could handle (Collini et al. 1983: 332).

At the outset, I suggested that the relative accessibility of Wallas' text should have given him a natural advantage over Marshall in shaping the future of politics as a field of inquiry. And, indeed, *Human Nature in Politics* was reviewed in a wider range of periodicals than *Principles of Economics*, including journals in disciplines outside of politics (e.g., *Ethics, Psychological Bulletin*) and even some highbrow magazines (e.g., *The Bookman, Saturday Review, Yale Review*). Moreover, whereas the general tone of Marshall's reviews was muted and respectful even when critical, Wallas' work elicited the entire gamut of responses, from adulation to sarcasm. Herein lies a hint of why Marshall succeeded and Wallas failed. Marshall's is a formidable book that took great pains to accommodate potential opponents, even if by significantly reconstituting them in terms of his own project. As such, the book did not lend itself to either easy dismissal or easy appropriation. This forced critical readers of Marshall's *Principles* into a dialectical corner. Since their own research agendas had probably been incorporated into Marshall's disciplinary vision, they were confined to disagreeing on matters of detail rather than of overall conception. However, at the same time, the systematic cast of the book meant that they could not simply pick and choose the bits that suited their own purposes. Whoever wanted to deal with Marshall had to deal with him on his terms–though they were terms designed to be reasonably hospitable to the likely reader. In this regard, it is significant that when Talcott Parsons (1937) made his initial stab at unifying the social sciences, he devoted the first part of his effort to an examination of Marshall's architectonic. By contrast, Wallas would seem to have written *Human Nature in Politics* in a way that made it

Two: Interpenetration at Work

all too consumable and malleable by its readers. To take a subtle point that recurs in the book reviews, reviewers tended to present the evidence that Wallas cited for his claims, which was often drawn from recent newspaper stories, as striking in its own right, but without giving a clear sense of the claims that it was meant to support. And, as for the reception of Wallas by other disciplines, the authors of these reviews respond enthusiastically, largely because they believed they had found in *Human Nature in Politics* independent corroboration for positions that research in their own discipline had already established.

If we were to rely solely on such traditional epistemic criteria as explanatory breadth, predictive accuracy, and problem-solving effectiveness, economics and political science would probably not seem so far apart–both relative failures in comparison to the natural sciences. However satisfying this judgment may be to philosophers of science, it willfully ignores some brute facts which suggest that the real value of bodies of knowledge lies in some entirely different set of considerations. For starters, though their epistemic credentials would indicate otherwise, physicists are taken to have a much more restricted range of expertise than economists. Yet it is no secret that the two disciplines are unrepentantly "academic," that is, both trade on ideal conditions and closed systems that bear little resemblance to the complexities of the actual world. Moreover, it will not do simply to argue that economists deal with what are in fact the ultimate determinants of the public sphere–the bottom line of cost-benefit analysis–since experience repeatedly shows that money matters do not enjoy any special ontological or epistemological status as a foundation for public policy. The authority of economics is not diminished by its repeated failures to analyze costs and benefits accurately, nor is the integrity of economic reasoning undermined by the ingenious ways in which even accurate cost assessments can be finessed, namely, by deficit spending and inflated currency.

All that having been said, it may nevertheless be reasonable to trust the judgment of economists over that of physicists simply on grounds of subject matter. But that does little to explain the authority that economists exercise in government, especially as policy consultants, where one would expect a preponderance of political scientists, who are, after all, the avowed experts in the area. Moreover, the aspiration of political scientists to capture the day-to-day complexity of the governing process, whether it be congressional voting patterns or public opinion change, stands in contrast, in what would seem at first glance to be a rhetorically advantageous way, to the abstract model-

Reflexion

building tendencies of economists. However, first impressions are deceptive, though they persist through the histories of the two disciplines. Although survey research methods from political science have influenced the way data are gathered for policymaking, the actual policymaking is put in the hands of the economists. Why? To simplify the story considerably, much turns on the economists' skill in rhetorically converting their epistemic liabilities into political virtues.

Economists have managed to portray their idealizations not as the false images of reality that, strictly speaking, they are (a point that physicists readily concede about their own idealizations; cf. Cartwright 1983), but as normative standards against which reality can itself be judged and toward which it can be corrected. Thus, the model of the free market becomes less an empirically inadequate picture of economic activity and more a desirable goal toward which policy ought to be directed. Goodwin (1988) has observed that when economists move from the academy to the policy arena, they signal this normative turn by exchanging metaphors from physics for those from medicine: abstract equilibria become symptoms of a healthy market. The self-assumed imperative of bending the real to the ideal is, for the most part, lacking from the assertions of political scientists, whose stress on the causal significance of locally varying factors makes them more guarded about giving any general advice (cf. Almond 1989: chap. 2). Consider even the rationality assumption that economists typically make of agents, which, over the last thirty years, has done the most to place political science on a scientific footing. Instead of using the assumption as the economist does, to measure the shortcomings of agents from a superior vantage point, the political scientist uses it to remind *herself* that she has yet to fathom the full rationality that belies the surface discontinuities in an agent's behavior (Keohane 1988).

Again, this difference between economists and political scientists occurs at the level of practice, as well as of theory. Unlike Wallas, Marshall refused to get caught up in the tariff policy debates of his day, which indirectly served to increase his political clout by enabling him simultaneously to claim the high moral ground and to avoid any immediate tests of his hypotheses (Collini et al. 1983: 336-37). A lesson that economists have learned from Marshall's practice is that the more long-term one's empirical perspective is, the more it becomes a de facto normative perspective. One's rhetoric thus shifts from talk about what will happen next to talk about whether the time is right to assert what will happen next. Needless to say, talk cannot temporize forever, but economists have been generally good at dictating the times and terms in which they pronounce on policy (cf. Galbraith 1988). This ability to

Two: Interpenetration at Work

manage the circumstances under which one exercises power—perhaps more than the actual power itself—is a sign of the economist's "expertise" (cf. Fuller 1988a: chap. 12). Rhetoricians have long recognized this phenomenon as *kairos* (cf. Kinneavy 1986; C. Miller 1992a, 1992b), and social psychologists have more recently focused on manipulating the moment of decision as an effective tool for a minority to use to determine an outcome (Levine 1989). In the economist's own terms, such metamanagement allows her to minimize the hidden "process costs" of making policy statements that turn out not to work, namely, a loss of credibility when making future pronouncements (cf. Sowell 1987: chap. 4).

Moreover, the rhetorical deftness just described is not merely part of the tacit practice of economists, but actually built into some of the core concepts of economic science. Consider two such concepts: *scarcity* and *unintended consequences*. Of course, the very need for economics arises from the phenomenon of scarcity, which is portrayed as the ultimate reality indicator, the material equivalent of logical contradiction (Xenos 1989). Two wants cannot be satisfied at once. One must either change the logical space in which the decision is made—for example, by expanding the time frame or range of resources involved in the decision—or, as economists more typically advise, hold the space constant and make tradeoffs between the conflicting demands. In their more metaphysical moods, economists have been prone to argue that scarcity is the source of real choice, for a world of pure abundance would generate an unmanageably large number of possibilities for satisfaction. Free will can be experienced only if the range of available options is sufficiently restricted that the agent feels that deliberating on them can determine her course of action. Informed by the economist's sense of metamanagement, this becomes a rhetorically powerful idea, as the economist is able to "structure" a decision in such a way that policymakers feel that they are freely choosing from among an array of options, even while the economist knows that the outcome of any of those choices is likely to fall within a very narrow range of events. By contrast, whereas scarcity closes the logical space of action, the appeal to unintended consequences opens the space up at a crucial juncture, when the economist needs to temporize. Typically, the appeal is made at once to both negative and positive consequences. Here the economist would show concern that hasty policy intervention may unwittingly disrupt the self-corrective processes of the market that take place over the long term, even though the signs of recovery remain, at best, ambiguous at the moment.

Reflexion

In considering the arguments that were made for and against capitalism prior to its eighteenth-century triumph, Hirschman (1977) has identified what may be the original bifurcation of attitudes on the part of economists and political scientists toward the rationality of agents, attitudes that stem from Plato and Aristotle, respectively. Capitalism was clearly predicated on a view of human psychology that enabled agents to calculate their "interests" and act on them accordingly. However, it remained to be seen whether such interests were a philosophical fiction or some suitably sublimated version of the "passions" that were known to motivate human behavior, normally in quite irrational ways. The proto-political scientists took the Aristotelian line that the wayward nature of the passions made them ontologically ill suited to numerical representation, which meant that agents could not be expected to make consistent utility assignments in their thinking. By contrast, the proto-economists took this numerical intractability as a moral failure, specifically as grounds for agents to engage in certain forms of self-discipline so as to transform their passions into mathematically focused interests. With this micro-Platonism, the groundwork for a "moral science" was laid. Over the next two centuries, it forged links between the determinate outcomes of mathematical procedures (e.g., cost-benefit analysis), the correspondence of those outcomes to real world events, and the decisiveness of the course of action dictated by those procedures. And as we have seen in this section, such a difference appeared in the psychological arguments that Marshall and Wallas used to stake claims for economics and political science as sciences of the political.

Conclusion: The Rhetoric That Is Science

Canonical histories of the social sciences must walk a fine line between subverting their own epistemic status and reinforcing that of the current scientific exemplars. The fineness of the line may be measured in terms of the awkwardness, if not downright silence, with which these disciplines confront the possibilities of, respectively, an anthropology of science, a sociology of science, a political science of science, an economics of science, and a psychology of science. By way of example, consider the psychology of science.

Two: Interpenetration at Work

Generally speaking, psychology is interesting because, more than any other social science, it has had to negotiate its disciplinary identity between the sciences and the professions. At the outset, some notable experimentalists, such as Hugo Munsterberg and Oswald Kuelpe, did not see such a sharp line, and the latter even welcomed the possibility of housing psychology departments in medical schools, as the "normal" counterpart to psychiatry (Ash 1980). However, the proposal has continued to be controversial over the last hundred years, as defenders of a pure "scientific" psychology have worried that the premature deployment of psychological techniques could do more harm than good. Without passing judgment on the groundedness of these suspicions, it must be said that psychology's easy ability to generate fear is no doubt connected with its techniques–borrowed as they often have been from medical practice–which have a greater capacity for leaving traces both *about* and *on* individual human bodies than, say, the abstract aggregative techniques of economics, even though the latter arguably have had a more lasting (albeit more hidden) impact on people's lives.

Psychology claims to offer authoritative knowledge about the nature of the mind. Two preconditions for exercising such authority are that the psychologist can explain her own mind better than a nonpsychologist can, and that the psychologist can explain people's minds better than the people themselves can. Although psychoanalysis, to its credit, has been singular in its efforts at meeting the first precondition (i.e., by requiring that the analyst be analyzed before analyzing others), every school of psychology has claimed to have met the second precondition. These claims typically assume that scientific reasoning is especially good at explaining things. But have psychologists explained the source of science's explanatory power? If science were like any other field psychology might study, its practitioners' success at explaining the reasoning of scientists would depend on their having regular access to the scientific community. This would invariably require at least temporary control over what scientists do. For example, scientists might be routinely sequestered like jury members to undergo various psychological tests and experiments for purposes of eliciting salient patterns of reasoning. The institution of such a procedure would signal the power that the psychologist had to draw scientists away from their work and into her own.

But, as a matter of political fact, science is *not* like any other field, and so the psychologist of science can mobilize little more than undergraduates on a regular basis–and then at universities (e.g., Bowling Green State [Tweney et al. 1981] and Memphis State [Gholson

Reflexion

et al. 1989] Universities) far from the centers of major scientific research. Given this state of affairs, it is common to think of psychology of science as a "fledgling" or "marginal" specialty. However, if one looks closely at the events that the canonical histories of psychology (e.g., Boring 1957) take to have been pivotal in the founding of the discipline, they turn out to be moments, such as the discovery of the "personal equation" in astronomical measurement, in which science is needed to counteract the *shortcomings* of scientists. Indeed, the methodological control that psychologists have subsequently gained over mental phenomena–via sophisticated experimental designs and inference strategies–has come from learning how to systematically check their own mental biases and limitations. Indeed, if, as has often been suspected, psychology is more advanced in the range of methods it offers than in the results it has reached by those methods, then that would suggest that most of psychology has been psychology of science. But that would be to tell the story of psychology as the development of more clever compensations for the cognitive weaknesses of scientists, not as the cumulative growth of knowledge by exemplary cognitive agents (cf. Campbell 1989).

Emplotting the history of psychology as itself a psychologically involved activity heightens certain tensions in the scientific status of psychology that would remain neglected if psychology were treated as merely "about" minds. Yet, with the exception of Merleau-Ponty (1962, 1963; cf. Rouse 1987), there has been surprisingly little recent philosophical treatment of the implications of the proposition that of knowledge is both *about* and *in* the world. Indeed, there has hardly been any treatment at all among analytic philosophers, in whom the post-Kantian tendency to regard matters of epistemology as separate from matters of ontology runs deepest. In that case, before attempts to reveal the embodied character of knowledge are likely to be generally persuasive, it will be necessary to go back to a point in history when a highly familiar form of knowledge–one that we routinely treat as being about, but not in, the world–came to be embodied. I take this embodiment to be a twofold process that is intimately connected with the *internalization* and *externalization* of standards. By "internalization," I mean that people are capable of making judgments about new cases even without the presence of standards in the environment. Drawing on the social behaviorism of George Herbert Mead and Lev Vygotsky, Rom Harré (1979) has usefully called this process "privatization," which is what I believe most of cognitive psychology is trying to get at in its characteristically reified way. By "externalization," I mean that people come to forget that earlier encounters with standards have struc-

Two: Interpenetration at Work

tured their later judgments, which leads them to confer an "independent reality" on what is, in fact, a subliminally standardized world. Patrick Heelan (1983) has given the best phenomenological treatment of these realist intuitions, while Donald Campbell offers the best social science explanation for them (Segall et al. 1966).

The familiar form of knowledge in question is experimental knowledge, the basis of the distinctive power currently wielded by the natural sciences. Shapin and Schaffer (1985) present the seventeenth-century debate between Thomas Hobbes and Robert Boyle on the epistemic authority of experiment as the originating myth of the modern knowledge-power nexus. For their own part, Hobbes and Boyle were clear that the ultimate objects of prediction and control were human beings, but Hobbes still regarded Euclid's geometric proof as the exemplar of science. As a result, he explained the source of knowledge's power in terms of the explicitness of its reasoning process, which enabled the scientist to communicate directly with his audience. The audience could either agree or disagree with the scientist, but everyone could be clear, as there was no asymmetry between the scientist's and his audience's knowledge of the situation. Of course, the big drawback of such an "ideal speech situation" was that it did not guarantee the resolution of any disagreements that arose, especially if the scientist's audience is large and heterogeneous. Indeed, to Boyle's mind, this was enough to explain why scholasticism had failed to advance the frontiers of knowledge. And, alas, Hobbes himself used the inability of reason to resolve such disputes as grounds for introducing a strong sovereign to "channel" the discussion in a more "productive" direction.

Needless to say, Boyle leapt on the authoritarian turn in Hobbes's thought, citing it as a *reductio* of the dialectical approach. Boyle thought that "the way of talk" promoted disagreement at ever finer levels of analysis–until the heavy hand of the sovereign issued its final judgment. Instead, Boyle appealed to a coarser medium of epistemic exchange, one which could elicit assent from people who have verbally different accounts of the situation. Experimental observation did the trick, once the audience was instructed on viewing the experiment in the right way. For Boyle could secure not only mass assent but also an assent that seemed uncoerced, given the ease with which observation could be made automatic through instruction, and hence made to appear "natural."

In this regard, Boyle's "experimental method" is a version of the rhetorician's method of places, in which the experimenter is taught how to embody knowledge in the world (i.e., how to code the concrete situation in theoretically significant terms) so that it can be later re-

Reflexion

covered (e.g., when observation is compared with theory after the experiment is done). For his part, Hobbes fully realized the rhetorical character of Boyle's experimentalism, and indeed did not himself object to the use of rhetoric to secure agreement on matters of opinion. However, Hobbes wanted the use of rhetoric to be employed self-consciously on the people whose agreement was being sought. Indeed, he tended, in a Clausewitzian manner, to see the explicit exchange of reasons for positions and the naked use of force to stop debate as two ends of the same continuum. In both cases, the audience could trace the source of the power being wielded and thereby respond to it in an appropriate manner. In that way, the source became *accountable.* By contrast, Boyle's trained observer forgot all such traces, treating his own response as natural under the circumstances, unmoved by the artifice of rhetoric. And while Boyle was interested simply in embodying experimental observation as natural knowledge, an entire system of rhetorical associations have come to make the cluster of practices we call "science" appear part of a seamless whole with transparent access to some natural reality.

For example, we rather automatically presume that there is a mutual correspondence between the words and deeds of science, with the two operating as convergent indicators of some "natural kind" that is the putative subject matter of that science. Moreover, the sui generis character of this natural kind is mirrored by the autonomous character of science as an institution that exists in an environment that sets it apart from, and makes it unanalyzable in terms of, other social institutions. However, the mounting anthropological and sociological accounts of science reveal that these mirrored wholes are more perceived than real, and that in fact the laboratory is simply a point of confluence for structures and practices found in other, normally unrelated parts of society (cf. Latour 1987). Each structure and practice is the proper study of one of the social sciences, but it is paradigmatically studied in a nonscientific social setting. It may be, for instance, that the self-interestedness of scientists crucially contributes to the growth of knowledge (cf. Hull 1988); yet the proper analysis of that trait will be found not in a study of science but in a study of business behavior, on which scientific behavior is parasitic.

As I have argued here and elsewhere (Fuller 1989a: chap. 2), science does not have an essence but is rather simply the sum of disparate strands of society that are mutually reinforced in specific places, both by the behavior of scientists themselves and by our learned perceptual bias to ignore the disparateness, and to see instead these strands as embodying a common form of knowledge that is a source of

Two: Interpenetration at Work

worldly power. Because the social sciences continue to be perceived as only partially autonomous from the societies that support them, their histories provide a special opportunity–matched only by the emergence of experiment as a legitimate source of natural scientific knowledge–to examine the processes by which knowledge tries to be *about* the world without drawing undue attention to its existence *in* the world. And until we take seriously the thesis that knowledge inhabits the same world as its putative objects, we will not fully appreciate the implications this point has for the legitimation of our knowledge enterprises.

5

SUBLIMATION, OR SOME HINTS ON

HOW TO BE COGNITIVELY REVOLTING

Artificial intelligence (AI) wreaks havoc on anyone who likes clean boundaries between science and the public, as well as between the natural and social sciences. (The very name "artificial intelligence" immediately suggests something that is neither natural nor social.) Our story begins with an overview of how rhetorical impasses can develop as scientific knowledge circulates in and out of scientific circles. After enumerating the impasses that are relevant to debates over AI, I focus on one celebrated recent debate in which I and other STS researchers participated: Can computer models of scientific discovery refute the claim that science is socially embedded? In considering the line of argument pursued by the leading AI advocate, Peter Slezak, I focus on the way he mobilizes the historical record to portray a "Cognitive Revolution" already in full force. However, I then go behind the scenes to see whether the people enrolled in Slezak's holy war would all admit to being on the same side. As it turns out, Herbert Simon and Noam Chomsky, two alleged allies, do not sit very well together–especially now that they no longer have behaviorism to fight. This point reveals the extent to which "cognitive" is an umbrella term that obscures as much as it reveals. In particular, the term obscures the social character of things. I compensate for such obscurantism by seriously entertaining the idea that cognitive machines, computers, are virtual social agents, for whom a new political economy is needed, and to which both AIers and sociologists should have an interest in contributing.

Two: *Interpenetration at Work*

Introduction: Of Rhetorical Impasses and Forced Choices

Scientific controversies often reach rhetorical impasses, especially when differences of opinions solidify into mutually exclusive groups of followers who perceive themselves as bound to a common fate. Under these circumstances, there is simply no way of comparing, and hence no way of negotiating, the positions in question. When a scientific controversy is transferred to the public sphere, such an impasse can be created where none existed in the scientific sphere. Participants in scientific debate are typically encouraged to treat major theoretical positions as ideal types to which one would appeal in varying degrees, depending on the particulars of the case under consideration. For example, when the so-called nature-nurture debate is conducted in the scientific literature, all sides tend to admit that an organism's behavior is determined by some combination of genetic and environmental factors, the specific ratios of which are to be a matter of the specific behavior under study. However, once the debate goes public, participants tend to mobilize into two fairly rigid camps which come across as holding that *all* behavior must be *exclusively* determined by either genetics or environment (cf. Howe and Lyne 1992). This simplification of the scientific debate is due largely to the fact that the possible outcomes in the public sphere and those in the scientific sphere are defined quite differently. Whereas success in the scientific sphere is often no better measured than by the vicissitudes of journal citation patterns, success in the public sphere is announced by such unsubtle indicators as votes, appointments, and funding decisions, all of which involve forced choices of one sort or another. In other words, rhetorical impasses emerge in scientific controversy as a result of the restricted media available for expressing opinions, which do not allow for the nuances and fluctuations normally registered in the scientific setting.

I do not mean to bemoan the above state of affairs. After all, in a crucial respect, the mass media play much the same practical role as philosophy has aspired to play at a theoretical level. Both aim to unify and focus scientific debate by eliminating surface differences in the context of inquiry. However, philosophers and journalists have typically focused the ends of inquiry in different directions. Whereas the journalists aim toward integration with other strands in contemporary public debate, philosophers ultimately want convergence with the histories of the other sciences. Thus, the rhetorical impassability of scientific controversy explains much of how the subject matter of the history and philosophy of science comes into being. Left to their own devices, scientists can entertain a variety of incompatible theories for an indefinite period of time, trading off between the alternative accounts as empirical cases present themselves. Yet, the history of science that typically most interests the philosopher–the "internal" history of science–consists of a canonical series of great decisions that the scientific community supposedly made between rival theories. Most philosophers who study these decisive moments–as when Copernicus finally trumped Ptolemy–have presumed that the weight of the evidence, or some other "methodological" criterion, made the difference between accepting one theory and rejecting the other. However, the question that goes unaddressed during such considerations is why the moment of decision was when it was, and not earlier or later. As Larry Laudan (1984) has observed, scientists seem to be able to converge on a theory *when they agree that it is time to make a choice*. And as Serge Moscovici's research in the social psychology of group decision making has shown, those who control *when* the decision is made control *what* decision is made (Levine 1989).

If rhetorical impassability is as robust a phenomenon as I am suggesting, then one ought to look not to some internally generated decision-making process, or "methodology," to explain the sequence of theory choices that have been made in the history of science, but rather to the restriction on cognitive expression that results from the translation a scientific controversy into the more coarse-grained currency of the public sphere. To recall one famous thesis to this effect (Forman 1971; cf. Fuller 1988a: chap. 10), German physicists were able to argue back and forth about the merits of quantum indeterminacy without any felt need for resolution until the pressure to survive in an irrationalist culture dictated that they plump for the indeterminist interpretation. It would be too easy to say that the ambient culture simply "determined" the scientists' response. For if we take Moscovici seriously, then a claim of that sort would unwarrantedly presume that closure *had* to be reached on the governance of quantum particles in Weimar culture, when in fact one of the things that a minority opinion group can do effectively is to prevent the "moment of decision" from ever arising by arguing that more evidence needs to be gathered and analyzed. Thus, in some places and times, there is no orthodox opinion on an issue, not because there is no majority view, but because the minority blocks that view from finally closing discussion. Once again, as the rhetorician would have it, time (*kairos*) is of the essence.

In what follows, the ideological wellspring of STS, known as the

Two: Interpenetration at Work

Sociology of Scientific Knowledge (SSK), or the "Strong Programme," constitutes a minority in just the above sense, as against the emerging alliance of interests around Artificial Intelligence (AI) research. But does SSK function as any more than a spoiler? I will argue that interpenetrative rhetoric can act so as to transform SSK from *paralyst* to *catalyst*, once SSK shifts the argument to a new plane. Speaking as the participant social epistemologist, I want to make the next move in this debate by showing how the nature of Slezak's intellectual project is very much bound up with the way in which he mobilizes allies and distances opponents. In other words, *AI is an instance of SSK in action.*

Some Impasses in the Artificial Intelligence Debates

One of the most intriguing features of the swirl of controversy that has surrounded AI research is the extent to which highly abstract debates in the sciences are so permeable to public interpretation that subsequent research reflects some of what was previously regarded as the public's conceptual coarseness and confusion. Unlike sociobiology, in which the media intervened only after "gene talk" had taken root in the scientific arguments of biology (Howe and Lyne 1992), AI research has been invested with public import from day one.

Can computers think? The rhetorical impasse surrounding artificial intelligence may be epitomized by the multiple interpretations that can be given to the following two answers to this question:

> 1. Computers will continually improve their cognitive capacities until they surpass humans in intelligence.

> 2. Computers will never demonstrate real intelligence because of the unprogrammable character of human expertise.

With some interesting exceptions that will be discussed later in this chapter, most SSKers stand for (2) whereas advocates of "Strong AI," like Slezak, defend (1). What makes the difference between (1) and (2) impassable is that the terms of the debate defined by these two positions can be understood in a number of alternative ways, yet the

Sublimation

coalition needs of the groups associated with each position make it imperative that those alternatives remain suppressed Here are three questions in which unresolved ambiguity is put to strategic advantage in public debate:

> a. Is the "intelligence" to which computers are held accountable defined by the range of behaviors that is normally criterial of human intelligence, or by the mechanisms that produce such a range in humans?
>
> b. Does "computers" literally refer to particular machines taken in isolation, or to machines functioning in some suitably normal environment (which may include interfaces with humans and other machines)?
>
> c. Is the thrust of the major theses descriptive (i.e., a statement of fact about machine capabilities) or prescriptive (i.e., a statement of value about the possibilities for interpreting machine behavior)?

For each of these three sets of alternatives, it is common for (1)-inspired defenses of one option to be met by (2)-inspired attacks from the other option. Thus, in debate (a), if (1) is defended by arguing that a given computer can produce certain behaviors, a proponent of (2) will reply that the computer cannot produce the behaviors in the way humans can, which frequently amounts to the claim that computers lack human bodies. Similarly, in debate (b), if (1) is defended by pointing to the computational power of an individual machine, the proponent of (2) will respond by pointing to features of human intelligence that require an ambient social world in order to be recognized. Finally, in debate (c), if support for (1) is mustered by noting the tasks in which computers actually outperform humans, the supporter of (2) will challenge the wisdom of letting computers handle such tasks. In the case of each of these three debates, the responses from the camps representing (1) and (2) are in fact not in contradiction because they are not addressing the same issue. However, the social epistemologist's task here is not over because to show the compatibility of two sides to a controversy is not to end the controversy, but to diagnose two incommensurable positions on a yet to be specified subject of common interest that can be better pursued in collaboration than in either spurious opposition or splendid isolation.

Two: Interpenetration at Work

Drawing the Battle Lines

AI and SSK are moving targets which over the past two decades have charted orthogonal courses in the study of scientific reasoning. Peter Slezak, the head of the first Cognitive Science Unit in Australia (at the University of New South Wales), has recently tried to find a conceptual intersection, but has succeeded only in staging a verbal collision. In 1989, the SSK journal, *Social Studies of Science*, devoted a special issue to the first head-on confrontation between the two heirs apparent to the throne of epistemology, SSK and AI. Leading the offense was Slezak (1989a), who claimed to have refuted the signature SSK thesis that science has an ineliminably social character. Specifically, he argued that computers could be programmed to reproduce at least some of the discoveries made by scientists of the past without reproducing their social context.

It is important to observe at the outset that Slezak is a philosopher in disguise. His doctoral dissertation tackled no less than Descartes, and no less than the philosopher-turned-psycholinguist-turned-cognitive scientist Jerry Fodor (1987) acknowledges Slezak as an influence in the writing of *Psychosemantics*. Slezak himself does little to hide his philosophical colors. Although (or perhaps, because!) he styles himself the avenging angel of AI, Slezak's argument is cast in nonnegotiable terms that are generally alien to AI research, the rhetoric of which tends to resemble that of the early Royal Society in its "show-and-tell" approach ("See what my machine can do!"). Here are some of the telltale philosophical signs of Slezak's argument:

> 4. The argument is set up as a zero-sum game, which is rigged so that AI wins as long as SSK is not *necessary* for modeling science.

> 5. What it is to model science is defined in terms—namely, the internal history of science—that make the evidence assembled look most persuasive.

> 6. Evidence for the likelihood that the claim can be supported—that is, evidence once removed—is taken as sufficient grounds for demonstrating the claim itself.

Sublimation

The inimitable mark of Slezak's philosophical gamesmanship appears in (6), the idea that the bare prototype of an alternative account—the idea that a machine *could* be designed to capture what SSK *already* can—can shift the burden of proof onto an already existing account. After all, none of the classical philosophical conundra—especially the problem of skepticism and of the existence of other minds—would have ever gotten off the ground had philosophers respected the burden of proof that their radical queries bore. (Of course, philosophers more than made up for their rhetorical insensitivity by conjuring up the specter of grave risk if their queries were *not* pursued.) Slezak manages to scare up a possible case by mustering an assortment of opportunistically chosen pronouncements from practitioners and theorists of AI, and, more substantively, by citing Herbert Simon's ongoing series of BACON programs (Langley et al. 1987). These programs ultimately aim to derive the largest number of discoveries from the history of science—in the order in which they occurred—from the smallest number of heuristics. (Each heuristic is basically a set of nested "IF you're in this situation, THEN do this" statements.) The BACON programs are philosophically interesting in that the computer makes its way through the history of science by "learning how to learn" (Shapere 1984). That is, each solved problem and new discovery is added to the knowledge base for future reference, sometimes in the aid of arriving at new heuristics. The bulk of Slezak's original article is thus devoted to appeals to authority and other people's computer sketches.

The defense of SSK was mounted by an assortment of philosophers (Steve Fuller, Ronald Giere), sociologists (Augustine Brannigan, Harry Collins, Steve Woolgar), and a psychologist (Michael Gorman), many of whom, it must be said, were more concerned with pointing out the weaknesses of AI (as presented in Slezak's paper) than with actually defending SSK. As it turned out, the most explicit target of Slezak's original article, David Bloor, a founding father of SSK, refused to enter the fray. Slezak had one overt sympathizer among his interlocutors, a fellow computer modeler (Paul Thagard), who struck a conciliatory note, reassuring SSKers that the social context of scientific work was not just eliminable "noise in the system" but a genuine anomaly that future computer programs will be able to solve. After that initial skirmish, Nobel Prize-winning super-social-scientist Herbert Simon (1991a) entered the fray, indulging in some schoolmasterly finger wagging at the SSK defenders, as it seems that each had failed, in some way or other, to fathom the epistemologically crucial features of the programs he and his followers have designed.

Two: Interpenetration at Work

For my own part, I am actually rather well disposed toward AI work, including Simon's. However, my reasons for approving of AI work are different from those of Slezak or the ordinary AI enthusiast. I believe that the rise in the estimation of computer models in epistemological discussions marks an important sociohistorical renegotiation of what we take "science" and, more generally, "epistemic authority," to mean. AI has disrupted any easy notions we might have had that either of the scare-quoted expressions is uniquely human. In this regard, the success of AI may be counted as evidence in favor of the basic SSK thesis that all concepts, even the ones that pertain to the concept makers themselves, are conventional. What I cannot promise, though, is that empirical corroboration of this thesis will have salubrious consequences for society at large. (This tends to be the tenor of SSK's hostility to AI: e.g., H. Collins 1990, especially chaps. 13-14). Indeed, it almost certainly will not, if we remain oblivious to the often subtle changes that the computer revolution has wrought in our ordinary self-understanding. Generally speaking, Slezak refuses to take seriously the proposition that the incompatibility of AI and SSK is an artifact of their sociohistorical circumstances. Where Slezak sees an essential difference, I see a lack of communication—one that has persisted for so long (at least twenty years) that AI and SSK have become incommensurable. This is in keeping with my general view that, more than we would like to think, conceptual difference is born of communication breakdown (cf. Fuller 1988a: xiii).

AI as PC-Positivism

Slezak rightly associates the epistemic orientation of AI with the accounts of rationality implicit in "internalist" histories of science, which, following David Bloor (1976), he calls the "teleological model." Positivist philosophers wedded to internalism can easily see a natural successor to their own projects in the advanced BACON programs, which generate a plethora of physical principles by applying a few well-chosen heuristics. Indeed, it remains an open question whether AI—especially the research on which Slezak relies most heavily—has surpassed old-fashioned positivist ingenuity in devising efficient methods for selecting hypotheses, an issue that ought

Sublimation

not to be confused with the obvious fact that over the years machines have been designed to apply these methods to more problems more quickly. A striking case in point is Thagard (1988), who had introduced all of his "problem-solving strategies" in mainstream philosophy journals long before he displayed the effects of prolonged exposure to a computer. The very names of these strategies, "inference to the best explanation" and "maximizing explanatory coherence," betray a philosophical lineage that goes back at least to Charles Sanders Peirce and perhaps even Sir Isaac Newton himself. Indeed, by the end of *A Computational Philosophy of Science*, the reader learns that Thagard's alleged philosophical breakthroughs were made in an effort to develop an automated logic tutorial for undergraduates–that is, PC-Positivism! (And, I mean "PC" in the late-eighties sense of "Personal Computer," not in the early-nineties sense of "Political Correctness.") At times, only a sexier rhetoric and a bigger machine seem to separate the positivist's "logic of justification" from the AI researcher's "logic of discovery." After all, what BACON and other such programs simulate is the selection of not *any* hypothesis, but rather the one that turned out to be *right*, which is to say, justified. But to his credit, Slezak has cut through the AI Newspeak to recognize his natural affinities with the positivists and other internalists.

However, Slezak should have stuck with his allies when he came to framing the internalist-externalist dispute. Unfortunately, he instead turned to his opponent Bloor for guidance. Bloor generally makes it seem as though the internalist and the externalist are offering competing answers to the same question: Were "cognitive" or "social" factors primarily responsible for the acceptance of a given scientific claim? Despite my agreement with Bloor on a great many issues, I am afraid that his archenemy, Larry Laudan (especially 1977: chap. 7), probably has a much better grasp of what is at stake in the dispute. In particular, Laudan recognizes that the internalist and the externalist use the history of science somewhat differently. The internalist's aim is to test a *normative* theory of scientific reasoning. For example, the fact that Heisenberg would not have argued for quantum indeterminacy had he not been sensitive to the rise of irrationalism in Weimar culture does not undercut the fact that the arguments themselves were "good" ones, that is, sufficient (in the internalist's eyes) to make the case for quantum indeterminacy. Whereas the externalist is concerned with the occasions that brought about the need to make indeterminacy arguments in physics, the internalist is concerned with the soundness of

Two: Interpenetration at Work

the arguments (given what was known at the time) regardless of their cultural timeliness. This difference in attitude toward time is especially acute in the case of Herbert Simon, who distances his computer simulations of scientific discovery from the actual history of science in exactly the same way as Laudan distances internalist from externalist historical interests.

For Simon, "time" is nothing but an abstract sequence of events, and the sooner with which it can transpire on the computer the better. Simon expresses this most vividly when he wonders how it could have taken Kepler so many months to discover his three laws, when BACON can do it in a few minutes. In effect, Slezak and Simon presuppose a strong content/context distinction, in which the role of context is either to impede or to facilitate the transmission of content. Ideally devoid of all context, the computer is made to appear as the proverbial frictionless medium of thought. Any delay between posing the problem and stating the solution is explained solely in terms of the time it takes to go through operations expressly designed for reaching the solution. As context is added to this process—which is to say, as content is distributed among finite human beings with differential access to one another—the knowledge enterprise is impeded to ever greater degrees. Being students of context, SSKers are assigned (by Slezak and other AIers) with the task of detailing the various ways in which humans have lagged behind computers in their cognitive performance. In short, it would seem that Slezak fails to observe the official party line of his own camp, which was voiced by Thagard in this symposium: namely, that the internalist and the externalist, or AI and SSK, are doing different but compatible things.

Moreover, Slezak is not the only one who could benefit from seeing internal and external histories of science as asking different questions. Bloor and his SSK defenders could as well. In particular, SSKers have all too readily embraced Quine's thesis that theory choice is always underdetermined by the available data (i.e., there are never any strictly empirical grounds for supporting one theory over another: for the ramifications of this point, cf. Roth 1987). As SSKers rightly see, if the thesis is true, then factors other than those sanctioned by the scientific method—especially "social factors" broadly construed—play a decisive role in theory choice. Unfortunately, Quine (1953) and other philosophers mean to allow these extra factors *only* because underdetermination is true. In other words, there is an implicit pecking order of explanations for science, which gives pride of place to internal factors and then, once those have been exhausted, turns to external factors to

Sublimation

make up the difference. Thus, Quine differs from, say, a logical positivist only in *how soon* he believes one will need to countenance social accounts of science. In Laudan's terms, the underdetermination thesis implies the *arationality assumption*, whereby sociology takes over from philosophy to account for the arational residue (however large) of knowledge production. By acceding to the idea that they are offering an alternative–yet "external"–account of the same phenomena, SSKers tacitly accept second-class epistemological status. This is an ironic dialectical error for SSKers to commit, given the great pains that Bloor (1976: chap. 1) initially took to stress the need for what he called "symmetry," namely, for the social sciences to use the same principles to explain *all* episodes from the history of science *equally*–the good, the bad, and the ugly.

At first glance, the dialectical disadvantage in which SSKers find themselves reflects a perverse way of dividing up the intellectual labor. After all, human beings appear to occupy the dregs of cognitive inquiry. But shouldn't computers be held accountable to human cognitive performance, not the other way around? Here, too, Slezak is onto something, for over the past twenty years the identities of the "modeler" and the "modeled" in AI research have been subtly exchanged, representing a shift in the balance of power within the cluster of computer scientists, neurophysiologists, experimental psychologists, linguists, and philosophers who define "cognitive science" (cf. Pylyshyn 1979). Back in the 1950s and 1960s, when Simon, Alan Newell, and Marvin Minsky were first plying their trade, terms such as "artificial intelligence" and "computer simulation" were meant to be taken literally. That is, computers were seen as trying to model human intelligence and as succeeding most notably in the relatively narrow range of "formal thought" that was tractable to linear programming. Computers were then taken as simplifying and amplifying something whose complexity could be fully fathomed only by studying humans directly. However, as various bigwigs became increasingly impressed with the prosthetic reasoning powers of the computer (e.g., in medical diagnosis, missile tracking, mathematical problem solving), the object of AI inquiry was subtly reconceptualized. What had been previously regarded as the rich complexity of human intellectual life that could only be feebly modeled on a computer was now portrayed as a "mechanical defect" that prevents humans from matching the cognitive efficiency of computers. Indeed, cognitive science's most sophisticated recent theorist, Zenon Pylyshyn (1984), has christened the new object of AI inquiry "cognizers," Descartes' *res cogitans* rendered computational, an essence

Two: Interpenetration at Work

that is instantiated equally in humans and in certain high-level computers—though more purely in the latter case.

Given this turn of events, the term "artificial intelligence" has become something of a misnomer, for it would now seem that computers manifest intelligence in a pure, natural state, whereas human intelligence is corrupted (by error, emotion, and other context sensitivities) and, hence, is at best a first approximation of the ideal form. This shift in the balance of ontological power partly explains the ease with which Slezak and other AI researchers tend to ignore the relevant experimental work on human reasoning, and the eagerness with which psychologists have retooled to include computer programs as relevant test sites for models of human reasoning (e.g., Anderson 1986; Johnson-Laird 1988).

Slezak is absolutely right that SSK must take this trend seriously, as cognitive science promises to be the strongest pitch yet made by the combined forces of Platonism, Cartesianism, positivism, and other forms of internalism to command political and economic resources. Ironically, this pitch comes at a time when these forces are perceived within the humanities and social sciences as having been intellectually discredited. However, SSKers would be foolish to suppose that the critique launched against the internal/external (content/context) distinction in AI by, say, Weizenbaum (1976) and Dreyfus and Dreyfus (1986)—let alone the more global critiques in Rorty (1979) and Latour (1987)—has trickled down to the science policy boardrooms. Despite the valiant efforts of social constructivists, AIers are still the primary authorities on how their research is interpreted. Consequently, opportunities for an SSK-style critique are strategically suppressed, as AIers focus the policymaker's attention—much as a magician focuses the attention of her audience—on what the computer does "by itself" (i.e., as an "automaton"), minimizing any awareness of the extent to which this framing of the situation depends on constant intervention by the AIer on behalf of her machine. Moreover, these interventions occur not only at the beginning and the end of programming: i.e., with the initial selection of the data from the historical record which the computer processes as input and the selection of a code by which a target audience will be able to make sense of the computer's output. In addition, the AIer must typically intervene in the middle of the program to supply needed data that the computer cannot get on its own or perhaps even a promissory note in cases where actual steps in the program have yet to be worked out.

Sublimation

How My Enemy's Enemy Became My Friend

From the black-and-white presentation of "AI versus SSK" that has so far been made, it would be easy to conclude that we have got two monolithic movements on our hands. While this is certainly false in the case of SSK, it is even more strikingly false in the case of AI, if we take AI to include all those who call themselves cognitive scientists. For within these ranks may be found influential advocates of SSK-like theses, such as Stich (1983) and Dennett (1987). Of course, for polemical purposes, it is convenient to collapse ideological nuances and to streamline a tortuous history. But Slezak does so to such an extent–and in the service of showing up Bloor's alleged ignorance of recent events in the history of psychology–that it is necessary to set the historical record a little bit straight before proceeding further. (For the record, a subtext of Bloor 1983 is a revisionist social history of psychology that portrays the field as struggling to define itself as a science that studies more than just accurate or inaccurate physical intuitions, which nineteenth-century philosophers would be inclined to locate as the purviews of physics and sociology, respectively. Thus, Bloor reinterprets the analytic strength that Slezak and other cognitive scientists claim to derive from psychology's commitment to studying the isolated individual as a desperate attempt to avoid social or physical reductionism.)

Slezak follows the canonical histories of the Cognitive Revolution that place Simon and Noam Chomsky on the same (winning) side of the battle against the behaviorists. This by itself is hardly a cause for criticism. After all, Simon and Chomsky knew each other and participated in the conferences that would later be taken as having founded "cognitive science"–many of which were officially on "signal detection." Indeed, both have had occasion in interviews to cite each other as contributors to a common cause (Baars 1986). Now, to philosophical ears, this last sentence emits a curious resonance: Does "contributors to a common cause" refer to the participants in the Cognitive Revolution or to corroborating evidence for the existence of the Cognitive Revolution's primary object, originally known as "the information processing system," but nowadays simply called "cognition" or "intelligence"? In fact, I mean a little of both, and probably the SSK analyst will be able to see why more easily than the AI enthusiast.

Two: Interpenetration at Work

Popular Whig histories of the Cognitive Revolution (e.g., Gardner 1987) dazzle the reader with an array of victories in the laboratory but obscure the overall strategy that won the war. As typically happens during scientific revolutions, the revolutionaries leave their most lasting imprint, not with particular findings or theories, but with a sensibility about the sorts of findings that are worth having and the sorts of theories worth testing. Thus, the Cognitive Revolution imparted to the study of the mind a legacy of *scientific realism* marked by the search for underlying mechanisms that explain how a seemingly disparate range of phenomena could have much the same structure, especially under ideal conditions of observation (which may be so ideal as to involve computer simulations of human output). It would be difficult to overestimate the gestalt switch caused by this turn to realism. Try talking methodology with a normal practitioner of cognitive science. You will find that she has a hard time imagining how Skinner and other behaviorists could have thought that they were doing science *precisely because* they failed to postulate mechanisms that were not susceptible to direct empirical test. In fact, quite sophisticated cognitive scientists have been known to forget what kept behaviorism afloat for fifty years in America. Despite persistent moral and intellectual objections, behaviorism's dogged commitment to "the methods of science" helped it prevail. But back then, these methods were primarily associated with the prediction and control of observable behavior, and *not* with the search for underlying mechanisms, which, absent the appropriate operationalization, would just dissolve into a species of metaphysics. (For instances, cf. the open peer commentary surrounding Skinner's "Behaviorism at Fifty": Catania and Harnad 1988: chap. 5.) Chomsky, who plays a crucial role in Slezak's argument, contributed decisively to the reversal of this sensibility back to realism.

However, scientific realism has historically been a tricky business to pull off, and the cases where it has seemed to work are ripe for a deconstruction of the actor-networks that had to be built along the way. Bruno Latour (1988) has been especially struck by the actor-network known as *explanation*. To follow Latour, we must conceive of explanation as a form of political representation in which the explainer must represent a diverse constituency, spokespeople for particular sorts of phenomena. According to "the politics of explanation," then, the so-called efficiency of a covering law (or universal generalization), the standard form that a scientific explanation takes, lies in its ability to minimize the resistance of the disparate elements that are subsumed

Sublimation

under such a law. For example, in the face of Newton's laws, celestial and terrestrial motions are no longer two things pulling in opposite directions (up and down, as Aristotle would have it) but are rather reduced to two versions of the same thing. Not surprisingly, then, scientific realism has taken the key to science's epistemic power to lie in its explanatory function. The charm that the explanation holds for the subsumed cases is in the prospect that the underlying reality implicated in their alliance will confer greater power on them collectively than they would have individually. As in any political situation, the trick is to make the representative accountable, which, to revert to philosophy of science terms, is to tread the fine line of "corroboration" between portraying the constituencies as pursuing largely *parallel* research trajectories and portraying them as pursuing largely *convergent* trajectories. However, neither option is completely welcomed.

The Scylla that awaits the parallelist is a stack of idle analogies that reveals a common structure only to the impressionable historian or philosopher. Thus, the realist does not want to look like a mere analogy monger, a throwback to the period immediately prior to the Scientific Revolution when Man and Nature were fraught with mystical "correspondences" that were ultimately underwritten by a Divine Emanator of Forms (cf. Foucault 1970: 17-25). Thus, even if the cognitive scientist believes that the structure of the *mind* is isomorphic in all of its embodiments (e.g., à la Lévi-Strauss, that the individual mind is the microcosm of some collective mind), she typically does not believe that the structure of the mind is isomorphic to the structure of *nature*, in all of its embodiments. In short, we have Jerry Fodor's (1981) doctrine of *methodological solipsism*: study the mind as if Descartes' worst suspicions were borne out, and the world presented to you is a complete illusion. Admittedly, Rom Harré (1986) is one scientific realist who embraces analogy mongering wholeheartedly, including its Aristotelian implications. But he rejects "cognition" as a proper object of inquiry independent of the environments in which embodied cognizers find themselves. Harré thus turns from Fodor to J. J. Gibson and his ecological orientation to psychology.

But drawing a boundary between what is inside and what is outside the mind is not quite going to do the trick for the scientific realist because there are still too many loose analogies available. Consider this heterogeneous group of people who, on both standard European (De Mey 1982) and American (Gardner 1987) accounts, are said to have been on the same side of the Cognitive Revolution:

Two: Interpenetration at Work

    1. Herbert Simon (political scientist-turned-computer simulator)
    2. Noam Chomsky (theoretical linguist-turned-psycholinguist)
    3. George Miller (communications technologist-turned-experimental psychologist)
    4. Marvin Minsky (mathematician-turned-computer scientist)
    5. Jerome Bruner (Gestalt psychologist-turned-educational theorist)
    6. Jean Piaget (child psychologist-turned-genetic-epistemologist)
    7. Claude Lévi-Strauss (structural anthropologist)
    8. Thomas Kuhn (historian of physics-turned-philosopher of science).

Given this veritable Chinese Encyclopedia of "Cognitivists," how might one characterize—not to mention explain—the relevant sense of resemblance between Simon's problem-solving heuristics, Chomsky's competence grammars, Miller's information processing stages, Minsky's modules, Bruner's principles of perceptual integration, Piaget's developmental sequence, Lévi-Strauss' cultural maps, and Kuhn's paradigms? This question finally suggests the Charybdis that awaits the convergentist, namely, the image of the Cognitive Revolution as a matter of elaborate collusion. Did the principals listed above secretly meet somewhere in Cambridge, Massachusetts, in the late 1950s to concoct the roles they would play in the planned intellectual coup and then, afterward, periodically meet to straighten out each other's lines in the unfolding drama? Clearly, nothing quite like this actually happened, although the contact that most of these people had with each other in metropolitan Boston during this period was much greater than their disciplinary differences would suggest.

Here are just some of the subtler connections that have earned Boston the title of "Hub of the Universe" (an expression presciently uttered by Ralph Waldo Emerson a century before the Cognitive Revolution). Kuhn was inspired to isolate "the structure of scientific revolutions" by Piaget's dynamic structuralism, which Bruner was promoting in the late 1950s, and indeed, a decade later, when De Mey came to Harvard to study with him. An even more direct infusion of French structuralism (including the teachings of Lévi-Strauss) occurred with Roman Jakobson's accession to a chair in linguistics at MIT at that time. Jakobson's principal colleague was Morris Halle, who early befriended Chomsky. (On the psychology side, of course, George Miller was instrumental in converting Chomsky's formal apparatus to testable hy-

Sublimation

potheses.) Gardner, himself a fixture at the Harvard School of Education, wrote his first book on the French influence on cognitivism (Gardner 1973). And although it is true that Simon's career has been most closely associated with the cities of Chicago and Pittsburgh, Gardner reassures us that Simon (and Newell) was present at Dartmouth College, New Hampshire–a Boston satellite–in the summer of 1956, when the Cognitive Revolution was officially declared. Moreover, the cognitivists were also in close proximity to the hub of their opponents, including the behaviorists B. F. Skinner and W. V. O. Quine, both of whom taught at Harvard.

None of these facts, well known as they are, would bother the behaviorist who tried to account for the Cognitive Revolution, since the behaviorist would be the first to argue that mutual reinforcement of the revolting parties was crucial to the maintenance of their collective behavior. However, the historian sympathetic to cognitivism would want to avoid any whiff of collusion on the part of the revolutionaries. Consequently, she would prefer to see them as inadvertently running across each other after having pursued parallel courses in which they had managed to survive (what they learn, after the fact, to have been) a common foe.

At this point, Sir James Frazer's early anthropological classic, *The Golden Bough* may come to our interpretive aid, especially his discussion of *sympathy* and *contagion* as principles by which "savages" explain change in the world. The historian of the Cognitive Revolution would have us believe that the principals knew of each other's work largely from a distance, so as not to be conspiratorial, yet close enough so as to enable them to see their points of commonality. Thus, by striking the right distance from each other, the cognitive revolutionaries can mutually implicate the independent objectivity of their respective viewpoints. The behaviorist would, of course, read these developments less charitably. She would explain a world seemingly fraught with ideational sympathies in terms of verbal contagion. (Indeed, that is exactly the kernel of truth in epidemiological models of conceptual diffusion, cf. Hull 1988.) That is, when a piece of language–such as "cognitive" or "information processing"–does the trick for someone in one setting, then interested onlookers try to see if the words will work the same magic for them. For example, "cognitive" continued to work some *negative* word magic in the behavioristically dominated clinical circles long after it had become ascendant in experimental psychology (cf. Mahoney 1989). Whether anything else about the onlookers' research practice changes is an open question, which the Whig historian is inclined to pass over in a tactful silence. Baars (1986: 138-64) implicitly raises this issue by styling an entire set of Cognitive Revolutionaries, including George Miller and George Mandler, as "adapters."

Now, so far I have been suggesting that there is some sense, be it ontological or sociological, in which the likes of Simon and Chomsky are rightly cast as being on the same side of the Cognitive Revolution. But I really want to argue for a much weaker thesis, namely, that this unity in arms runs no deeper than the fact that Simon and Chomsky (and the rest of the cognitivists) shared a common foe, the behaviorists. For ever since the foe was put safely out of dialectical range, the cognitivists have roamed wherever they pleased on the conceptual map. Not surprisingly, this has also led the principals to distance themselves from one another in interestingly asymmetrical ways, reflecting their respective senses of how the cognitive should be bounded–now that the coast is clear of The Behaviorist Menace. I will confine my remarks mainly to Chomsky and Simon, since Slezak focuses on them.

But Now That the Coast Is Clear

The indefinite continuation of the "Cognitive Revolution" (Gardner's term) into the "Cognitive Paradigm" (De Mey's term) provides an interesting case study in consensus formation and deformation, which are the processes to which philosophers have most recently turned in order to "sociologize" their accounts of science (cf. L. Laudan 1984; for a critique, see Fuller 1988a: 207-32). The most important implication of the present discussion for consensus theories is that a consensus emerges only if there is a reason (an external force, as it were) for it to do so; otherwise, the constituent individuals will move off in their own directions, with the consensus language (in this case, cognitive talk) becoming semantically diffuse. In terms of classical political theory, then, a scientific consensus is better regarded as a relatively short-term convention to act in mutual aid than as a group deeply committed to the same long-term goals and values: i.e., more a *societas* than a *communitas* (for more on this distinction in the history of social theory, cf. Manicas 1986: 24-37). In particular, Herbert Simon and Noam Chomsky would no doubt be surprised to find themselves fighting on the same side in Slezak's holy war against SSK, as each places quite a different value on continuing the alliance that originally

Sublimation

enabled them to vanquish The Behaviorist Menace. Simon continues to be the covering cherub, given to abstracting the form of "intelligence" from as many disciplines as he can just in order to get the analogies to stack up right, but Chomsky will not let linguistics be coopted into Simon's scheme so easily.

There is a tendency among AI enthusiasts like Slezak to presume that the computer models a self-sufficient Cartesian reasoner–perhaps because of the lingering folk association between computers and robots. However, it takes only a second thought to realize just how alien a program like BACON is to Chomsky's radically Cartesian sensibility. Chomsky (e.g., 1980: 76), after all, is notorious for claiming that humans are endlessly creative creatures, able to generate new sentences without any obvious prompt from memory or the immediate environment. What "discoveries" BACON makes are not new and are highly dependent on programmer intervention. In contrast, and much more in keeping with the spirit of BACON, is Simon's (e.g., 1981: chap. 3) equally infamous assertion that the complexity manifested in human behavior is entirely a function of the complexity of the environments in which humans manifest their behavior. In other words, according to Simon, when we try to solve a problem, we already know in vague terms what an adequate solution would look like. The "problem" lies in realizing that solution within the means at our disposal, broadly construed to include anything we can use in the environment. Thus, the human condition is so inordinately complicated only because our means are typically so ill suited to our ends that we are forced to concoct backhanded solutions. Progress is made as more efficient means are designed to realize more solutions. Not surprisingly, then, Simon-influenced AI work has shown little interest in modeling the actual cognitive processes of scientists, since, from Simon's standpoint, these processes are little more than clumsy way stations on the road to completely efficient thought. It is an attitude that recalls the early behaviorist view of deliberation as hesitation prior to response, a process that warranted not further study but elimination through efficient conditioning.

To put the point as a dilemma: Slezak must *either* accept Simon's programmed discoveries and reject Chomsky's creationism *or* accept Chomsky's creationism and reject Simon's programmed discoveries. Slezak (1989b) has responded that only an equivocal use of "creativity" drives the wedge I would place between Simon and Chomsky. Here Slezak wins the point at the cost of losing the entire argument. He observes (1989b: 691) that by "creativity" Chomsky really means, not the ability to produce something genuinely original, but the ability to produce a sentence from rules without having been previously exposed to

Two: Interpenetration at Work

the sentence. Slezak then correctly notes that this would seem to make Chomsky, no less than Simon, concerned with rule-based productive systems. Yes ... but what would Slezak then have us conclude about the ability to produce something genuinely original? I can see only two possibilities, in light of Slezak's clarification: (1) there is no such thing as genuinely original scientific discovery (there is only ignorance of the relevant rules), or (2) neither Chomsky nor Simon has anything of relevance to say on the topic of scientific discovery. Most SSKers can probably live with (1), and Slezak cannot live with (2), so maybe there is agreement, after all! In any case, Chomsky himself has increasingly dissociated his account of linguistic creativity from any larger generalizations about how humans acquire knowledge, *especially* scientific knowledge, which Chomsky has even sometimes suggested is too amorphous a phenomenon for us ever to expect to grasp (Botha 1989: 37-38).

In terms of the dilemma, then, I recommend pursuing the first (Simon's) horn as empirically the more fruitful one, especially in regard to potential linkages with SSK. I take Simon's analysis of the sources of human complexity to be a more generalized version of what actor-network theorists in SSK have to say about the totalizing tendencies of technoscience. For example, Callon, Law, and Rip (1986) present the technoscientist as establishing her credibility by "translating" the interests of an increasingly large number of others into her own work, as reflected in the textual constraints placed on journal articles and grant proposals, through which she must pass before being granted her point. In Simon's terms, these intervening interests constitute the "environment" in which the technoscientist must achieve her goals. However, whereas Simon's discovery programs are designed with an eye toward eliminating or simplifying as many of these "middlemen" (immortalized in Latour 1987 as "obligatory passage points") as possible, actor-network theorists see a trend toward engulfing all of society as middlemen in the production of scientific knowledge. Yet, both Simon and actor-network theorists seem to agree that there is nothing inherently complex or special about scientific reasoning as a cognitive process: it is simply ordinary strategic reasoning deployed in extraordinarily resistant environments. In a related move, Latour and Woolgar (1986: Postscript) have urged a ten-year moratorium on explanatory appeals to the "cognitive."

SIMON—THE COVERING CHERUB

Perhaps a word about the unity of Simon's career is in order (Simon 1991b is a rambling but informative intellectual autobiography). Simon

Sublimation

(1976) originally introduced the concept of "bounded rationality" in the mid-1940s to account for the adaptive character of corporate decision making, which occurred under conditions of much greater uncertainty than are assumed in the standard models for maximizing expected utility in neoclassical economics. This point is of more than idle historical interest because by the late 1950s bounded rationality had become the concept that unified Simon's forays into economics, organizational theory, cognitive psychology, and computer science—including, most recently, computer models of scientific discovery (cf. H. Simon 1981). From Simon's own account, then, it follows that the prototype for Slezak's computer refutation of SSK is something as "sociological" as an account of corporate decision making. (Readers are invited to ferret out the dead metaphors from organizational theory that infest Simon's informal descriptions of his computer programs: cf. Langley et al. 1987: especially 299-300, which describes an "integrated discovery system.")

In the short but synoptic sweep of *The Sciences of the Artificial*, we learn that intelligence is inherently artificial, in that it emerges once an "organism" (understood in that abstract systems-theoretic sense that is indifferent to biology) develops reliable ways of maintaining—and sometimes even enhancing—its identity against the resistance of its environment. Thus, we differ from the thermostat "only" in the variety of productive ways in which we adapt to change in the environment (H. Simon 1981: 15-19). Simon's main thesis is that the truly smart organism primarily tries not to seek wins, or avoid losses, or even stay in the game forever, but rather to pursue a bounded version of all three simultaneously, i.e., it tries to get the most from the least for as long as it can. Since this thesis goes against the conventional wisdom of virtually every discipline in the social sciences (especially formal philosophical models of rationality), Simon is able to use it as a pretext for reconstituting all of those disciplines into "sciences of the artificial."

By the time Simon conceptualizes the scope of the sciences of the artificial, it has become clear that his unit of analysis is not the lone administrator, *but the system of administration*, in which the administrator functions as a major node in the administrative network. Indeed, her behavior may well provide the richest symptoms of the system's overall state. But the administrator is only a *part*, not the whole unit of analysis. Unlike Chomsky's competent language user, she is not only embodied as an individual, but environmentally embedded as well. Of course, it does not follow that there is complete clarity as to how one exactly goes about individuating administrative systems. But that very ambiguity concerning criteria of individuation renders the "system" a contested terrain, and hence an apt SSK object of inquiry. In particular, whose response to what sort of feedback is relevant to telling one system apart from another? The introduction of computers into Simon's project complicates systems analysis, as well as the ensuing AI-SSK debate, since a computer can be treated in any of the following ways:

> 7. as a system in its own right, composed of machine parts or functions, depending on one's mode of analysis (cf. Dennett 1987: chap. 1)
>
> 8. as a model of a system, e.g., the BACON computer simulation of the scientific discovery process
>
> 9. as an individual in its own right, like Simon's administrator, embedded in a larger social system.

Although the AI-SSK debate officially transpires at level (8), SSK supporters, in fact, move quickly to level (9), culminating in the recent discussion of computers as *actants* (which will be considered at the end of this chapter), whereas AI advocates tend toward level (7). Simon himself has now bought into this perspective, which permeates what may be called Slezak's "rhetoric of testability." In other words, "intuitive" appeals to the social, as in Simon's administrative systems, remain "mere" metaphors until they have acquired a technology, such as the digital computer, which can then be used to reproduce the relevant nonsocial fact and thereby explain the fact's persistence in a variety of social settings. Slezak's argument has bite just as long as we can focus on the final stage of this operation and ignore all the previous ones, including the embarrassing fact that Simon's intellectual perambulations began with an interest in the environments in which harried bureaucrats worked.

CHOMSKY–THE REVISIONIST HISTORIAN

Since much of Slezak's argument against SSK, and especially Bloor, turns on a folk valorization of Chomsky's significance, something should also be said about what Reber (1987) has called "the rise and (surprisingly rapid) fall of psycholinguistics." We have only recently begun to get enough historical distance from the Chomsky phenomenon to understand what actually took place. While Chomsky's work remains vital in linguistics, it was subject to a meteoric rise and fall as a research concern in psychology between, roughly, 1967 and 1980. In this regard, it is telling that Slezak's main authority for Chomsky's signif-

Sublimation

icance is a history of *linguistics*, not of psychology (Newmeyer 1980). Why did Chomsky fade so fast in psychology? Two related reasons stand out which have import for Slezak's argument.

First, Chomsky refined his generative grammar only in response to anomalies arising from considerations in theoretical linguistics (e.g., the simplicity of rules in the grammar, the ability to parse intuitively grammatical sentences), while remaining unresponsive to the recalcitrant data raised by experiments on the grammar's psychological validity. Hence, the psychologists pulled out of the enterprise in frustration.

Second, whatever one makes of the fortunes of behaviorism, it must be said that it conformed to the "functionalism" implicit in all successful research agendas in experimental psychology since Wundt, to wit, behaviorism attempted to derive principles that explain an organism's behavior as a function of some change in its environment. Chomsky's work failed to meet this basic principle, for it supposed that linguistic competence remains invariant in spite of differences in language training and other environmental stimuli. Indeed, this very lack of interdependence between what is postulated as transpiring inside and what outside the organism posed a major obstacle to conducting decisive experiments on Chomsky's model. However, as the experimental psychologist Michael Gorman (1989) remarked during the debate, the typical AI research environment is quite unlike the normal working situation not only of the human scientist, but even of in-use expert systems. This lack of what experimentalists call "ecological validity" exacerbates the problems involved in computer testing Chomsky's model, given the computer's very limited capabilities to respond to its environments.

But Chomsky was more than just a shooting star in experimental psychology. He has also had lasting impact on how historians conceptualize the trajectory of psychology since World War II. In particular, Chomsky was most responsible for portraying behaviorism as a degenerating research program, one that had to be overturned for progress to be made in psychology, and one that could be successfully overturned only by importing a strong "cognitivist" or "nativist" orientation. In the rest of this section, I will begin to deconstruct this conception, the pervasiveness of which approaches that of Kant's division of the history of philosophy into "empiricists" and "rationalists" at the end of *Critique of Pure Reason*. But first, it is worth remarking that B. F. Skinner and other behaviorist targets were taken completely off guard by Chomsky's relentlessly negative portrait of their scientific status. When Skinner spoke of behaviorism's "success" as a research program,

Two: Interpenetration at Work

he tended to point to its applications, many of them in clinical settings starting in the 1960s—which is to say, *after* Chomsky had begun to sound the school's death knell. To this day, both historians and psychologists tend to overlook this fact because Chomsky has managed to persuade psychologists that the epistemic status of their research programs should be judged solely on the basis of their scientifically derived findings, not on the basis of their practical applications. Thus, the perceived decline of behaviorism was, at the same time, the triumph of psychology the academic discipline over psychology the liberal profession (cf. Baars 1986: chaps. 1-3). In short, Chomsky changed the standards of success in the field to his strategic advantage.

In light of this subtle historiographic coup, Bloor can only be congratulated for reviving SSK interest in Chomsky's original target, Skinner's *Verbal Behavior* (1957), since it is clear that Chomsky had hardly delved into the work when he wrote the review that launched him into stardom. It seems that Chomsky relied mostly on Skinner's remarks on language in his William James Lectures, which provided many of the basic ideas for the book but none of the detail about developments in other fields that supported Skinner's view. Instead, Chomsky fell back on the stereotypical behaviorist account of language that emerged from Clark Hull's simple Markov chain models of language learning. But by the late 1950s, Skinner and even more orthodox behaviorists had begun to distance themselves from that model. Lost on Chomsky was the comprehensiveness of Skinner's theory, which incorporated into the behaviorist repertoire a strong audience component (as the selection environment for operants) that was reminiscent of the early reader-response criticism proposed by I. A. Richards, as well as the emotivist theories of language use found in logical positivism and general semantics. In effect, Skinner had made great strides toward "socializing" the behaviorist bias toward the isolated organism by transforming the concept of a text's "meaning" into a network of operant responses that has the text as the nodal stimulus, and which in turn enables efficient communication between an author and multiple readers at once. Having judged Chomsky's review to be an uninformed attack by an upstart linguist, Skinner, not surprisingly, deemed it unworthy of timely response (Mac Corquodale 1970 is the first sustained behaviorist response). Needless to say, this turned out to be a tactical blunder, one which was admitted only after it was too late (Czubaroff 1989).

While Slezak is right that Chomsky succeeded in minimizing *Verbal Behavior*'s impact on psychology, the book did bear substantial fruit in analytic philosophy, via the theory of reference developed by Skinner's Harvard colleague W. V. O. Quine (1960), which continues to be very influential. In particular, from behaviorist premises about how one would draw up a translation manual for a language radically different from our own, Quine arrived at the "indeterminacy of translation thesis," which asserts that there is no fact of the matter about a speaker's mental states that could determine the correctness of a translation of the speaker's utterances. According to Quine, correctness of translation is entirely relative to the purpose for which it is sought and its fit with other of the speaker's utterances that have already been translated.

Admittedly, the thesis remains controversial, but it has served to shift the burden of proof among analytic philosophers in the United States to those who would maintain that mental entities such as "meanings" fix the reference of our words or, more generally, that something transpiring in a speaker's mind is the ultimate arbiter of what the speaker is talking about. (Wittgenstein's argument against the possibility of a private language has had the same sort of effect in Britain.) This line of thinking is also very strongly represented among cognitive scientists trained in the analytic tradition, especially among those who deny that regularities in human thought or behavior can be specified without reference to environmental variables. For example, on the one hand, Stephen Stich (1983) has argued that the empirical unreliability of appeals to reasons, beliefs, and desires in explaining behavior–items in the very ontology that Slezak holds to be at the foundations of AI–merits the elimination of these entities. On the other hand, Daniel Dennett (1987), a student of the philosophical behaviorist Gilbert Ryle, has argued that only the "interpretive stance" that one adopts to a computer always beaten in chess determines whether the computer is a bad chess player or simply a machine designed for some other purpose. Nothing intrinsic to the computer could alone decide the issue. Both Stich's and Dennett's views sit quite well with social constructivist accounts of AI (although both have flirted with evolutionary biological explanations that would make the constructivists bristle).

SIMON AND CHOMSKY:
THE FINE ART OF STRATEGIC POSITIONING

Simon knows that Chomsky will not be easily bought into his system. Simon's strategy here has been to resolve the differences between his own empiricism and Chomsky's rationalism at an appropriate level of abstraction from the phenomena of language. As is well known, Chomsky (e.g., 1980: 136-39) treats language as a self-contained, or

Two: Interpenetration at Work

"modular," organ whose fundamental workings are little affected by the vicissitudes of the speaker's contact with the environment. However, Simon attempts to soften this line by arguing that what is hardwired in the organism is simply the ability to learn from interactions with the environment. In that case, language may be distinguished by the efficiency with which humans learn it: small and simple input seems to elicit massive amounts of complex output (H. Simon 1981: 89-91). The British linguist Geoffrey Sampson (1980: 133-65) has tried to show how this sort of efficiency could have arisen from the evolutionary forces of syntactic variation and social selection, which shaped language into one of Simon's "nearly decomposable" systems.

Notwithstanding Simon's efforts at keeping the old revolutionary alliance intact, Chomsky persists in his wayward course. For example, when explicitly asked to comment on Simon's work, Chomsky admits to having never taken much of an interest, indeed, for some rather deep methodological reasons that make one wonder how the two could have ever been allies at all (Baars 1986: 348). Chomsky believes that Simon overestimates the significance of computers that have the ability to solve certain classes of problems as well as, or even better than, humans. Indeed, he thinks that in this respect, Simon repeats the errors of behaviorists who were overimpressed by the success of animal conditioning experiments. From Chomsky's standpoint, both Simon and Skinner, say, seem to focus mostly on modeling the sort of behavior in which humans are unlikely to outperform machines or pigeons (e.g., serial computations, simple motor skills), while spending relatively little time on trying to model behavior that humans would be especially good at. For Chomsky, this reflects the fact that Simon, like the behaviorist, is still principally concerned with the prediction and control of behavior (regardless of the relevance of the behavior to what one is ultimately interested in) rather than with the search for underlying mechanisms that would genuinely (e.g., neurobiologically) explain why *humans* have the distinctive capacities that they have, most notably, language. Thus, Chomsky reserves his approval of AI work for people like David Marr, who actually tried to model mechanisms (in this case, for visual perception) and not merely behavior.

It should be noted that Chomsky is not being entirely unfair to Simon, who, after all, is on record as claiming that what makes humans so good at science is not some special creative faculty but the variety of imperfect heuristics that we have come to be able to juggle to good epistemic effect (Langley et al. 1987: 7). As heuristics, no one of them is

Sublimation

foolproof, and each of them can be found in some combination with the others in all the realms governed by the sciences of the artificial. Skinner's follower, Howard Rachlin (1989), likewise denies any special status to human intelligence, but instead resolves the "cognitive" experience of humans into the complex networks of operants and reinforcement schedules. Moreover, even in the passage that Slezak cites to demonstrate Simon's distance from behaviorism, it is clear that Simon's principal criticism of Skinner is that the latter officially refuses to posit intervening variables–such as "programs"–in the prediction and control of behavior, even though (so Simon claims) such posits are necessary for Skinner's own project to get off the ground (Newell and Simon 1972). Here Simon is reflecting his own commitment to the "purposive behaviorism" of famed Berkeley psychologist E. C. Tolman, whose imputation of "cognitive maps" to maze-running rats provided some early clues to how the black box of thought may be scientifically pried open (Simon 1991b: chap. 12). Again, we see Simon trying to blur rather than build boundaries in dealing with his opponents, very much against the spirit of Chomsky's own starkly drawn antibehaviorism.

Clearly, Chomsky and Simon employ quite different rhetorical strategies in their pursuit of normal science in the "cognitive paradigm." On the one hand, as Rudolf Botha (1989) recently observed, Chomsky has managed to embroil himself in over one hundred separate debates since first proposing his model of generative grammar in 1957, with virtually all of his research designed to gain dialectical advantage in one or more of these encounters. Chomsky (e.g., 1980: 89-91) typically argues by shifting the burden of proof onto opponents: Why *shouldn't* language be thought of as a special module, seeing that we understand it so much better than our other capacities? More often than not, Chomsky uses the alleged cognitive superiority of linguistics to other social sciences as an argument for the distinctiveness of language rather than for the role that linguistics may play in reforming social science. On the other hand, Simon typically argues by juxtaposing a set of relatively simple studies from a variety of fields, no one of which is especially impressive, but which when presented together enable the reader to see a heretofore undiscerned pattern of intelligence at work. In principle at least, Simon treats all of the social sciences as cognitive equals. His current focus on AI reflects more the ability of computer programs to serve as a lingua franca for discussions of intelligence than any deep-seated belief on Simon's part in AI's superiority as a discipline. Indeed, when Simon (1991a) finally entered the fray, he immediately conceded the very point that Slezak claimed to have refuted in his

Two: Interpenetration at Work

original article. That is, Simon freely admitted that his new computerized scientists are just as social as the old-fashioned human ones. With friends like Simon, Slezak does not need enemies!

LANGUAGE AND THOUGHT: HORSE AND CART

From where, then, does Slezak get the idea that AI poses a direct challenge to SSK? Here the invocation to Chomsky provides a clue– Chomsky's own aversion to programmed discoveries notwithstanding. One view held in common by Chomsky and Simon is that there is a *language of thought*, an ideally efficient medium for the transmission of content. Chomsky and Simon disagree, of course, on how one reaches this ideal: for Chomsky, it is by recovering our innate linguistic competence from actual linguistic performance, whereas for Simon, it is by rendering our environment more tractable to our goals. In either case, however, it is presupposed that there is a way of determining the relative efficiency with which a particular content has been transmitted in, say, speech or a computer program. Such judgments presuppose, in turn, that two texts can have "the same content," with one text perhaps conveying this content more efficiently than the other. However, as Quine originally claimed and experimental psychologists have since shown, there is no empirical basis for such a presupposition: what is counted as having the same content is not only conventional but also contextually malleable. Given these points, it becomes radically unclear what a language of thought, such as the one embodied in BACON or in Chomsky's generative grammar, is supposed to be modeling. As a result, it is equally unclear what would constitute a proper empirical test of the model.

In light of the above problems with trying to render Slezak's position empirically testable, it is bizarre to find Slezak claiming that SSK cannot explain the fine-grained detail of scientific reasoning, by which he means why, say, Newton specified the laws of motion as he did and not in some other way. For we have just seen that the language-of-thought thesis makes sense only insofar as it is possible to collapse differences in detail in order to identify alternative formulations of the same content. Only then are Simon et al. justified in dropping out the historical specifics from their simulations. Indeed, so many details are collapsed that it is never clear whether BACON is modeling an individual, a collective, or a historical reconstruction. By contrast, SSK has at its disposal STS resources for explaining particular textual selections, namely, in terms of the reading and writing traditions with

Sublimation

which the author is familiar and with which she associates distinct audiences (or interest groups) and expected responses. Each of these traditions is transformed as it is combined with others in the course of their being jointly reproduced in a given text. It is difficult to see how much more fine-grained Slezak would want SSK to get. My guess, however, is that Slezak thinks that by building an enormous amount of contextual variation into the construction of scientific knowledge, SSK destroys the "realism" or "objectivity" of that knowledge—when, in fact, all that is challenged is its *universality* and *univocality*.

But from the fact that Slezak dismisses the SSK project, it does not follow that his own project deserves similar treatment. As might be expected, however, behaviorists and SSKers alike are prone to dismiss the entire language-of-thought project that underwrites AI as just so much reification: an illicit inference from the fact that we show agreement to the conclusion that there is something—some common content—on which we agree, which in turn is supposed to be part of some ideal medium for communicating content. However, I think we should take Slezak at his word and try to come up with some empirical tests for the "asocial" concept of content that he presupposes.

For example, could two cultures with radically opposed starting points, given comparable opportunities for collecting data and the like, reach the same correct solution to a common problem? If so, this would show that the initial cultural differences were overcome in the course of looking for optimal ways of relating conjecture to evidence. However, to do this experiment right, it is not enough for the experimenter to observe at some point that the two cultures have reached convergently correct results. In addition, she must see whether her judgments of convergence match what each culture thinks the other has accomplished, as well as whether the two cultures can agree between themselves on exactly what has been accomplished. Moreover, for the duration of the experiment, the experimenter should keep her own judgments private, but permit the cultures to monitor each other's activities so as to enable them to declare on their own that convergence has been reached. If there is a language of thought, then all these judgments of convergence should themselves converge, which is to say, both experimenter and subject cultures should be able to get beyond their particular perspectives and agree on the results of the experiment. My guess is that as the experimental task more closely approximates the rich environment in which science is done, such a harmonic alignment of opinion will be less likely, which will in turn highlight the empirical elusiveness—if not downright unfoundedness—of the concept of content on which the language-of-thought thesis is based.

Two: Interpenetration at Work

Three Attempts to Clarify the Cognitive

When Bruno Latour and Steve Woolgar (1986: Postscript) declared a ten-year moratorium on appeals to the cognitive, whatever else they might have been doing, they were shifting the status of the cognitive from an *explanans* to an *explanandum*, that is, from something that can be used to explain human action to something that is itself in need of explanation. In what follows, I draw on three general STS strategies for providing such an explanation. Given what we have seen so far, these will clearly have to be strategies for explaining the *variety* of accounts that travel under the banner of *cognitive*. The first, grid-group analysis, aims to plot the dimensions of this diversity as a function of social organization; in it "cognitive" defines what anthropologists call a "sacred space." The second, inspired by Marx's ideology critique, aims to demystify the "inherent" qualities in things deemed either cognitive or the proper objects of cognition by showing that they are systematically misappropriated features of society. The third strategy returns to my original diagnosis of systematic misunderstanding between AI and SSK. By focusing on the common image of the "black box," it is possible to trace the sources of this incommensurability.

THE COGNITIVE AS SACRED SPACE

The fact that Simon is much more hospitable to Chomsky than Chomsky is to Simon shows that how one defines the "cognitive" depends very much on who one takes to be a proper student of cognition, which, in turn, reflects how distinctive (or "sacred") an object one takes cognition to be. Not only is Simon more hospitable than Chomsky to would-be students of cognition, he also operates with a more flexible sense of what counts as a cognitive process. Thus, ontological and sociological space are bounded simultaneously. Slezak's obliviousness to this point in his attempt to mobilize intellectual resources against SSK serves only to muddle his defense of AI. However, his insensitivity to the rhetorical character of AI's own history may be remedied by placing his work in a context quite common to SSK and cultural studies generally (Thompson et al. 1990), namely, *grid-group analysis*. Grid-group analysis became part of SSK's intellectual armament when David Bloor used Mary Douglas' account of tribal responses to strangers in order to explain the different strategies that mathematicians used to manage anomalies raised against Euler's Theorem, as portrayed in

Sublimation

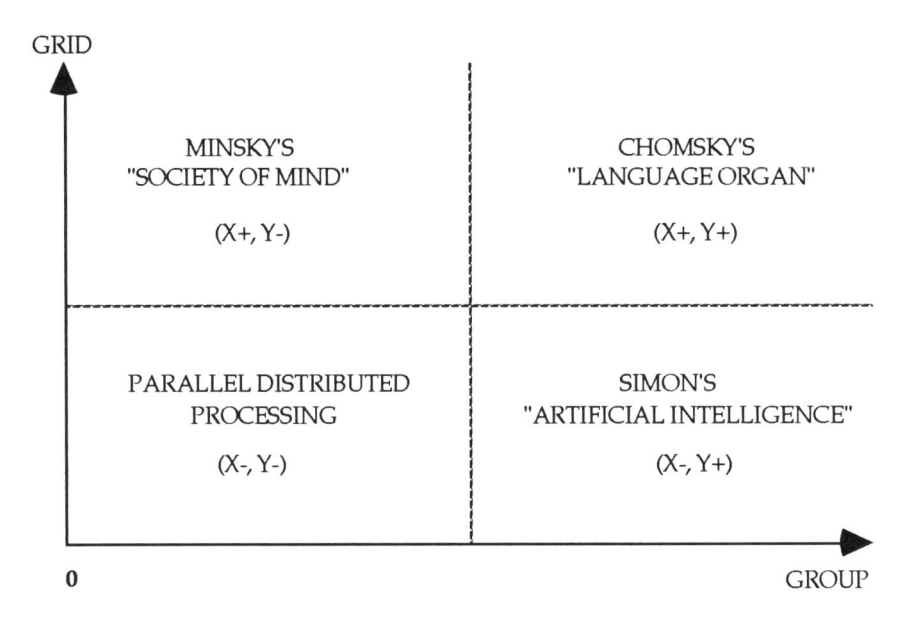

```
GRID
 ▲
 │     MINSKY'S            │    CHOMSKY'S
 │   "SOCIETY OF MIND"     │  "LANGUAGE ORGAN"
 │                         │
 │      (X+, Y-)           │     (X+, Y+)
 │                         │
 ├─────────────────────────┼─────────────────────────
 │                         │
 │  PARALLEL DISTRIBUTED   │      SIMON'S
 │      PROCESSING         │ "ARTIFICIAL INTELLIGENCE"
 │                         │
 │      (X-, Y-)           │     (X-, Y+)
 │                         │
 └─────────────────────────┴────────────────────────▶
 0                                              GROUP
```

Figure 3. A Grid-Group Analysis of Schools of Cognitive Science

Lakatos' *Proofs and Refutations* (Bloor 1979; cf. Bloor 1983: 138-45). I will now briefly sketch, in Tom Gieryn's phrase, the "cultural cartography" of cognitive science based on what I have said so far and offer some suggestions as to how the remaining grid-group quadrants may be interpreted (See Figure 3).

In grid-group analysis, "grid" refers to the internal organization of some body of knowledge-and-knowers, whereas "group" refers to the external differentiation of that body from other such bodies. A body of knowledge-and-knowers is plotted as either "high" or "low" on both dimensions. Thus, "high group, low grid" would mean that the body in question strongly differentiates itself from other bodies but manifests little internal organization. In fact, this is where I would place Simon, who identifies the essence of intelligence in the interface between organism and environment, yet who then stresses that the mark of intelligence in the organism is its adaptability to change rather than its execution of fixed procedures. Admittedly, the ability of, say, the business firm or the scientific discoverer to adapt to change in its situation is limited, but that does not imply that the firm's or discoverer's response to that situation must be rigid. As we have seen, this attitude also captures Simon's policy toward enlisting allies in the cognitive

Two: Interpenetration at Work

paradigm. By contrast, Chomsky should be considered "high group, high grid" in his highly formal and rigid manner of demarcating cognitive allies and objects from one another.

What is missing, then, is a sense of what the "low group" half of the cultural cartography would look like. These would be the people who do not postulate a great ontological and sociological divide between the cognitive and the noncognitive–or, as a follower of Simon might say, they are the ones who attenuate the interface between organism and environment. On the "high grid" side, I would locate the "society of mind" approach to AI, long championed by the founder of MIT's AI Laboratory, Marvin Minsky (1986). According to this approach, the mind is a collection of specialized modules that are indexed to situations in which expertise is required in everyday life. Although Minsky's modules are themselves well-defined, it is not clear whether they are meant to reflect the situation-specific character of social learning or, rather, the biofunctional preconditions for social learning to be situation-specific. So much of the work in Minsky's argument is done by metaphors drawn from organizational communication that it would be easy to conclude that individual thought is nothing but a microcosm of the social structures in which an individual performs functions–especially given Minsky's (1986: 38-46) basic constructivist tenet that the "self" is a mythical entity inferred from the fact that the diverse modules cohabit the same body. As might be expected, this view resonates with dramaturgically oriented theories of personhood in philosophy (Dennett, Harré) and sociology (Cicourel, Knorr-Cetina).

On the "low grid" side may be placed the recently popular parallel distributed processing (PDP, or "connectionist") models in AI which associate particular mental states with the spread of neural activation across the entire brain, thereby obviating the need for functionally specified modules. As an idea, connectionism originated over forty years ago in Donald Hebb's (1949) efforts to provide a neurophysiology that could underwrite the image of the maximally plastic organism presupposed by behaviorism. In the intervening period, several AI researchers designed connectionist models that for some simple motor skills and feature detection tasks outperformed more orthodox serial processors (Rumelhart and McCelland 1986 is the canonical presentation of connectionism; a good introduction is Johnson-Laird 1988: 174-94). However, because connectionism did not recognize the strong behaviorist-cognitivist split that fanned the fires of the Cognitive Revolution, it remained in obscurity until enough distance had been created from The Behaviorist Menace that cognitivists could afford to reintroduce it

Sublimation

through the backdoor, a suitably opportunistic fate for something as free-formed as a low grid-low group tribe!

THE COGNITIVE AS MISAPPROPRIATED SOCIETY

In the next chapter, I consider a set of historians and philosophers of science who have tried to stake the middle ground in the AI-SSK controversy by proposing a "cognitive history of science." The clearest philosophical attempt to legitimate this approach is probably Ronald Giere's *Explaining Science* (1988). But in what sense are Giere and his allies taking a "cognitive turn"? I would say that it is in a quite conservative sense, insofar as their turn is biased toward findings and interpretations that support the image of the scientist as a competent, largely self-sufficient human agent. Consequently, they downplay research pertaining to the cognitive limitations of individuals, especially the failure of individuals to appreciate the context dependence, and hence global inconsistency, of their thought and action. Moreover, our cognitivists underestimate the cognitive power that is gained via group communication and technological prostheses. But in the course of displaying these biases, the cognitive turn has brought to light important metaphysical issues that previously eluded philosophers of science. They pertain to the bearers of scientific properties: Where in the empirical world do we find *knowledge, theories, rationality, concepts*–to name just four philosophical abstractions hitherto left in ontological limbo? Our cognitivists are clear about arguing for the individual scientist as the relevant locus. Their focus is "cognitive" in the familiar sense of being concerned more with the individual's thought processes than with the products of her thought. That is probably because cognitivists do not challenge the idea that these processes produce the sorts of things that more traditional philosophers of science would regard as having "cognitive content," such as theories. As a result, while the cognitive turn tends to give us a full-blooded sense of what theorizing is like (e.g., a pattern of neural activation), we are still left with a rather pale, abstract sense of what theoretical output is like. For example, I suspect that different "styles" of theorizing radically underdetermine the types of theoretical texts that are written, yet it was those texts that initially led philosophers and historians to believe that there was something cognitively special about science.

My own perspective on the cognitive turn in the history and philosophy of science is very much like Marx's on the capitalist turn in the

Two: Interpenetration at Work

history of political economy. In capitalism, relations among people are mistaken for properties of things. What Marxists typically mean by this claim is that goods do not have an inherent value, or natural price, but only an exchange value that is determined by the social relations among the capitalist, worker, and consumer. Likewise, I believe that, in its attempt to locate abstractions in the empirical world, cognitivism mistakes (1) rational reconstructions for actual history, (2) properties of groups for those of individuals, (3) properties of language for those of the mind, and (4) properties of society for those of nature. Let me consider each in turn.

> 1. A vivid way of making this point is by examining one ambitious attempt at a cognitive historiography of science: *Patterns, Thinking, and Cognition* by Howard Margolis (1987). Margolis accounts for Kuhn-style paradigm shifts (especially Ptolemy to Copernicus) as the overcoming of cognitive barriers. But like Piaget's genetic epistemology, this makes for better pedagogy than history of science. In other words, teachers could use Margolis to get students to see beyond the shortcomings of their current framework to a more comprehensive one—but only once that next stage of comprehensiveness has already been achieved by the scientific community. His is a method for meeting standards rather than setting them. Margolis' confusion here probably stems from his insensitivity to the normative dimension of Kuhn's account of scientific revolutions. In particular, although in politics it makes sense to speak of "failed revolutions," all of Kuhn's revolutions are success stories. That is, the only cognitive changes that he recognizes as "scientific revolutions" and "paradigm shifts" are the ones that moved scientists closer to our current paradigms. Beyond that, Kuhn has little to say about how such revolutions occur, for that would involve accounting for a variety of individuals, most with interests quite distinct from those of the original revolutionary, but who nevertheless found that person's work of some use for their own. Thus, Margolis mistakes reconstructed history for the real thing because he typifies in one individual a process that is better seen as distributed across a wide range of individuals.
>
> 2. This last point is worth emphasizing, as it brings into focus the simplistic sociology that often informs the cognitive turn. Kuhn is more to blame here than any of the latter-day cogni-

Sublimation

tivists, especially his tendency to characterize scientists as having a common mindset or worldview, which, in turn, makes it seem as though, for a given paradigm, once you've seen one scientist, you've seen them all. Sociologists regard this typification of the group in the individual as a methodological fallacy, the "oversocialized conception of man [*sic*]" (Wrong 1961). The problem with the conception is that in attempting to account for the social dimension of thought, it actually renders the social superfluous by ignoring how the interdependence of functionally differentiated individuals makes it possible for a group to do certain things that would be undoable by any given individual. Philosophers are prone to an oversocialized conception of humans because of bad metaphysics. They tend to treat a *part-whole* relation as if it were a *type-token* one: to wit, society is an entity that emerges from the arrangement of distinct individuals, not a universal that exists through repeated instantiations. Indeed, as I argued in Chapter 3 of this book, the signature products of cognitive life–knowledge, theories, rationality, concepts–are quintessentially social in that they exist only in the whole, and not in the parts at all. For example, it is common for cognitive psychologists to treat conceptual exemplars, or "prototypes," as templates stored in the heads of all the members of a culture (cf. Lakoff 1987), when in fact they may be better seen as concrete objects that function as public standards in terms of which the identities of particular items are negotiated. It may well be that each party to such a negotiation has something entirely different running through her mind, but their behaviors are coordinated so as to facilitate a mutually agreeable outcome.

3. Continuing in the spirit of the last remark, if one is looking for an account of the brain that starts with minimal common capacities and then builds up quite different neural networks, depending on an individual's experience, one need look no further than the promising array of parallel distributed processing (PDP) models. However, contrary to what Giere seems to think, I believe that the extreme context sensitivity of PDP models implies that whatever sustained uniformity one finds among members of a scientific community is due not to any uniformity in their private thought patterns, but rather to some uniformity in the public character of their behavior, especially the language in which members of that community transact business.

Two: Interpenetration at Work

(In fact, that might be the point of scientific language.) For if PDPers are correct about the variety of neural paths that can lead people to say, do, and see roughly the same things, then I take that as an argument that the nervous system does *not* provide any particular insight into the distinctiveness of science as a knowledge-producing activity. (Of course, PDP would still say a lot about "how we know the world" in the looser sense of successfully adapting to the environment.) But even if one were to find this conclusion outlandish, it remains to be seen whether the cognitivistshave a story to tell about scientific communication, the means by which findings are ultimately judged to be normal, revolutionary, or simply beside the point. From works such as Nersessian (1984), which Giere cites approvingly, it would seem that communication is the process by which a later scientist reproduces an earlier scientist's thought processes in order to continue a common line of research. However, if thought is as context sensitive as PDPers suggest, then it is unlikely that this story could be literally true—especially if the relevant thought processes are defined in terms of what we now, only in retrospect, regard as a "common line of research." And even if a later scientist wanted to pursue an earlier scientist's work, it is not clear that either her means or her motives would involve the reproduction of that work (cf. Wicklund 1989). My guess is that the "concept maps" and other heuristics that cognitivists elicit from scientific texts are more formal analyses of scientific rhetoric that conveyed the soundness of the scientist's work than representations of "original" scientific reasoning that readers followed step-by-step in their own minds. This is by no means to demean the accomplishment, but simply to put it in perspective.

4. Finally, perhaps the grossest sociological simplification behind the cognitive turn may be termed its "visually biased" social ontology: to wit, social factors operate only when other people are within viewing distance of the individual; if no one is in the vicinity, then the individual is confronting nature armed only with her conceptual wiles. The solitary laboratory subject working on experimental tasks—the source of much of the cognitivists' background psychology—certainly reinforces this image. The biggest offense here lies in the failure to see that cognitive patterns are memories of socially framed experiences, which are resistible and replaceable only in socially permissi-

Sublimation

ble ways. The project of altering one's point of view (e.g., adopting a new theory), for the sake not merely of entertaining the alternative but of making the alternative the basis of one's subsequent research, involves the simultaneous calculation of what philosophers have traditionally called "pragmatic" and "epistemic" factors. This serves to bind "the social" and "the natural" in one cognitive package that cannot be neatly unraveled into, respectively, impeded and unimpeded thought processes. Relevant to this point is the Machiavellian Intelligence Thesis, recently proposed by two Scottish animal psychologists (Byrne and Whiten 1987). They argue that cognitive complexity is a function of sociological complexity, such that the organisms which respond to environmental changes in a less discriminating fashion tend to be the ones with a less structured social existence. One conclusion that Byrne and Whiten draw is that the complexity of nature distinctively uncovered by science may be little more than a reflection of the combination of people who must be pleased, appeased, or otherwise incorporated before a claim is legitimated in a scientific forum. A more simply organized science would, then, perhaps reveal a simpler world.

THE COGNITIVE AS BLACK BOX

Those who are sanguine about the possibilities of AI tend to regard humans and computers as partaking of the same substance at a certain level of abstraction: e.g., they are both "cognizers." By contrast, skeptics tend to regard humans as necessary complements of computers: e.g., a human must interpret computer output for it to make sense. This difference may be cast in terms of the two sorts of operations that Piaget (1971) has identified as essential to how people orient themselves in the world. The former involves enclosure in logical space (i.e., both humans and computers are members of the class of cognizers), whereas the latter involves separation in physical space (i.e., human and computer are distinct parts of one intelligent system). These two operations have precedents in the structuralist literature as, respectively, *metaphoric* and *metonymic* modes of linguistic analysis. Neither operation as such implies the superiority of either computer or human over the other. For example, in the metaphoric mode, cognizers can be defined so that either machine computability or human complexity is the norm against which the other is a degraded version; likewise, in the metonymic mode, either humans may confer sense on computers or computers may serve to discipline human judgment, as when a computer

Two: Interpenetration at Work

is treated as one of B. F. Skinner's programmed learning machines. Nevertheless, fights begin to break out once defenders and opponents of AI enter a prescriptive mode, which typically involves treating the computer as a kind of "black box." Consider two ways in which both sides deal with this image.

> *The image of "closing the black box"*: Those who are sanguine about AI close the computer's black box by trusting its output and adjusting their interpretation of the computer's design so as to render the output appropriate. This image reflects the tendency in the history of science for instruments to become "cognitively impenetrable," i.e., one trusts the readings from the instruments, even if it means discarding the theory one would like to see confirmed by the readings. Skeptics, however, close the black box by evaluating the computer's output by a standard external to the computer's design, such as human judgment, so that there is no intrinsic interest in the computer's operation, merely in the extent to which it simulates a predetermined understanding of what human beings can do. Whereas the former group, boosters of AI, adopt Daniel Dennett's (1987: chap. 1) "design stance" toward the computer, the latter group, critics of AI, adopt his "intentional stance." In the case of the design stance, the sense in which the black box is "closed" is that it operates as a final authority on epistemic judgments. In the hands of actor-network theorists in science studies, such as Michel Callon and Bruno Latour, the machine is made to appear to be a cynosure in terms of which many diversely interested parties must define (or "translate") themselves (Callon and Latour 1981). As in the analysis of commodity form in Marx, the computer gradually shifts from being a mere medium of exchange to being something consubstantial with the parties involved in the exchange. Thus, as scientists come to rely on the outputs of expert systems to test their hypotheses, these systems come to be endowed with genuine expertise. By contrast, "closing the black box" on the intentional stance means rendering the machine's epistemic authority finite, i.e., measurable by some external, especially human, standard of performance.
>
> *The image of "opening the black box"*: Boosters of AI envisage opening the black box as the process of discovering what it is about the computer (or the human being, for that matter) that

enables it to think. The assumption is that the answer will be given in terms of subsistent and essential properties of the computer mechanism. By contrast, AI critics imagine that the black box will be opened when a history of the interactions between the computer (or human being) and other things has been written. Whereas the concentration of intelligence in one enduring place marks a "cognitivist" orientation toward the computer, the diffusion of intelligence over time and space marks a more "behaviorist" orientation to what would be more properly called a "learning machine." Thus, for boosters, "opening the black box" means peering inside the machine to see how its hardwired program constrains the range of potential interactions it has with the environment. But for critics, the relevant sense of "opening" is in terms of letting the contents of the box spill out so as to reveal the sequence of contingencies that have determined the machine's applications. In Marxist terms, what in the first instance would be regarded as a property of things, in the second instance would be seen as purely relational. Thus, as Herbert Simon first pointed out with regard to firms, what appears at a distance to be a consistent decision-making strategy may, on closer inspection, be seen as series of ad hoc adaptations to environmental changes.

Clearly, then, some metaphors–such as "black box"–are too fertile for their own good. The "black box" was originally Skinner's way of saying that behavioral psychology could go about its business without waiting for neurophysiology to determine an organism's capacity for operant conditioning. In other words, the psychologist declared his disciplinary autonomy by putting physiology in a black box. Soon afterward, cyberneticians started using "black box" to indicate that they were really concerned with *any* machine that could generate the relevant input-output functions. In the cybernetic black box, then, could be found the set of functionally equivalent mechanisms, not just the physiology of a single organism. Latour (1987) has more recently gotten into the habit of using "black box" to refer to a situation in which differences (of opinion, interpretation, operation) have been suppressed for purposes of contributing to a united front. The Cognitive Revolution, as portrayed originally by its principals and more recently by sympathetic historians (e.g., De Mey 1982; Baars 1986; Gardner 1987) and Sons of the Revolution like Slezak, is a Latourian black box. As we have seen, the passage of time is itself a good tool for opening this

Two: Interpenetration at Work

particular black box and watching all the hidden differences spill out.

In the Slezak controversy, considerable rhetorical confusion was caused by the cognitivists, especially Ronald Giere (1989b), black-boxing (or is it black-balling?) Latour as a Skinnerian behaviorist. *Pace* Giere, Latour is just as much interested in opening his black boxes as Giere is in opening the ones that Skinner is willing to leave shut. But they are not the same sort of boxes. A relevant point for any future challenges that Slezak might want to make to SSK is that, however else they may differ, SSKers typically believe that with enough ingenuity, any Latourian black box can be opened. One of Simon's scientific discovery machines would turn into a Latourian black box if scientists routinely deferred to the machine's output as an instance of scientific discovery, much in the way that scientists already defer to the readings of their more technical instruments without wondering, in each case, whether the instrument is working properly. The clever SSKer could presumably destroy the trust invested in a primitive discovery machine by eliciting multiple interpretations of the machine's output from its users, which would then serve to raise more global doubts in their minds. However, someone like Slezak should believe that a successor machine could alleviate such doubts and thereby close the Latourian box. Why not try and see whether the SSKer can then wreak hermeneutical havoc on the users of the next generation of discovery machines? Even if Slezak does not like the theory that informs the SSKer's practice, he can hardly afford to pass up the opportunity for some free diagnostic service.

Now let us suppose that, after a time, one of Slezak's machines has become so sophisticated that the SSKer cannot penetrate the interpretive closure that the machine's users have reached. It might seem that the SSKer would have to admit defeat. However, a follow-up question first needs to be asked: Are the machines still regarded as doing science, or have the activities performed by the machines lost the status of "science" precisely because they have become subject to such routinization? The lesson, once again, is that AI researchers cannot simply presume that they are trying to simulate a stationary target. Rather, they need to take seriously what has become a near obsession in recent SSK theorizing, namely, the reflexive character of their enterprise: To what extent will progress in the manufacture of discovery machines cause the goal of computerized science to recede even further in the distance?

Of course, the elusiveness of successfully computerized scientific discovery may be due not only to machine achievements but also, and

Sublimation

more likely, to machine failures. This latter possibility is kept in play by such verbal hedges as "relevant" (as in "relevant information" for a machine discovering a scientific law) and "sufficient" (as in "sufficiency criterion" for a discovery program), which pepper the writings of Simon, Slezak, and their associates (cf. Langley et al. 1987: 32-34). These words make direct falsification of the AI project difficult, as they bob and weave between *describing* what some scientist actually thought and *prescribing* what would have been enough for her to have thought, had she had the benefit of a frictionless intellectual medium. (Downes 1990 documents Simon's gyrations on this matter.) To put the point vividly: Would Kepler or his contemporaries have been satisfied with the outputs of Simon's machines as constituting a scientific discovery? No doubt they would have been put off by the fact that machines were producing the outputs. But probably Simon would be unimpressed by an objection of that sort. In that case, what sort of objection could they raise that *would* cause Simon to admit his program's inadequacy? Needless to say, Simon's work will not be appreciated by SSKers, or STS practitioners more generally, unless he and his students stop mining the history of science opportunistically and start taking to heart that the divide between "real" content and "mere" context has itself been continually subject to social construction throughout the history of science. What sounds like background noise to Simon may be essential to the understanding that the original actors had of their activities. And if Simon does not want to be held accountable to local historical standards, then at least he ought to provide an intelligible explanation. All of this is to say that Simon's enterprise would be best served by actually including an STS researcher in his next grant proposal!

AI's Strange Bedfellows: Actants

After all the reservations that I have expressed about AI research up to this point, it may come as a surprise that I essentially agree with Slezak on including a metatheoretic requirement of computability in our accounts of cognition. This is simply the positivist point that you have not fully understood a phenomenon until you have developed an explicit procedure for reliably bringing it about. (As a metatheoretic requirement, it does not commit me to endorsing the adequacy of any of the procedures instantiated in today's computer simulations—a point where I probably diverge from Slezak.) I do not think it pure coincidence that the aspects of the human psyche which most resist computer simulation, such as moods and emotions, are also generally recognized as being the least understood. Moreover, in the spirit of SSK's own constructivism, I would hazard to say that our continuing lack of understanding about moods and emotions may result from a lack of agreement over what we mean by "moods" and "emotions." The "mystery" here may be simply logistical. My comments here are aimed against certain potentially obscurantist tendencies in STS, *practice-mysticism*, which hold that the holistic nature of the scientific enterprise makes it resistant to any systematic, procedure-based analysis. Practice-mysticism can be traced to Michael Polanyi's stress on the "tacit dimension" of scientific knowledge, which was meant to keep both methodologists (e.g., Popper) and commissars (e.g., Bernal) from holding scientists accountable to publicly scrutable standards. Ironically, despite their radical patina, ethnographic studies of scientists often reinforce this image of inscrutable competence by presuming that scientists do indeed know what they are doing, but that this knowledge can be gleaned only by becoming acculturated to their specific habitats, paying attention to what the scientists do, not to what they say. The advantage of computer simulations in this context is to remind us that complexity need not imply ineffability or inscrutability.

At this point, epistemology and ontology start veering into political economy. Is a computer *entitled* to know? Should we confer epistemic status on its outputs? But before broaching this question, the possibility of practice-mysticism must first be brought right to our own doorstep. One of the principal sources for contemporary work in the rhetoric of science has been academic programs in *technical communication*, a somewhat grandiose name for the ability to write instruction manuals that will enable ordinary people to do or use technical things (for an STS-inspired approach to this field, see Collier and Toomey 1994). Not only must these manuals be comprehensible to their intended audiences, but they must also enable people to use the accompanying gadget. The technical communicator takes for granted that anything can be explained to anyone, given enough time. Unfortunately, the time is not always available to articulate all that a particular person needs to know. But this is a problem of economics not ontology: no practice, in principle, escapes verbal instructability. From the technical communicator's standpoint, then, the practice-mystic misreads her own impatience as the unskilled's incompetence. One's own need to apply effort becomes a measure of another's cognitive liabilities. The same applies

Sublimation

to our own (un)willingness to interpret computers as having done enough for us to attribute certain thoughts and capacities to them. I may lack the time, imagination, or interest to interpret the computer as performing intelligently–perhaps because I have more important things to do and the computer is in no position to prevent me from doing them, or because I would have to end up interpreting the computer as doing something other than I would have expected or liked it to do. The political implications of this point (which resonate with much in the critical literature on colonialism: cf. Forrester 1985: pt. II) become clearer, as we now turn to a critique of *Artificial Experts*, in which Harry Collins (1990) offers perhaps the most sophisticated defense to date of practice-mysticism.

Collins argues that computers will probably never be recognized as "members" (or "peers") in a scientific community, if the community's local standards hold sway. From Collins' detailed accounts of computer ordeals, one is reminded that, in certain respects, the "computer" stands for anyone who can pass all the regular examinations but does not come from the right background. Ever stiffer tests are set–usually ones that members with the "right background" would themselves be hard-pressed to meet–and ever less charitable readings are given to the individual's responses. Indeed, Collins signals what, in a more politicized context, would be called a "prejudice" against computers by admonishing that "our" humanity may be endangered by allowing machines too quickly into the fold of intelligent beings. Of course, "our" doesn't mean all of humanity, but only those members of Homo sapiens who, in the unabashed language of the eighteenth century, have the appropriate "taste" or "sensibility" to evaluate those others who now also lay claim to some humane qualities. Today "skill" and "expertise" are the preferred terms of art (cf. Bourdieu 1986), yet whole classes of people are still just as eligible for exclusion as classes of machines. In other words, although Collins's book is ostensibly about distinctions between humans and computers, it is really about distinctions that already exist among humans, but whose coverage, in recent years, has been extended to computers.

Two sorts of strategies uphold the political economy of expertise. We have already discussed the first, namely, strategies that make one's initiation into a community of experts difficult and hence relatively rare. Were everyone thought to be competent in some sphere of action, then it would probably lose its status as "expertise." Instead, one would probably start assimilating it to the debased epistemic currency of "habit," "routine," or "common sense." The second sort of

Two: Interpenetration at Work

strategies maintain a double standard of evaluation for "experts" and "nonexperts" (or students, for that matter). Once you are presumed to be expert, the level of scrutiny drops considerably, as the extent of your discretionary judgment rises. Thus, actions that might seem anomalous if performed by a nonexpert are allowed to pass and perhaps are even taken as innovative in the hands of an expert.

Collins' ethnographic proclivities may have made him oblivious to these important differences in evaluative standards. In particular, his repeated appeal in *Artificial Experts* to the nuance and unpredictability of human behavior plays to an artifact of human cognition. That is, in everyday life we presume that if people's behavior does not violate a certain range of tolerance (or "civility," to use another eighteenth-century word), then it passes. Such default standards can serve to reinforce a variety of behaviors whose variety can give the impression that people are many-splendored things, when in fact it only shows how coarse-grained our behavioral standards normally are. Unfortunately, the impression of many-splendoredness can be conveyed to someone who is trained to attend to behavioral differences on the presumption that each deviation is meaningful. In other words, "nuance" and "skill" may be ethnographic overinterpretations of behavioral variation that normally escapes the notice of the natives. And as for human "unpredictability," that can probably be explained as an artifact of the imperceptible shifting of our standards of behavioral scrutiny from context to context, such that the context in which we initially predict someone's behavior is typically different from that of the standards by which we later evaluate the prediction. Of course, in terms of keeping evaluative standards constant, machines are at a distinct disadvantage because their performance is typically scrutinized under conditions that more closely resemble those of a laboratory experiment.

The enforcement of double standards is a far cry from Collins' previous incarnation as an SSKer, in which he foreswore any allegiance to either a computer or a human essence. In those days, he would not have been swayed by–dare I say, "merely" philosophical–arguments that rely on meticulously comparing computer responses to what some hypothetically competent human would do. Rather, he would simply, say, plop a computerized expert system in a medical habitat and treat it as an open-ended empirical question whether the medical professionals follow the system's advice more often than that of any of the ambient humans. No doubt it would be hard to determine the follow-up behaviors that constitute "following advice," but no less so for the

Sublimation

humans than for the computer. The level of expectation, the closeness of scrutiny, and the degree of charity conferred on a piece of advice will depend enormously on whether the prospective advisor–human or computer–is taken to be already part of the fold or someone who has yet to prove her worth. For standards are typically relaxed for anyone in the fold to such an extent that they may become difficult for the outsider to identify. Indeed, insiders may become confused about who gave the advice they are following. And it is probably this looseness that enables the standards to acquire the illusory depth that generally passes for "tacit knowledge."

If you are under the impression that I regard the injection of the computer into the scientific workplace as a kind of democratizing move, then you have picked up the scent of my argument. At this point, it pays to draw a distinction and note an unwitting convergence, in light of that distinction. The distinction in question is between AI and what the philosopher of science Clark Glymour (1987) calls *android epistemology* (AE). Both are machine-centered pursuits, but the former treats the computer as a (better or worse) representation of human thought processes, whereas the latter treats the computer as a thinker in its own right, though perhaps one that thinks in ways radically different from human ways. Needless to say, the same machine can be treated either way, or even both ways. In other words, contrary to the tenor of AI-SSK debates, there need be no a priori decision between making the human or the computer the standard for measuring the adequacy of thought. After all, actual scientific practice incorporates both entities, and one does not hear scientists making blanket pronouncements that everything–or, for that matter, nothing–a human or a computer says is cognitively significant. Thus, Glymour's call for an AE alongside AI is only true to the social facts. Indeed, most procedure-based theories of rationality, be they derived from economics or from epistemology, work better on computers than on humans. And insofar as humans have been willing to evaluate their own thought and action in light of these theories, we would seem to be hard-pressed not to credit the computers with some measure of cognitive ability. How much, of course, depends on how often we change our behavior in light of what the computers say. The unwitting convergence arises because Glymour, a practicing PC-Positivist of the sort mentioned at the beginning of this chapter, is in agreement with the recent theory of *actants* in SSK, which would ascribe such properties as "agency" and "cognition" to all manner of things, in part as a means of opening up aspects of those phenomena that are not properly appreciated because they can never speak for

Two: Interpenetration at Work

themselves or establish their own standard of evaluation. To redress the balance, Steve Woolgar (1985) has thus proposed a *sociology of machines*.

In conclusion, I agree with much of the tenor of AI work that treats the computer not as merely a model, but as a virtual agent in the scientific enterprise. I take the substance of my position to be close to Woolgar's sociology of machines, except that whereas he is agnostic about the changes wrought on the concept of the person by the integration of computers into the human lifeworld, I am more openly enthusiastic (cf. Fuller 1989a: 88-102). Being an only slightly reconstructed modernist, I believe that as you become more conscious of the mechanisms of conceptual change you can change your concepts more freely (cf. Dolby and Cherry 1989). This view would also seem to reflect the implicit position of most AI practitioners, who want to grant the computer at least some epistemic authority: i.e., there are certain cases in which we should trust the computer's judgment over our own. Ironically enough, that much-battered behaviorist B. F. Skinner bears some credit for the view I am espousing on behalf of an enlightened attitude toward computer agency. Skinner's original programmed learning machines were designed to shape the behavior of students who wanted to learn linguistic and mathematical skills by subjecting them to the principles of operant conditioning, with the machine doling out the appropriate reinforcement for each student response. In a world where knowledge of, say, mathematics is valued largely for its abstractness and precision, why wouldn't one of Skinner's machines be the ideal entity to do one's apprenticeship under?

Some SSKers may be disturbed by these developments, but I think that any loss of sleep over this matter would be just the result of SSK's failing to follow through on its own message. In other words, if SSKers generally hold, say, that the meaning of one's actions is what the community takes them to mean, then why should this not also apply to whatever a computer does? To put it in terms of the Turing Test, the classic criterion for judging the intelligence of artifacts: If you can confuse the machine with a scientist, then it simply *is* a scientist. Given the great lengths that SSK has gone toward conventionalizing even the slightest hint of a human essence, it would be only consistent to argue that convention alone causes us to trust human over computer utterance. Indeed, we already defer to the epistemic authority of the calculator over our own or some other human's computational efforts. Admittedly, arithmetical computation is not the most esteemed form of cognition, but perhaps that is due precisely to its being a task conventionally delegated to machines. If so, imagine the implications for the ordinary conception of science if scientists routinely trusted the output displays of, not only calculators and meters but hypothesis-testing machines as well! I envisage that an interesting unintended consequence of our coming to accept BACON and its successors as competent prosthetic reasoners may be to remove the cognitive functions that these machines perform from the valorized realm of "science." In short, in trying to understand scientific reasoning, AI may unwittingly end up drastically altering the social definition of science itself.

6 EXCAVATION, OR THE WITHERING AWAY OF HISTORY AND PHILOSOPHY OF SCIENCE AND THE BRAVE NEW WORLD OF SCIENCE AND TECHNOLOGY STUDIES

Whatever one may think of Kuhn, he did a marvelous job of making History and Philosophy of Science (HPS) appear to be a disciplinary blend of utmost importance to the intellectual community at large. Moreover, Kuhn did the job unselfishly, in that he also drew attention to his positivist, Wittgensteinian, and Popperian antagonists, many of whom won adherents of their own. Indeed, it has now become commonplace, especially in the social sciences, for debates about methodology to involve worries about where one's discipline is on the road to becoming a "paradigm" and how a "revolution" may be staged to set the discipline aright. However, the remarkable ability of HPS to establish spheres of influence in other disciplines is no indicator of the fate of the Kuhnian revolution at home, where the last fifteen years have witnessed a steady retreat behind disciplinary boundaries. In short, historians of science have succeeded in pulling in a few philosophers to examine the details of past science, perhaps with a greater sense of the institutional character of science than before the Kuhnian revolution, but in a way that is still studiously atheoretical and nonprescriptive. And with the latest round of the scientific realism debates (cf. Leplin 1984; Churchland and Hooker 1985), philosophers of science have returned to a quasi-transcendental mode of arguing that betrays their roots in classical epistemology and metaphysics. Thus, we have realists proffering just-so stories about what "must have happened" in history to enable science to be so "successful." Instead of raising historical counterexamples, antirealists tell simpler versions of the same story. Often it seems that the homeliness of the scenarios imagined is made to take the place of critical historical scholarship. This retreat from the Kuhnian revolution is significant, as it reflects an ambivalence on the part of HPS toward breaking new theoretical ground, specifically, *an ambivalence toward making the transition from the humanities to the social sciences*–a reluctance to make the transition from HPS to STS.

Positioning Social Epistemology in the Transition from HPS to STS

Where does social epistemology stand, such that it can accuse HPS of dragging its heels along the inevitable path toward STS? In brief compass, my answer is this. Like HPS, social epistemology also starts off "humanistically," in the sense of using the language of science as the entry point for understanding the nature of science. What becomes immediately evident, however, is that the descriptive and prescriptive functions of scientific language are in tension with one another, and hence require rhetorical management. In short, certain things can appear in the world only because certain other things have been made to disappear. Only once the normative conditions enabling this rhetorical management are uncovered, can one then envisage that alternative normative conditions would produce alternative forms of knowledge. Thus, the social epistemologist quickly moves from deconstructing texts, to surveying the material bases of power relations, to designing experimental utopias: from humanities to social science! Now, let us look at this transition a little less breathlessly.

Ordinary language is ill suited to any of the usual philosophical conceptions of epistemic progress, largely because the relatively unscrutinized level of ordinary usage serves more to maximize a sense of group identity and historical continuity on the part of the language users, and less to establish the exact extent or even presence of some common objects of agreement. This is what the American rhetorician Kenneth Burke (1969) has called the "consubstantial" quality of discourse, which is

Two: Interpenetration at Work

the proper object of rhetorical study. It enables large numbers of people to move in a common direction without ever having to reach explicit agreement on a label for that direction. For example, a call to "patriotism" may unite many citizens in war, even though, if asked, they would probably give divergent opinions of what they are defending when they defend their "country." The positivist account of language as tool, typified in A. J. Ayer's (1936) emotive theory of ethics, is sensitive to this point. That is, unless special institutional arrangements are made—say, the introduction of a verificationist semantics—language functions primarily to move people to act, speak, and feel in certain ways. Nothing as fine grained as the distinction between truth and falsehood is required for these functions to be performed. Here the positivist parts company with the pragmatist, who holds that instrumental success and long-term survival are prelinguistic surrogates for truth found throughout the animal kingdom. In siding with the positivist, I am admitting that the search for truth is quite an artificial inquiry, one that is directly tied to the regimentation of linguistic practice. Such an inquiry cannot simply be reduced to brute pragmatic utility.

From the scientific standpoint, the consubstantialist tendencies of ordinary language foster only miscommunication and cognitive stasis by minimizing the opportunities for raising latent differences to the surface. Such opportunities are opened up once utterance is held to stricter standards of accountability, even if that means simply asking more follow-up questions. The Socratic dialogues illustrate this move, whereby two people who originally assented to some seemingly simple proposition are asked to articulate the reasons for their opinion, which turn out reveal a deep disagreement that then requires philosophical assistance for its resolution. Applied systematically, such assistance aims to reconstitute ordinary linguistic practice into one that can be scrutinized on a regular basis and thus be rendered an appropriate vehicle for epistemic progress. In this way, truth and falsehood become institutionalized as properties of utterances. However, this institutional arrangement, often called "representation" or "reference," is rather expensive to maintain and goes very much against the efficiency of language as a prod to action. For a variety of procedures and products—repeatable experiments, canonical methods, final examinations, pure samples—need to be established to which specific utterances can then be shown either to correspond or not. This variety embodies the process of "standardization." We are able to say that standards are subject to "determinate" readings because of the control that is exerted over who

Excavation

can speak for them. This, in turn, confers an "objectivity" on the utterances that are held accountable to those standards. What I have just described is the verificationist theory of meaning expressed as a piece of sociology (cf. Porter in press).

In everyday life, an utterance is presumed to move its audience unless explicitly challenged. But once the utterance is challenged, the utterer will often justify it by invoking standards that, indeed, would test the validity of the utterance if construed representationally, but which, under normal circumstances, simply serve to terminate discussion of the issue and to move the conversation to some other topic. Hence, it is important to distinguish the *representational function of language* from the *rhetorical function of representation*. The former involves a vast deployment of human and material resources for what are, essentially, surveillance operations, while the latter simply involves granting an utterance the same warrant for action as one would the surveillance operations that would ideally stand in its place. A common example of this representationalist rhetoric occurs whenever one scientist incorporates another's results into her own research without feeling a need to reproduce the original study.

Here is a piece of philosophical shorthand that epitomizes the way in which the social epistemologist combines views on the nature of knowledge that are typically seen as antagonistic. Am I a *scientific realist* ? A *logical positivist?* Or, a *social constructivist?* The answer is that I am all three. My realism is predicated on positivism, which is, in turn, predicated on constructivism. The difference between the three positions is that the social dimension of knowledge is least evident in realism (which, as in Peirce, always alludes to the theoretical language of a community in the indefinite future), somewhat more evident in positivism (which makes the possibility for knowledge relative to a currently available language), and completely self-conscious in constructivism (which relativizes knowledge still further to an extension of the language currently in use).

In a sense, the social epistemologist wants to beat the positivists at their own game by envisaging what it would be like to implement their account of language. The whiffs of Burke and Foucault are meant to vivify a point that can be traced to Frege and Carnap: truth and falsehood are properties of sentences in a language that has been designed to represent reality; prior to the construction of such a language, there is neither truth nor falsehood. The twist is that, unlike the positivists and their logical forebears, I take account of the diachronic dimension of language, especially the fact that most speakers of a language will

Two: Interpenetration at Work

have interests and understandings quite different from, and often at odds with, those of the originators of the language. From that I infer that as verification practices become routinized, they become more susceptible to consubstantiality effects, as similarly trained individuals come to take for granted that others mean what they mean when they say certain things. Thus, while routinization bespeaks a certain efficiency of practice, it also opens the door to incommensurable conceptions that rise to the surface only during a round of critical inquiry, as, say, happens during a "crisis" in one of Kuhn's paradigms. In that sense, the success of scientization (that is, routinization of scientific standards) sows the seeds of its own destruction.

Another way of seeing this point is in terms of what marks the conceptual transition from an instrumental to a representational approach to language, namely, when one's personal ends are no longer sufficient to justify the linguistic means used in their pursuit. For once enough misunderstandings, deceptions, and failed ventures have been acknowledged, people will come to realize that it is in their own interest to make their usage first satisfy some mutually agreeable end–a standard–before it can satisfy more personal ends. This makes one's pursuits less direct, but also less arbitrary, as everyone will have an interest in catching violations before they contaminate activities of the entire community. The first systematic effort to make this transition occurred during the Scientific Revolution of seventeenth-century Europe.

Although the Scientific Revolution is generally seen as transforming attitudes toward the natural world, it is better taken to have introduced a new attitude about ourselves–namely, as *imperfectly embodied standards of knowledge* (Sorell 1991: chap. 2). Thus, Francis Bacon expressly presented the experimental method as a form of self-discipline designed to counteract cognitive liabilities, or Idols of the Mind. The evolution of that method over the next 350 years has probably been the most important contribution that *psychology* has made to social epistemology, namely, a series of proposals for institutionalizing inquiry so that the whole of human knowledge may consist in something more than the sum of the participating human knowers. And this is what I mean to endorse by a *naturwissenschaftlich* approach to knowledge in *Philosophy of Science and Its Discontents*. Contrary to how such matters are normally understood (cf. Rouse 1987), I see the experimental method more as a means for *macro*-reproducing the lab in the world than for *micro*-reproducing the world in the lab.

The cultural distinctiveness of the Scientific Revolution has often been noted by historians and anthropologists, who nevertheless disagree about what exactly constituted the epistemic "takeoff" that led

Excavation

the West to surpass China, India, and the Islamic world in knowledge production after 1700. What crucial "factor" or "idea" was absent in the East which was present in the West? Although this is not the place to settle a historical dispute of such magnitude and complexity, social epistemology's sense of the history of science offers up an answer that may be pursued as a hypothesis. Rhetorically speaking, the Scientific Revolution enabled the translation of theoretical speculation into experimental practice. To give the point less of a linguistic spin, and more of a perceptual one, Western scientists came to see experiment as not merely an instantiation of theory, but as a test for the well-foundedness of theory. The trick here was the realization that because we are inherently imperfect knowers, our reasoning processes are unlikely to reach the truth simply of their own accord. Experimental techniques and apparatus, then, become both prosthetic devices to extend our cognitive capacities and standards against which those capacities are evaluated. The computer's dual role as the extension and the measure of rationality in the modern era is a clear case in point. By contrast, while the East had the technology and the speculation, the two were pursued independently of one another. No matter how certain or fallible our reasoning processes were taken to be by the Eastern philosophers, those processes were treated as self-contained, or at least not enhanceable or revisable by technological mediation. Generally, this was because the human soul was held to already contain the essential ingredients of reality, much as Plato and Aristotle had thought in Western antiquity. In the end, then, the difference between Occident and Orient, *circa* 1700, boiled down to ontology: the former portrayed humanity as an incomplete part of the natural world, while the latter portrayed it as a micro-instantiation of the entire world order. Only the former was suited to modern science.

The long prehistory of Mill's Methods of Induction testifies to the existence of experimental ideas in the West before the seventeenth century. However, earlier attempts to isolate necessary and sufficient conditions were speculative, and hence largely consubstantialist in their effects. Thus, medieval arguments about some factor's being the sine qua non of some state did more to elicit a sense of group identity between author and reader than to open the claim to empirical scrutiny. Indeed, this rhetorical appeal to thought experiments is very much alive today in the definitions of knowledge proposed by analytic philosophers, which turn on test cases that have become so well rehearsed in "the literature" that they enjoy the status of "intuitions" among the cognoscenti (cf. L. J. Cohen 1986). I would also put in this category most historians' narrative attempts to isolate causes, in that it is not clear

Two: Interpenetration at Work

whether one is persuaded by the general familiarity of the historian's plotline or its particular relevance to the case under study. This is the *geisteswissenschaftlich* approach to knowledge that is criticized in Fuller (1989a). Yet, insofar as historians take themselves to be knowledge producers, the general principles that provide the implicit warrant for their particular causal analyses are open to scrutiny from the social science disciplines whose business it is to generate and test such principles. These principles consist of the assumptions that historians make about people's motivations, collective tendencies, and cognitive horizons. Indeed, the most fruitful way of understanding the positivist "unity of science" thesis is as reminding inquirers that if they are inquiring into roughly the same subject matter, then they are accountable to each other's epistemic standards, from which they may then negotiate a common standard–ideally (so held Carl Hempel, e.g., 1965) a formal language of confirmation and explanation. Thus, historians open themselves to criticism by psychologists if they borrow outmoded theories, just as much as psychologists open themselves to attack from historians who question the generality of their experimentally derived principles. This, in short, is the implicitly scientific character of historical explanation.

The Price of Humanism in Historical Scholarship

As a discipline, history is both admirable and atavistic. It is admirable–at least from the social epistemologist's perspective–for its ability to court a wide readership that often extends beyond the academy. However, history is atavistic in that it reaches this wide public by continuing the nineteenth-century Rankean practice of portraying itself as a relatively neutral resource for finding out what actually took place, whose claims can be assessed simply in terms of their conformity to the available evidence and not in terms of their conformity to general explanatory principles put forth by, say, the social sciences. Both philosophers and sociologists of science would seem to err in their use of history, insofar as they insist on trying to isolate certain decisive "factors," be they "internal" or "external" to the knowledge enterprise, that are responsible for determining the course of science across a variety of sociohistorical settings. Historians have been inclined to look upon, say, the infamous exchange between

Excavation

philosopher Larry Laudan (1977) and sociologist David Bloor (1976) over the *rationality assumption* (sociology explains only the arational parts of science) vis-à-vis the *asymmetry principle* (sociology can explain both the rational and arational parts of science) as purely ideological, perhaps a mere cross-disciplinary turf war (J. R. Brown 1984 recaps the debate). After all, hasn't the historical record shown that *both* sorts of factors are at work all the time? Indeed, doesn't the inconclusiveness of the Bloor-Laudan debate prove the bankruptcy of any attempt to infer generalities from the history of science? A good case in point would seem to be the chilly reception given to Laudan's own attempt to stage a "crucial experiment" between the two viewpoints.

Under the rubric of "normative naturalism," Laudan (1987) has proposed a research program whereby the central claims made by internalists can be put to the historical test on, so to speak, a case-by-case basis (L. Laudan et al. 1986; R. Laudan et al. 1988). The fruits of this project would earn Laudan a place alongside Bacon, Bentham, Galton, Sorokin, and the other great tabulators of our times. A bureau or institute could be entrusted to collect the data, periodically publishing the latest tallies in a handbook: e.g., "Do scientists wait for a new theory before giving up an old one plagued by anomalies? In 10 cases this happened, but in 8 cases not." A good way of looking at this project is as an empirical test of Laudan's arationality assumption, which supposes that there are norms of sufficient transhistorical purchase to count as rational grounds for theory choice in science. Laudan (1987) quite deliberately chose enumerative induction–the idea that each confirming case counts in favor of a hypothesis–as his guiding meta-norm, since it clearly is a method that virtually every philosopher (Popper being the exception) has taken to lend credibility to a knowledge claim. Laudan hopes that the project will come up with less intuitive ones as well. However, if it turns out that none, or very few, of the 300+ norms under consideration are decisively accepted or rejected by the annals of science, then that would indirectly lend support to Laudan's externalist foes, who argue that theory choice is primarily determined by local social factors in which methodological appeals figure willy-nilly.

Now, despite the great care that Laudan has taken to formulate the norms and to justify his inductivist testing procedure, his project has been resisted from all quarters. Is this resistance simply a case of theorists not wanting to see their pet theses falsified? The main objection that has been so far voiced to the Laudan program is this: even if every philosopher has endorsed enumerative induction as a necessary part of

Two: Interpenetration at Work

the scientific method, certainly no philosopher has endorsed it as the entire scientific method; indeed, the various other methods proposed by philosophers have generally been designed to counteract the irrational consequences that would follow from the strict pursuit of inductivism (cf. Nickles 1986). I would argue that, Laudan's explicit appeals to history notwithstanding, his selection of enumerative induction as the method for his own project reveals more of his true, *nonhistorical* interests. In effect, Laudan has abstracted a lowest common denominator from the views of various philosophers of science and then reified it as the essence of the scientific method. In this light, we can read Laudan's critics as arguing that he has mistaken an "accidental universal" (i.e., a feature common to a set of particulars that fails to define their real nature) for a "natural kind." But this essentialist strategy is not unique to Laudan, or even to philosophers. Indeed, I propose to show that when philosophers and sociologists debate the merits of internalist versus externalist histories of science, they are really contesting a point of *ontology*: Does science have a transhistorical essence of its own ("Is science sui generis?" as Durkheim might have asked), or is it reducible to a historically persistent combination of some other, specifically social, sorts of essences? But first we need to diagnose and evaluate the reluctance of historians to take sides on this issue.

Some historians erroneously believe that an ontological debate of the sort just described could not be adjudicated by historical means, and, in a positivist spirit, they conclude that the debate should be terminated. True, a simple reportage of history "as it actually happened" will not do the job. But then historians are not the only ones who make use of historical evidence—everyone does. In fact, the territorial claims that historians are inclined to make over historical evidence bear an unfortunate resemblance to the claims that philosophers have traditionally made for their expertise over what is rational. When speaking in this territorial mode, one acts as if a discipline intruding on her turf has only two courses of action. In the case of historical turf, the intruder has the option of either submitting to the scrutiny of historians or admitting that she is using historical evidence in a (probably less literal) way that evades the standards of historical scholarship. In neither case is the intruder made to feel like she is doing something intellectually worthwhile. For its part, history, as the turf protector, reveals the extent to which its autonomy is grounded in xenophobia, specifically a fear of being held accountable to the claims made by other disciplines that draw largely from the same body of evidence.

While some philosophers have alleviated their xenophobic tendencies by making their theories of rationality accountable to the

Excavation

findings of economists (especially rational choice theorists) and psychologists (especially cognitive scientists), historians have been much more reluctant to admit officially that the validity of their research is affected by the findings of other disciplines. However, the practice of historians reveals a less consistent stance, one captured by the following observations: (i) historians maintain that they range over an intellectual terrain that is distinct from philosophy and the social sciences, though members of those disciplines periodically wander into the historian's turf; (ii) historians reinvent parts of these disciplines as a matter of course in their studies, even though more clearly articulated and tested theories on these matters could be found in the other disciplines; (iii) when it suits their purposes, historians will sometimes rely on the research of other disciplines, but when such research does not suit their purposes, they either ignore it or criticize it on methodological grounds—even if the methodology employed (e.g., controlled lab experiments) was the same as that of research on which, on another occasion, they had relied. In most of these respects, historians are no different from other specialists—and, indeed, one of the primary academic functions of the social epistemologist is to compensate for these liabilities. Nevertheless, they tend to be especially trenchant in the humanities, where the presumed generality of "human nature" has traditionally licensed casual sampling from the literatures of the special sciences. Indeed, the "liberality" of the humanist's general education is said to be displayed in such bibliographic forays.

The inconsistency that I have just noted in the behavior of historians suggests that a double standard is afoot, one that may be brought out by examining the latest version of the internal history of science. This new version is generally called "cognitive history." It is a species of intellectual biography that is concerned with reconstructing the thought processes of great scientists, usually on the basis of private notebooks and with the aid of the conceptual apparatus of cognitive psychology. This work tends to be done by people who have had substantial training in "cognitive psychology" broadly construed (i.e., including not only recent lab and computer work but also Gestalt and Piaget) and who openly support the HPS movement (e.g., Holton 1978; Gruber 1981; Nersessian 1984; Tweney 1989). These histories have an uncanny tendency to provide "independent corroboration" for internalist theses. If what I have been saying about double standards is correct, this so-called corroboration should be traceable to the suppression (or ignorance, as the case may be) of countervailing considerations that could easily be found in the psychological literature on which these

Two: Interpenetration at Work

historians draw. Let me now turn to three obvious instances in which this occurs: (1) the fixation on genius; (2) the presumption of scientific competence; (3) the analytic significance of individuals.

THE FIXATION ON GENIUS

Historians influenced by developmental psychology are prone to argue either that geniuses (e.g., Einstein) achieved a sixth stage of cognitive development after having exhausted Piaget's normal run of five stages, or that near geniuses (e.g., Poincaré) failed to make the big discovery (e.g., relativity) because they were stuck at stage five (cf. A. Miller 1986). In effect, this line of reasoning supposes that the relative significance of individual scientists to the scientific enterprise is an implicit acknowledgment (or at least a reliable indicator) of the relative quality of the scientists' minds. It is as if the principle of progress in science is that the entire community should try to approximate its most intelligent member. This principle may well have functioned as a regulative ideal during the Heyday of Humanism, the sixteenth-century Renaissance, when scholars saw themselves as recovering the pristine wisdom of the ancients that had become vitiated through repeated cultural transmission. However, it would be anachronism at its worst to portray the spread of relativity physics in the first three decades of this century as a matter of scientists playing catch-up with Albert Einstein. Even on her own terms, the cognitive historian would have a hard time explaining how rank-and-file physicists could come to reproduce routinely a discovery that originally took incredible mental powers (though perhaps she could argue, with a little help from scientific realism, that Einstein just needed fewer clues to arrive at relativity, which then enabled him to set down the additional clues that the rank-and-file needed). Even reasonably sophisticated inquirers interested in improving the social conditions of knowledge production (e.g., Root-Bernstein 1989) continue to focus more on how individuals may generate better ideas than on how ideas may circulate better in the scientific community, despite the fact that the most systematic psychological study of scientific discovery to date suggests that quality of mind is *not* what separates the geniuses from the also-rans of science (Langley et al. 1987).

More to the point, we know enough about the psychology of intellectual reception and appropriation to suppose that scientists would try to understand Einstein only as much as they needed for their own purposes, and, moreover, that they would use his theory with the understanding that it provided only an incomplete or partly erroneous view

Excavation

of things that their own contribution would be especially well suited to correct (Wicklund 1989; cf. Fuller 1988a: chaps. 5-6). Indeed, a wide range of studies suggests that a group working on a common set of problems is much more effective than any of its members in eliminating error and almost as good as its most insightful member in arriving at correct solutions. Together these traits make groups consistently better than individuals in most forms of problem solving (Clark and Stephenson 1989). These findings already start to explain why relativity theory was adopted and extended in quite a variety of ways, many of which were unexpected and even unsatisfactory to its creator–though better than if it had been merely transmitted intact in its original conception. Notice that in the course of presenting this alternative psychological account, I have granted the cognitive historian the controversial point that to understand an author is to reproduce for oneself the author's original thought processes. If this point is *not* granted, then the possibilities for even more radical critique are opened up, one that drives a wedge between the skills involved in *doing* (writing, performing) and in *recognizing* (reading, evaluating) good science (cf. Fuller 1989a: chap. 3).

THE PRESUMPTION OF SCIENTIFIC COMPETENCE

Even when the cognitive historian does not go so far as to treat the great scientist as a genius, it would seem that the great scientist can do no wrong, or at least not for very long. On this point, I will be briefer, as the argument has been developed elsewhere (Fuller 1989a: chap. 3). Cognitive scientists are now generally agreed that heuristics are liabilities on borrowed time, which is to say, mental shortcuts that work well in a limited domain but disastrously outside of it. We should thus expect that a heuristic-based account of a scientist's thinking over a span of several years would illustrate a great many cases of cognition run amok, perhaps never to be resolved properly in the scientist's own mind. Unfortunately, nothing of the sort is to be found. Instead, cognitive historians tell us a story of, say, Michael Faraday as someone who just so happens either to access the right heuristic at the right time or to correct a misapplied heuristic by the time the story is over (cf. Tweney 1989). The probability that this would capture the thinking of a real human being, given our best theories of cognition, is minuscule.

I am tempted to say that the humanist demand for a well-told story–one where the hero wins in the end–has undermined the scientific credibility of the cognitive historian's account. Not surprisingly, the Faraday case is greatly aided by meticulous notebooks that Faraday

Two: Interpenetration at Work

deliberately kept to assist himself in developing a continuous line of thought. However, our cognitive historians make reference to this fact, apparently without realizing that this "metacognition" of Faraday's may render the notebooks a more opaque, overwritten record of his actual lab work than their meticulousness might at first suggest. And notice here that I have yet to call into question the reliability of Faraday's memory, even though that would be the first thing that a psychologist would do, given the time lag between the events in Faraday's lab and his recording them in the notebooks. Ironically, although there is now an entire subfield of psychology devoted to this issue ("protocol analysis," cf. Ericsson and Simon 1984), it was routinely raised in the manuals on historical inference published at the turn of the century (Dibble 1964).

THE ANALYTIC SIGNIFICANCE OF INDIVIDUALS

Finally, we come to the very idea of making the individual scientist, genius or otherwise, the unit of historical analysis. Again, this seems to be a harmless practice, since in the interest of thoroughness, the historian will be forced to take in the scientist's cultural context and thereby sweep up ambient social factors that might otherwise be lost. That historians take this to be enough of a response to our concern underscores the extent to which they still think of themselves as akin to novelists for whom the choice of subject is largely a matter of personal taste. However, there is more at stake here. One remarkable point of convergence between ordinary language philosophy, experimental cognitive psychology, and cross-cultural anthropology is that our concepts are normally calibrated to fit our visual horizon. It thereby becomes the default level of ontological analysis. And so if, as Kahneman (1973) has argued, people tend to regard a freely moving, foregrounded object as causally determinative of its surroundings, then it would be fair to conclude that the historian's choice of subject represents an implicit causal judgment about the events and entities that have made a difference in history. Thus, even when the cognitive historian portrays Einstein and Faraday as "socially situated reasoners," this phrase serves only to frame a portrait, as the historian makes sure that the scientists are portrayed as having transformed their contexts in ways that are interesting for the subsequent development of scientific thought. In that way, they also satisfy the Kuhnian sense of "paradigm" as exemplar. By contrast, what we never find is a cognitive history in which the scientist's most distinctive contributions are based

Excavation

on misunderstanding or some other sort of error (perhaps on the part of an influential later reader). In short, by virtue of being invested with such self-possession, the great scientist is conceptualized as an "agent" (cf. Harré and Secord 1979).

Now, why do historians of science continue to focus on great individuals as causal agents (in the manner of political historians) and not on more aggregate notions of institutions, cycles, and trends (in the manner of economic historians)? It is certainly *not* because more detailed historical work has shown that individuals matter more than groups in determining the course of science. On the contrary, each new sophisticated history of science seems to uncover crucial social factors that change one's entire sense of what transpired. However, these social factors—such as Max Weber's triad of class, status, and power—have an ontological diffuseness that renders them unwieldy tools with which to think about the mechanics of historical change in science. It comes as no surprise, then, that historians who freely wield these Weberian entities are often criticized for endowing them with agent-like qualities, as if a class, say, were itself a kind of purposeful superindividual who presses ordinary individuals into its service. If there is a clear case of our natural modes of thought (or "cognitive biases") working against what we are trying to think, then this would be it: Our attempts to overcome the fixation on individuals by treating social factors as themselves new individuals (cf. Tilly 1991, which raises this critique against history in general).

My own proposed solution is to tell the history of science as the history of an object whose nature it is to be distributed, namely, the sort of thing, of which copies can be mass-produced and hence inserted into many situations, thereby generating a dispersion of effects. The obvious candidate is the *book*—taken not as the captive essence of a great mind, but as a commodity whose value is negotiated in a variety of local exchanges (including the exchanges it took to concentrate the capital and labor needed for producing the original copies: cf. Fuller 1991b). If we do it properly, without smuggling vestiges of authorial intent as an invisible hand, we should be presented with a rather chaotic history of the book, a pattern of diffusion in which the artifact shaped behavior quite differently in different settings, with many parallel and interactive effects in tow. The unwieldiness of this pattern would probably lend itself more to spatial than to linear presentation: messy tree structures more than neat lists. In any case, the result would be to disrupt the mnemonic compulsion to collapse the history of science into a sequence of great discoveries by great people through which the World-Historic

Two: Interpenetration at Work

Spirit has happened to pass. For only sheer memorableness lulls historians into continuing to center their narratives around individuals, often in spite of what they know about how the history really works.

We should not lose sight of the legitimatory function performed by telling the history of science as a series of heroes. The sequence of Aristotle, Copernicus, Galileo, Newton, Laplace, Maxwell, Einstein betrays the hand of the textbook tradition, whose principal aim is to present the welter of past discoveries in a pedagogically tractable form. *Pace* Laudan (1990b), it is not clear that this sequence of great physicists requires a "methodological" explanation of how one hero could have laid the groundwork for the project of the next (though the impulse will be diagnosed below as symptomatic of an "overdetermined" historical sensibility). If cognitive historians were more self-conscious about the pedagogical psychology that makes the trail of geniuses an easy way to envisage the history of science, they would probably not be so quick to associate *mnemonic* and *causal* significance in their selection of what to write about.

A Symmetry Principle for Historicism

Common to the three points just examined is a sense that the new wave of cognitive historians of science are still locked into a humanistic frame of mind, which, for all its pretense to being scientific, unwittingly serves to reproduce the biases of the "prescientific" internal history of science. A good way to appreciate the atavistic quality of this humanism is to focus on the strategic use that historians make of *historicism* as a methodological doctrine. I take historicism to be a family of positions that involve the claim that the epistemic differences between times and places are more important than their similarities for understanding why people think and act as they do. Such differences may matter for a variety of reasons, depending on the version of historicism that is endorsed. For example, Comte and Hegel are historicists who take the radical epistemic differences between times and places to constitute a directed sequence of changes, whereas Dilthey and Popper (1957) are historicists who take these differences to preclude the possibility of any such sequence (Ironically, after flagging this distinction in historicisms in 1957, Popper went on in the next fifteen years to become the sort of teleological historicist he originally condemned; cf. D'Amico 1989, on Popper's emergent "World Three"; also

Excavation

Fuller 1988a: chap. 2.) In any case, it is probably fair to conclude from the history of philosophy that historicism has been the most fruitful way of getting skepticism's critical edge without suffering skepticism's more self-debilitating consequences.

However, historians often undercut historicism's own critical advantage by applying it *asymmetrically* to the past and the present. On the one hand, the historicist is supposed to demystify the tendency of today's philosophers and scientists to stress superficial continuities with the past that serve to suppress deep differences which, once revealed, generally show just how little we contemporaries understand about our own historical situatedness. But, on the other hand, the historicist is also supposed to recover the self-understandings of past figures, who are portrayed as having had a keen sense of their historical situatedness, to the point of transforming the available traditions in arch ways by investing even the most ordinary of objects with scads of "cultural meaning." Very much in the spirit of Renaissance Humanism, this asymmetrical application of historicism makes the people of the past our cognitive superiors, i.e., people whom the historian strives to understand, in large part, because they understood themselves better than we understand ourselves. Of course, the *prisca sapientia* that today's historian valorizes in the "ancients" is not quite the same as was valorized in the sixteenth century. Although some historians of political theory seem to think that John Locke and his predecessors had a better grip on Human Nature than any of his successors, most historians of science defer to the great scientists on the more modest grounds that they had a culturally (or at least cognitively) integrated understanding of their inquiry, next to which today's scientists seem either alienated or simply shallow. I wonder: How could we have fallen from such an Age of Heroes?!

To their credit, thoughtful humanists have been dissatisfied with the temporal asymmetry exhibited in this application of historicism. Unfortunately, in moving toward a more symmetrical historicism, many of these "postmodern" humanists and semioticians have been led to overcharitably read the thoughts and actions of members of our own culture—as if the solution lay in elevating the present to the mythic levels of the past. We are now led to believe (by quite an ideologically diverse group of inquirers that includes radical social constructivists and reactionary followers of Michael Polanyi) that unspeakable amounts of expertise are built into routine laboratory practices and that the entire scientific workplace is an enchanted realm of deep meanings (cf. Knorr-Cetina 1981). If the reader is inclined, as I am, to wince at the implausibility of a world superabundant with competence, she

Two: Interpenetration at Work

should save some of her cringing for claims of an analogous sort that are normally accepted without notice when the lab in question is that of Faraday or some other ex post facto notable.

For my own part, I prefer to achieve the desired temporal symmetry by a social scientifically informed cognitive egalitarianism that brings people from the past down to the realistic level of shortsightedness that both historicists and experimentalists have been so good at detecting in people from the present. Among other things, this strategy drives home the point that knowledge is *necessarily* a social accomplishment that cannot be completely understood by adopting the perspective of any one of society's members; hence, the need for a *social* epistemology (cf. Faust 1985). Without attempting to evaluate any specific claims they make about the history of science, there are several extant models for the sort of historiography that I am calling for. Structural Marxism and Freudian psychohistory are perhaps the most explicit in their symmetrical historicism, largely because false self-understandings, as either ideologies or ego mechanisms, are granted powerful roles in explaining what historical figures do. However, historiographies that simply postulate the inability of agents to predict what other agents will make of their work will do for my purposes. In this regard, diffusionist accounts in the history of *technology*–in which an artifact takes root in ways unanticipated by the original inventor– would stand as good models for the history of science (cf. Basalla 1988).

But why the resistance of even cognitive historians of science to a social scientifically informed historicism? I must turn frankly speculative at this point and suggest that historians continue to act as if the configuration of academic disciplines has not changed since the late nineteenth century–just before the emergence of the social sciences–and that, as a result, history remains (in the minds of many historians) the final authority on human affairs. Seen in a generous light, the debate between philosophers and sociologists over the ontology presupposed by the history of science serves to counteract this inertial tendency of historians to think that all they need are good archives, common sense, and some opportunistic reading in other disciplines.

At first glance, it may seem that a philosopher like Laudan is doing the historian's ontological bidding, since he shares with historians of science the most basic assumption that science (at least in the broad sense of *Wissenschaften*) is a distinct object of study that is defined primarily by the accomplishments of great scientists like Faraday and Einstein–all of which would seem to be denied by a sociologist like Bloor. However, historians have complained as much about Laudan's own positive program as they have about Bloor's. It is here that they

Excavation

show their true ontological colors. Not surprisingly, because historians are generally unreflective about these matters, their colors turn out to be somewhat mottled.

Historicism's Version of the Cold War

If what I have been saying about the discipline of history is correct, then it is ironic that Richard Rorty should have assumed the guise of the historian in *Philosophy and the Mirror of Nature* (1979) to cure philosophers of their imperialistic tendencies. But, of course, what Rorty wanted was the historicism of the historian to subvert the essentializing tendencies of philosophers and social scientists. That is, he wanted to show that beneath the superficial long-term regularities that philosophers and scientists detect lie deep changes in thought and action. But while this move may capture the historian's self-understanding, it probably does not capture much more than that. Here I will resort to the little known rhetorical tactic of *argumentum ad disciplinam*. I submit that there is a deeper disciplinary motivation for the historian to be fearful of essences. For to be true to the historical method, one must engage in an exhaustive study of the documents of the time and place that one is writing about. Indeed, very often more time and energy are devoted to this task than to preparing an analysis of the materials for publication. Under these work conditions, the most psychologically satisfying thing for the historian to believe is that the causal significance of the things discussed in these documents is proportional to the amount of time and energy spent in wading through them. In other words, it would be hard for a historian to admit that after spending several years in an archive, she arrived at basically the same conclusion about the mechanics of some episode as someone who never visited the archive and, in fact, conceived of an alternative account by bouncing off some secondary sources.

Yet the challenge posed by essentialism is precisely that the quality of evidence is not proportional to its quantity, and that consequently much of the actual historical record may be incidental to what has really mattered in the course of history. Presumably, then, someone with a higher sense of cognitive efficiency than the historian–a Comte or Hegel perhaps–could penetrate the surfeit of texts to glean the defining patterns of history. Call this challenge–that there might, after all, be something like a faculty of intellectual intuition–the *Platonic Plague*. I

Two: Interpenetration at Work

submit that the Platonic Plague is the historian's biggest epistemological nightmare, one that is realized on both sides of the Laudan-Bloor debate. It is also one that is realized in the experimental method of science, i.e., in the possibility that an abstract system of interacting variables can model the complexities of the phenomenal world. In a moment, I will draw some explicit connections between "historicism ontologized" and the experimental method. But first, let me outline what I take to be the best way for the historian to counteract the Platonic Plague.

This counterargument starts with the classical nominalist account of our knowledge of universals, an account that still has some psychological validity: namely, what Platonists and others have wanted to cast as our ability to intuit universals is really our *in*ability to remember the manifold differences among particulars. However, armed with the canons of the inductive method, even nominalists have believed that this adversity can be turned into a virtue, as we learn to focus our forgetfulness by retaining only those differences that we think might make a difference for bringing about the sorts of things that interest us. Experimental controls provide one environment that enables this mental discipline to work. Now notice the highly pragmatic character of all this talk—as if the pursuit of knowledge were only a matter of carving out an epistemic niche from within the welter of unmanageable data. Given that such a pursuit might involve systematically ignoring and compensating for data that our minds are incapable of handling, we could well be left with a seriously skewed picture of the nature of reality, one that would perhaps never be penetrated *unless we explicitly set out to do so*. Herein, then, lies the epistemic importance of the historian's practice not to leave any page unturned in the achives—as an antidote to the modes of convenient and pragmatic thinking that the search for universals invites. However, it is also clear that to fully realize this role, the historian must embrace the symmetrical historicism that I advanced above, so as not herself to succumb to the Platonic Plague. In other words, she must be open to the possibility that what is least suspected (or recalled) turns out to be most significant.

The upshot of the above argument is to recommend that historians switch their exemplars from Gadamer to Foucault, i.e., exchange their familiarizing posture as the keeper and dispenser of practical wisdom for the more alienating one as archaeologist of knowledge (cf. Fuller 1988a: chap. 6). In this game of epistemic bluffsmanship, the savvy historian—the one who has made the turn toward Foucault—can now issue her own counterthreat to the Platonic Plague. This counterthreat

Excavation

would come in handy in dealing with philosophical and social scientific attempts to ontologize historicism, as in the case of the Laudan-Bloor debates. I dub this counterthreat the *Idiographic Incentive*. It effectively answers the question, Why should we bother sifting through all the data of history if what really interests us is discerning long-term trends and other such essential motions?

The answer is based on the now classic experimental findings of Tversky and Kahneman (1974), which showed that people tend to take the *availability* of a memory as a sign of its statistical *representativeness*. In other words, the easier it is for me to recall an item, the more likely that I will take it to be typical of the class of items to which it belongs (the relevant class here being dictated by what the experimenter asks the subjects to recall). Tversky and Kahneman call this tendency the "availability *heuristic*," to underscore the fact that it works enough of the time so as to discourage people from investigating the many other times in which it fails to work. The heuristic has also been aided by the captivity of common sense to Aristotle's wax tablet view of the mind, according to which more frequent encounters with an object leave a more lasting mental impression. However, given the ease with which the structure of human memory can be altered by seemingly incidental factors, there are no good psychological grounds for thinking that the statistical and mnemonic qualities of things are so directly correlated.

I submit that a version of the availability heuristic is at work in the case of accessing historical evidence. For various reasons, some planned and others not, it is easier to get at certain kinds of evidence than at others. Usually it is easier to access a scientist's journal articles than her private notebooks, though if (as arguably happened in the case of Darwin) the notebooks are widely publicized and celebrated as literary works in their own right, they may become more readily available than the works that were originally designed for public consumption. Among the diverse factors that affect one's cognitive access to historical evidence are the availability of translations in one's own language, the substitutability of original sources by glosses, not to mention the historical figure's sensitivity to the chance that future generations might want to eavesdrop on her conversation (cf. Fuller 1988a: chap. 12). If Tversky and Kahneman are right, then there should be a tendency to think that most of the story is told by the evidence that is readily available, and that the less available evidence would not appreciably alter the story (though it would undoubtedly fill in the details). No doubt such a tendency may be found among such ontologized historicists as Bloor and Laudan, and perhaps myself. After all, don't

Two: Interpenetration at Work

we already know enough to decide whether (or under what circumstances) scientific theories are selected on the basis of "internal" or "external" criteria?

By contrast, part of the professional training of historians is to unlearn the availability heuristic by taking seriously the possibility that the next bit of evidence to be uncovered may radically reconfigure all the previous evidence; hence, the incentive to pursue the idiographic method. In this way, the historian inhibits the economizing tendency of the heuristic, which presumes that a principle of diminishing marginal utility exists for the epistemic value of evidence. Now, clearly, when speaking of the "professional training" of historians, I am idealizing somewhat, since historians are probably unaware of the relevant biases in human psychology that their inquiries are useful in counteracting and, as a result, do not counteract their own biases as often as they might if they were made to see the psychological significance of their practices. Indeed, an important research project could be undertaken to drive home this point in the history of science (or of anything else): to wit, a *Critical History of Access*. One could trace how the documents on which historians most heavily rely (including translations and secondary sources) came to be made so readily available, and the effects that this ready availability has had both on the sorts of facts that figure prominently in the histories written and on the historians' search for other documents. Such a history would reveal the biggest fallacy plaguing humanistic thinking, namely, *the unwarranted inference from the disposition of the evidence to causal dispositions*. This fallacy appears in many guises, the subtlety of which is a tribute to their commonplaceness. Here are just three.

> *The amount of evidence available is often taken to be a measure of the causal significance of the thing evidenced.* Thus, if most of Newton's manuscripts pertain to theological matters, then the humanist is prone to conclude that theology was the driving force in Newton's work. However, some causes may be documented well out of proportion to their significance, either because of the literary conventions of the time or because of the survival patterns of the documents over time. Nevertheless, the humanist is motivated to commit this fallacy in order to avoid having to admit that effort has been wasted in poring through the archives. It shows a failure to pass one of the classic tests of the distinction between science and superstition. For whereas the scientist sees no a priori reason why the mere presence of a piece of evidence will turn out to have significance,

Excavation

and hence readily concedes that most of the data gathered will be trivial or misleading, the humanist cannot quite face this possibility, and hence is more likely to invest whatever she finds with spurious significance.

The self-referential features of the evidence are often taken to indicate the type of causal role played by the thing evidenced. Thus, it should come as no surprise that the fields of history in which most of the evidence is personally signed—namely, political (i.e., treaties) and intellectual (i.e., treatises) history—are said to be about a succession of personalities, whereas the fields in which most of the evidence is left unsigned—namely, economic and social history—are said to be about impersonal forces. Of course, a hospital's accounts and health records are done by particular people, and the pronouncements of a politician or an intellectual are subject to linguistic constraints beyond her control and awareness. Yet, these truisms are easily forgotten when the humanist insists on being so evidence-driven.

Historical events often become exclusively associated with the canonical locations for finding evidence about them. This shows that the deceptiveness of historical access is, in a sense, a by-product of the scarcity of the material world: to wit, the present and future are recycled versions of the past. Yesterday's events are reconstituted and preserved as tomorrow's archives. Indeed, this scarcity may be seen as a form of the Platonic Plague: every particular is typecast for posterity as one of the universals that participated in its production. In less metaphysical terms: while any event is clearly part of the intellectual, economic, political, and other currents of its time, after the event has transpired, traces of these currents are distributed to various archives, only one of which becomes typically linked to the event. Consequently, the historian does not really make a "free choice" in her selection of facts for interpreting the event, since she will be immediately drawn to the stereotypical archive. For example, if the historian is interested in the proceedings of an academic conference, she will probably be drawn to an academic library and completely neglect the receipts that were taken in funding the conference, since that evidence is probably located—if at all—in some obscure place like a university's business office. The difficulty in obtaining this alternative source of evidence is then unwittingly taken by the historian to indicate its diminished relevance for understanding the event. In short, if you store the intellectual and economic records of an event in separate locations, then the two locations will soon be taken to symbolize separate causal "factors" that combined to bring about the event.

The depth and subtlety of the above fallacies ensure that it will not be easy for the humanist to adopt the mindset that is appropriate for doing a Critical History of Access. However, the experimental psychology literature cited above is not the only source of refuge here. Feminist historians routinely incorporate a Critical History of Access in whatever they write about because of the long-standing systematic efforts to alienate this half of the world's population from recorded knowledge. Moreover, the form of oppression involved here is distinctive, in that women have been excluded, more often than not, without any of the oppressor's rhetoric of evil and mystery that typically accompanies, say, racial discrimination. Rather, women have simply been passed over in silence, as men render women's lives and works part of the taken-for-granted background conditions of everyday life. Thus, students of women's knowledge are professionally alert to potential discrepancies between the amount and the significance of evidence. In this way, feminists also tend to be especially sensitive to the major ontological and epistemological issues surrounding incommensurability, which more traditional historians might be inclined to underestimate. As originally defined in *Social Epistemology*, the ontological issue surrounding incommensurability is how conceptual differences arise from communication breakdowns; the epistemological issue is when a failure to communicate implies a conceptual agreement or disagreement. If the former captures the process by which autonomous bodies of knowledge emerge, the latter captures the process by which these bodies are related to one's own.

Feminism has anticipated social epistemology's two-pronged probe of incommensurability. On the epistemological side, feminists have counteracted the bias imparted by Leopold von Ranke's historiographical maxim of recalling the past "as it actually happened." As the cornerstone of professional history, this maxim encouraged a document-driven inquiry, in which causal significance was assigned on the basis of the size of the paper trail one left, where "one's" identity was determined by the signatures left on the particular pieces of paper. It was thought, quite in line with Ranke's empiricist-inductivist sentiments, that any truly important event would be recorded. Not surprisingly, on

Excavation

this view, a major historical event transpired whenever a few heads of state met in the same room long enough to sign a prominently placed piece of paper. Once historians moved away from the national archives to less obvious repositories for documents, other sorts of people—merchants, priests, scientists–started getting their due. However, causal significance was still measured by the ability to leave permanent traces, usually written ones, to which the historian, at least in principle, could gain access. Given that women both did not write and were often not written about, any comprehensive history would have to transcend the historian's standard interpretive techniques–indeed, perhaps to the point of entertaining the possibility that those techniques are themselves complicitous with the male-dominated culture that the historian studies (cf. Scott 1987; Nielsen 1990).

I have labeled this potential for radical critique in our understanding of the past, the *inscrutability of silence* (Fuller 1988a: chap. 6). But once the critique has been made, and women's voices are heard, will they sound much different from men's? This is the ontological side of incommensurability: Is there anything *more* to conceptual difference than communication breakdown? If not, then maybe the articulated voices of women should sound like those of men. This is certainly the hope of Enlightenment liberals like Habermas who equate increasing the sphere of freedom with enabling the disenfranchised to air their views in the open forum. In that case, feminist appeals to a specifically nondiscursive "intuitive" orientation to the world may be simply an artifact of the traditional prohibitions on women's speech. But this is not the only possibility. For even if communication breakdown is all that is involved in alienating women from the public sphere, it does not follow that emancipation will come when women are brought into the open. After all, men are equally alienated from women, which explains the peculiar form that masculinist domination has taken, in particular a tendency toward radicalizing the difference between a self-contained active self and a passive nature that can be understood only in terms of its responsiveness to the self. Plato, Aristotle, Bacon, and Kant cloak this form of domination in rather different metaphysical trappings, but the metaphors that seep through their abstractions strongly suggest that the male-female relation is the analogue through which the self-world relation is understood (Keller 1985). This is not the place to proffer new metaphors, since my only point here is that, unlike the usual explanations of conceptual difference in terms of a creative leap or a normative infraction, the communication breakdown account supported by social epistemology implies a mutual loss and a mutual gain that squares with feminist ontological sensibilities.

Two: *Interpenetration at Work*

Under- and Overdetermining History

I do not want to give the impression that historians provide a foil for the Platonizing tendencies of philosophers and sociologists simply by transcending considerations of material scarcity and cognitive limitations in the course of inquiry. Actually, the issue is more complicated. The Platonic Plague and the Idiographic Incentive are ultimately alternative viewpoints about the amount of *evidence* that is needed before making historical inferences. However, this debate is often made to stand in place of an important subterranean debate about historical *causation* that is often fought between rationalist philosophers of science and constructivist sociologists of science. Whereas rationalists tend to presuppose that historical events–at least the exemplary ones that interest them–are causally *overdetermined*, constructivists presume that they are causally *underdetermined*. The distinction simulates the two sides of the metaphysical debate over how tightly the world is held together. The overdeterminationist simulates *determinism*, while the underdeterminationist simulates *voluntarism*. Although the latter often advertises itself as more "empirical" than the former, we shall see that they both appeal to what can only be regarded as occult notions of causation. Indeed, both historiographical stances are primarily *normative* positions, evincing certain attitudes that philosophers and sociologists have toward history.

An *overdeterminationist* view of history postulates that there is only one world order, which consists of certain nodal events through which all possible histories would have had to have passed, though not necessarily as a result of all the other events that actually fed into these nodes. The "nodes" in question are, for a rationalist philosopher of science like Lakatos, the sequence of correct theory choices in the history of science: they had to have happened in a certain rational order, though not necessarily in the length of time it actually took. In particular, the sequences could have transpired more efficiently (cf. Langley et al. 1987 and Chapter five of this book). Overdeterminationism is also consistent with the idea that history could have transpired *less* efficiently, yet nevertheless transpired in the requisite order. Recently, philosopher Stephen Downes has proposed a striking hypothesis that raises this possibility.

Contra the logical positivists and Quine, Downes argues that

Excavation

formal logic was not necessary for the development of science, in that had logic never been formalized, the same sequence of theories would still have been chosen in the history of science, though perhaps the formulation of these theories would have been somewhat inelegant. Downes's thesis can be subjected to the test of counterfactual history. Going back to the latest period prior to the formalization of logic (clearly one needs to specify whether Aristotle, the Stoics, the Scholastics, Leibniz, Boole, or Frege is meant here), imagine that formalization had not occurred, and then see whether the crucial events in the history of science would have still taken place, assuming that the absence of formal logic had the *smallest* possible collateral impact in the course of history. The idea would be to presume that history is generally overdetermined, so that other factors could have brought about events close to the actual history by compensating for the factor removed *ex hypothesi*. Such a presumption makes sense if we further suppose that every event can be identified primarily in terms of the *function* it served in bringing about some other event; hence, the overdeterminationist would search for a "functionally equivalent" combination of events (1984: chap. 1; McCloskey 1987: chap. 4, on the use of this principle in econometric history). I would not pretend that an adequate test has been made of Downes's thesis, but a couple of points are worth mentioning. First, given the nature of the thesis, what matters, strictly speaking, in the counterfactual history is that the relevance of formal logic to the conduct of inquiry remain obscure, not that logic fail to be formalized per se. (For example, formalization may be introduced as a pedagogical technique, much as how Descartes and Hobbes regarded experimentation.) Second, if it turns out that Downes is right, and the history of science is overdetermined with respect to formal logic, then that would be a good empirical argument for denying that science has an a priori component.

By contrast, historical *underdeterminationism* says that a given event need not have occurred, but once it did occur, everything that followed did so by necessity. Thus, instead of the Lakatosian account of inevitable progress, the constructivist sociologist paints a picture of the history of science governed by "turning points," such as Robert Boyle's successful exclusion of Thomas Hobbes from membership in the Royal Society–a triumph of the new experimentalism over the old scholastic rationalism, according to Shapin and Schaffer (1985). Indeed, it would be hard to cast constructivist accounts as involving "free choice" by the historical agents if the agents' perceived options did not turn out to have significantly different consequences over the course of time. If

Two: Interpenetration at Work

experimentalism would have become the most esteemed form of knowledge even with Hobbes's admission to the Royal Society, then his actual exclusion could hardly have been a "turning point" (Lynch 1989). In that case, we would have an overdetermined history of experiment's epistemic ascendancy: several alternative trajectories with the same endpoint—"equifinality," as the economists say.

Thus, with regard to causation, the sociologists economize just as much as the philosophers, insofar as Shapin and Schaffer, say, imply that the experimental method of today is little more than the latter-day reenactment of a decision that was originally made in the seventeenth century. Another way of putting the point is to say that Shapin and Schaffer seem to think that the seventeenth century continues to "act at a distance" on twentieth-century science, as if nothing occurred in the intervening three hundred years to sublimate the resolution of the Boyle-Hobbes debate. It would seem, then, that the Royal Society was able to construct its reality only at the expense of our being constrained by our own reality. In history, as in physics, underdeterminationism presupposes an occult sense of causation, indeed, no less so than the overdeterminationist account, which tends to presume that the robustness of the sequence of theory choices in the history of science, under a variety of counterfactual conditions, implies that there is some hidden logic, or "method," that orders the choices, which, in turn, suggests a quasi-deductive structure to history. Perhaps both the promise and the peril of underdeterminationist historiography is best captured in Immanuel Wallerstein's "world-system" approach, which stipulates that some fairly local changes in the organization of agriculture in medieval Europe triggered a series of dispersed effects that have since stabilized as the capitalist world-system. The point, then, would be to locate the next chaotic episode—the functional equivalent of a revolution—that will destabilize the existing world-system and ultimately reconfigure a new one (cf. Shannon 1990).

The occult causal sensibilities of the two historiographies betray the fact that their real interests lie elsewhere. In the case of overdeterminationism, the implicit normative agenda, "Whiggism," is fairly evident: the sequence of theory choices in the history of science is no fluke, but destined to triumph, in spite of the variety of ways it may be locally resisted or even temporarily delayed. However, the normative agenda of underdeterminationism is considerably subtler, yet still present.

Shapin and Schaffer, for example, express disappointment that Hobbes did not persuade more natural philosophers, for had he done so, the scientific community would today probably be conducting its activities in a much more dialectically responsive environment. Two features of this attitude are worthy of note. In the first place, underdeterminationists typically ground their sense of historical regret in exactly the same set of "internal" norms of science–valid reasoning, true premises, and the like–that the overdeterminationist espouses. The difference between the two historiographies turns on whether the "good guys" really won. Whereas Lakatos would say that the history of science has borne out the correctness of Boyle's position, Shapin and Schaffer strongly suggest that, once properly understood in its original context, Hobbes had "better arguments" than Boyle, but Boyle had greater political and rhetorical savvy. The second point is that historians who confer great significance on relatively rare turning points will tend to underestimate the causal efficacy of their own sense of goodness. They are likely to veer between a feeling of relief (when the good guys win) and regret (when they lose). Correspondingly, Whig historians overestimate the efficacy of people who share their values, and hence straddle hope (when the good guys are losing) and triumph (when they win). Thus, in terms of their normative sensibilities, an interesting analogy between the histories of science and politics may be drawn, with overdeterminationism corresponding with political utopianism and underdeterminationism with political realism.

When in Doubt, Experiment

I have been portraying the practicing historian as engaged in a battle of wits with other historicists over the proper use of historical evidence. However, the Critical History of Access conjures up a specter that could make all this thrust-and-parry beside the point. The specter is inspired by the Cartesian Demon of classical epistemology, but with a more restricted scope and, hence, with amore realistic chance of being true. Call it the *Diltheyan Demon*. Instead of a superhuman intelligence with the ability to create a world with all the evidential cues needed to cause humans to hold a seamless web of false beliefs, imagine a quite human intelligence with the ability to plant all the evidential cues needed to cause future historians to believe exactly what she would have them believe, regardless of its correspondence to what really happened. Of course, historians are very alive to the efforts that people have made to perpetuate a certain image of themselves and their accomplishments, but often historians unwittingly contribute to this perpetuation by focusing their inquiries on people who

Two: Interpenetration at Work

have already succeeded in self-perpetuation by making their work indispensable for our own.

For example, no matter how many books are written revealing Galileo's counterfeit experiments and philosophical bluffsmanship, Galileo's demonic wiles worked long enough on the generations immediately following him to make it very difficult now to write a history of science that gives him a diminished role. It no doubt could be done, but it would require a fundamental reassessment of the significance attached to the most readily available evidence as well as a concerted search for evidence that has not already been focused through Galilean lenses. And to what end? In asking this question, I do not mean to understate the importance or even the ultimate feasibility of disentangling "what really happened" from "what they would like us to think happened." However, it is not clear that such an enterprise is the most efficient way of trying to decide the ontological issues raised by the internal-external history of science debates.

Suppose we were to check out the historical track records of the methodological norms that Laudan has identified. It would neither be sufficient nor even necessary to operate within the confines of current historical scholarship: not sufficient, because historians are at best only obliquely concerned with taking sides in the battle that Laudan is fighting; but also not necessary, because historians in practice vary tremendously in their cognitive captivity to the "available" evidence. More methodologically sophisticated historians treat both the things that people do and the evidence produced for making reference to those things as historical events of equal standing, both equally in need of interpretation and explanation. Such a historian takes seriously that people always have one eye on the future when they talk about what they have done. However, it is hard to tell just how arch historians generally are about these matters, since the nonspecialist reader often cannot determine the sort of evidence that the historian takes as her license to make a certain claim about the past. Moreover, this is not a matter of the nonspecialist being unschooled in the relevant jargon, since historians are generally among the most jargon-neutral stylists in the academy. Rather, the problem is one that is endemic to their humanistic heritage, namely, that historians tend to use evidence, usually extended textual quotation, either to make narrative transitions or to provide occasions on which to speak authoritatively on the events surrounding the evidence. While these uses are hardly sins in their own right, they can be if they impede the reader's ability to assess the epistemic warrant for the historian's conclusions, especially on such

Excavation

tricky issues—crucial for Laudan's project—of whether a given scientists *really* did use a given method.

Of course, one answer to all of this history bashing is to challenge Laudan et al. to do the history better themselves. Sometimes this charge is portrayed as a matter of "getting into the details of the case," the idea being that this practice will bring to light nongeneralizable local factors that were necessary for the case to be what it is. Under these circumstances, the sort of abstract, free-standing norms that interest Laudan and other philosophers will simply be beside the point. When historians are in an ontologically generous mood, they may say that "implicit norms" emerged in response to unique features of the situation. However, there is no reason—*especially* no good historical reason—for historians to have such an inductivist attitude toward norms. Historians are no less susceptible to mystified conceptions of their own history than the rest of us, and the role of "case studies" in historical scholarship is, as it were, a good case in point. There is an increasing tendency to run the history of the case study approach through the idiographic tradition in the human sciences: namely, hermeneutics, ethnography, and other methods that let the specificity of the case dictate the methods appropriate to its understanding. Unfortunately, the idiographic tradition is rather alien to the sort of people who originated the history of *science*, namely, philosophers and physicists. Even George Sarton, the person most responsible for institutionalizing the history of science as a field by founding the journal *Isis*, advocated Henri Berr's positivistically inspired "synthetic" history (Stern 1956: 250-55). Yet, these proponents of principles, norms, and laws were not insensitive to cases, either. Buxton and Turner (1992) have recently unearthed this alternative tradition of case studies by examining its diffusion through the faculties of Harvard University from the late nineteenth to the mid-twentieth century.

At the near end of this alternative tradition is Harvard's president James Bryant Conant, who pioneered the case study approach to teaching science by recapitulating the design and interpretation of historically important experiments, specifically seeing where each major scientist continued and departed from his predecessors. From this came Gerald Holton's famous physics text that introduced the central concepts of the field in rough historical sequence, and ultimately Thomas Kuhn's portrayal of science as an activity in which innovators must struggle with how much of the textbook tradition (or "paradigm") needs to be carried over in solving an outstanding problem. But Conant and his associates in science education were only latter-day converts to

Two: Interpenetration at Work

an approach that had flourished in the Harvard Law and Business Schools for over a half century. The use of case studies to teach law is quite familiar, but its point is radically different from what one finds in today's idiographic appeal to cases. The law professor wants to display how an exemplary judge tailored a general principle or precedent to fit the case at hand, thereby revealing something of how the legal mind ideally works. This practice presupposes that there are, indeed, general principles of legal reasoning, but that one needs to survey a wide body of cases to discern their overall pattern. In this way, the law student gets a sense of the limits on the applicability of principles to cases and the degree of flexibility one has in interpreting previous cases and statutes. Thus, the cases are interesting only as illustrative devices, not in themselves.

Wallace Donham adapted the case study method to the Harvard Business School by having students examine the consequences of introducing innovative management techniques into a variety of workplaces. Notice that we have here a sense not only of generalizable norms, as in legal case studies, but also of norms being prescribed so as to increase the likelihood of some preferred outcome. Whereas, in the law, any decision by an authorized judge is ipso facto a potential exemplar of legal reasoning, the soundness of management thinking is ultimately borne out in the marketplace, where slight differences in technique can make substantial differences in productivity and sales. Thus, case studies of business successes and failures are of equal pedagogical importance. The moral of Donham's adaptation of the case study method would seem to be that the more one is inclined to alter the normative structure of a situation, the more the validity of the new norm should be judged in terms of the consequences of that intervention.

Following from Donham's example, then, the problem with traditional philosophy of science is that its treatment of historical cases straddles between the law school way and the business school way of using case studies. On the one hand, like the law professors, philosophers want to confine their attention only to exemplary episodes in the history of science—whether they believe such episodes to be rare or frequent. On the other hand, like the business professors, they want to take seriously the possibility that even the best science could have been done better. The tension between these two tendencies issues from the philosopher's desire to meddle in the conduct of inquiry with impunity. Indeed, "meddling with impunity" may neatly capture the legal-economic conditions under which the Platonic Plague can take place. In other words, a philosopher can seem to be extracting, rather than merely imposing, the normative structure of some situation just insofar as she is able to intervene in that situation without leaving a

Excavation

trace of her presence. (The scientists can't fight back!) This inversion of appearance and reality—the invisible philosophical hand, so to speak—is exactly of the sort that led Nietzsche to demystify the hidden power structure of abstract ethical systems, thereby providing the model for subsequent deconstructionist projects (Culler 1982: chap. 1).

It is not surprising, then, that historians and others have disapproved of a project like Laudan's. However, the grounds for resisting philosophical norms are ultimately not empirical and methodological, but, as we have just seen, moral and political ones: namely, their lack of accountability to the phenomena subsumed and governed by them. Indeed, the constructivist sociologist Bruno Latour (1988) has gone as far as to denounce the very offering of explanations for individual cases in terms of general principles as complicitous in such unsavory political practices. But we hardly need to go this far in order to address what is wrong with the traditional philosophical approach. In fact, we can start by looking at the histories of science that Laudan (1981) and his followers have written. It turns out that they reveal that the normative impulse begins at home. In the course of keeping one eye on the future, scientists end up saying, in their official writings, what they themselves think they ought to have done to get the right result. This tends to make for better methodology than the mess the scientists actually made in the lab, which charitable historians might dignify as exhibiting "implicit norms." Admittedly, this strategy does not immediately dispel the Platonic Plague, but rather evokes a Cartesian image of the solitude—away from the maddening lab—needed to think through a scientific conclusion from first principles. However, in this context, the Cartesian image carries certain strategic advantages for the normative philosophical project.

First, it makes a virtue out of the social constructivist charge that the scientific research report is a rationalization detached from the original scene of activity. Second, it nimbly avoids the need to judge the reliability of Faraday's memory when he writes of what transpired in his lab: if the events reported would have led to an epistemically desirable outcome, then that is good enough for the Laudanian historian. Third, by not having to hang on the actual practice of scientists, the Laudanian historian also avoids the "cult of science" mentality associated with followers of Michael Polanyi, who would make the scientists themselves the final authorities on how science should be done. But more than these strategic advantages, the most interesting feature of Laudanian history of science is the concession it tacitly makes, namely, that it is more important that a methodology *would* have worked, had it been used, than that it was used frequently, or perhaps even ever. In other words, the historical character of Laudan's project is

Two: Interpenetration at Work

really quite incidental to the normative conclusions he wants to draw. If that is the case, then there is a better way to get the sort of evidence he needs: *psychological experiments* (cf. Fuller 1992c).

Historians are not the only ones who retard the development of HPS by sticking to humanistic approaches. Philosophers indulge in their fair share of methodological backwardness when they openly embrace "rational reconstructive" approaches to the conceptual (Carnap, Reichenbach) and historical (Lakatos, Laudan) aspects of science, but then avoid functionally equivalent empirical approaches: respectively, ethnosemantic surveys of scientific discourse and controlled experiments on the efficacy of various methodological norms. I have already defended the importance of the social history of language to a systematic understanding of knowledge production (cf. Fuller 1988a: pt. II). Now I will limit my discussion to the role of experiments.

Consider the case of falsification as a methodological norm of science. Experiments showing that subjects taught to falsify hypotheses are better problem solvers enable clearer thinking on the counterfactual questions that typically concern rational reconstructionists, such as whether (or by how much) the introduction of a falsificationist strategy would have hastened some major scientific discovery (Gorman and Carlson 1989). In designing experiments to test the efficacy of falsificationism, psychologists are forced to come to grips with issues that philosophers manage to sidestep because of the level of abstraction at which they normally pitch their claims. These are some of the issues: What is the measure of methodological efficacy, and how is it to be operationalized (e.g., how does one count "solved problems," à la Laudan)? Is the norm meant to govern each individual's practice, group practice, or some other unit of analysis? Is the norm meant to be representative of ordinary scientific practice, or is such a concern (i.e., for "ecological validity") beside the point because the norm is meant to improve, not merely reproduce, ordinary scientific practice? In this light, rational reconstruction is best seen as a first pass at a proper experimental design, one in which relevant variables are isolated for further refinement and testing in controlled settings.

A significant advantage of philosophers' having recourse to experiments in testing their normative claims is that they are forced to become clear about the exact object of their enterprise. To say, as many realists and positivists often do, that we need to explain the "remarkable success" of science is to be tantalizingly vague about the terms in which we are to judge this alleged success. Attempts to historically specify such a standard quickly face resistance from the evidence. In the fullness of time, all scientific theories are eventually shown to be false. Indeed, arguably, the longer entrenched theories are the ones that turn

Excavation

out to have had the deepest flaws (which, in part, explains why the replacement of these theories is so long in coming). Insofar as "improvement" can be discerned in a sequence of theories, that is usually because they are all taken to be solving roughly the same set of problems. However, as Kuhn and his successors have emphasized, that set is subject to change, often because the problems simply lose their urgency, which makes any overarching sense of progress elusive. Even the technocratic criterion of success in terms of enhanced prediction and control will not sustain scrutiny as a metric for the success of *science*, since any technique can be explained by a variety of incompatible scientific theories, which makes it impossible to credit any of those theories with the technique's success. Yet those who question the generalizability of experimental results will wonder whether having groups solve problems in artificial settings will bring us any closer to determining the sense in which science is "successful." My response is that, at the very least, reflexively speaking, the deliberations required for designing an experiment will enable philosophers to become more self-conscious about the sorts of situations and effects that they are prone to term "successful."

Given both the pro-science stance avowed by philosophical supporters of an internal history of science and the extent to which these philosophers take the experimental method to be definitive of science, it is ironic that they have been among the most vocal opponents to experimental approaches to the study of science itself (e.g., Shapere 1987; Brown 1989). Perhaps even more ironic is the fact that their grounds for objection are essentially the ones that Aristotelians raised against the legitimacy of generalizing the experimental method throughout the natural sciences–the very objections that had to be overcome before the Scientific Revolution could take off (cf. especially Harré and Secord 1979)! These objections are raised and answered in detail elsewhere (Houts and Gholson 1989; Fuller 1989a: chaps. 2-3). For purposes of the argument here, I would just stress that the need for experiment arises naturally from the sorts of questions that philosophers (and their sociological interlocutors) tend to pose, which involve the extent to which a given factor (usually a method) contributes to a generalizable outcome. As we have seen, history as normally practiced is ill suited to dealing with these questions, which is why they never are resolved. Critics of experimental approaches, though ostensibly sophisticated in their attitudes toward science, seem to have a stereotyped view of the possibilities for experimental design, one modeled on, say, Galileo's (alleged) inclined plane experiments–this despite the fact that the biggest innovations in experimental design have come from psychology, precisely because of the tricky nature of its subject matter, namely, hu-

Two: Interpenetration at Work

man beings (e.g., Campbell and Stanley 1963; Campbell 1988: pts. I-III. The debate over the viability of experiments as a testing ground for the philosophy of science has been continued by Kruglanski 1991 and Tweney 1991, taking the pro- and anti-experiment stance, respectively).

A conceptually deeper worry for the critics of experimental approaches is the Scylla and Charybdis that awaits the internal history of science once the experimental study of science is granted legitimacy. In their search for generalities, philosophical defenders of internalism are methodologically compelled to turn from the anecdotal evidence of history to the nomothetic approach of experimental science. However, once they agree to experiment, these philosophers will probably find that insofar as the methods of science are generalizable, they can also be characterized in fairly abstract terms–that is, without having to make reference to the content of particular sciences. (Of course, such abstractness also becomes a practical necessity when the pool of experimental subjects is confined to undergraduate students.) Indeed, it may even be that these methods can solve any of a wide variety of problems or facilitate any of a wide range of social actions. It should come as no surprise, then, that philosophers who have been attracted to the "problem-solving" model of science, such as Dewey, Popper, and Laudan, have also been quite liberal in what they will countenance as a science or "intellectual practice." But with such liberalism comes the threat that sociologists like Bloor may be right after all, in that there is really nothing epistemically very distinctive about science: if certain methods seem to make science work better, that is only because they would make any social practice work better. Thus, science would be shown to have no essence of its own. Were philosophers to stick to internal history of science, they would never be under an obligation to compare the workings of science with that of some other institution, and thus not be tempted to rise to the level of abstraction at which the distinction between "internal" and "external" to science no longer makes a difference. Experimentation, by contrast, imposes just such an obligation.

STS as the Posthistory of HPS

If interdisciplinary fields rarely become disciplines in their own right, that is only because their central problems continue to

Excavation

be defined in terms of the old disciplines. Take the difference between HPS and STS. Whereas the HPS person tends to blur positions that emerged *after* the Strong Programme in the Sociology of Scientific Knowledge (and hence all the sociologists sound like David Bloor), the STS person tends to blur those that had existed *before* (and hence all the philosophers sound like Larry Laudan). In this respect, HPS belongs to the prehistory of STS. As we have seen, HPS failed to make substantial progress because it became embroiled in disciplinary turf wars between philosophers and sociologists, who argued (subject to historical arbitration) about the relative contribution made by "internal" and "external" factors to the growth of knowledge. By contrast, the central problems of STS are not principally defined along such disciplinary lines. Rather, there are signature problem areas on which inquirers with a variety of disciplinary backgrounds work, whose differences of opinion are no longer predictable simply on the basis of those backgrounds. These include the thick interpretation of experimental practice, the mapping of the circulation patterns of scientific artifacts and interests in society, and the deconstruction of artificial intelligence programs. Consequently, the Laudan-Bloor debates of ten years ago, pitting "*the* philosophical" against "*the* sociological" perspective, seem very dated and uninformative to the STS practitioner of today.

Even identifying STS with the triumph of "sociological" approaches can be misleading if the term is meant to suggest the discipline of sociology. For the training and work of most of the leading STS researchers bear slight resemblance to what is normally found in professional sociology journals. It is, rather, a case of STS researchers being *sociologistic*, which is to say, they tend to presume that an ontology of social entities is needed for explaining science. This commitment should be taken as analogous to the traditional scientific presumption of *materialism*, which is primarily a metaphysical position that is neutral with regard to the particular theory that has the best grasp on the nature of matter.

Despite STS's professional distrust of disciplinarity, it must be said that the *Realpolitik* of academic survival dictates that STS move toward becoming a discipline—or die. This leaves open the question of which sort of discipline STS should become. In particular, should it acquire some of the trappings of the liberal professions and applied fields? We have already seen that the normative motivation of the philosophy of science can be fruitfully understood as straddling that of law and business. Indeed, in order to prevent STS from losing its radical potential by becoming just another "normal science", perhaps it should be housed in a professional school—alongside the clergy, law, education,

Two: Interpenetration at Work

business, engineering, and medicine—rather than in the liberal arts division of universities. Although more traditional defenders of the university are loath to admit this, professional schools have been much more "liberal" in the types of relationships they have forged both inside and outside the academy than the so-called liberal arts. For starters, most professionals work in environments where their financial survival depends on enlisting the support of lay people, who usually rely on their expertise. And while by no means do I wish to endorse the cult of expertise, neither do I want to lose sight of the public-spiritedness that accrues to a field that is defined in terms of members whose practices are oriented more toward non-members than toward each other. As a matter of fact, the conference program of any annual meeting of the main STS professional association, the Society for Social Studies of Science (4S), reveals that STS researchers have already found their way into virtually every kind of knowledge-producing site. Unfortunately, they have remained as nonobtrusive in their participant-observation status as possible, packaging their insights in ways that could make sense only to other STS practitioners.

To be sure, a conception of STS centered on social epistemology would require such institutional liberality, so as to allow intervention in already existing knowledge practices. However, this intervention would not be in the form of second-order pronouncements worthy of a philosopher-king. Rather, a more apt model is the participant observer approach of the ethnomethodologist, that is, an a posteriori Socrates (a Popperian!?), who is both open and strategic in her probes. In that case, the social epistemologist would be required to spend much of her time not as a studious scholar in a department of her own, but as a catalytic agent in someone else's department—or, better yet, *between* departments. Research would consist of recording and analyzing the results of these interventions. At professional meetings, social epistemologists would trade techniques that worked (or didn't) in various interdisciplinary (for cases of science vis-à-vis science), interagency (for cases of science vis-à-vis government), or interconstituency (for cases of science vis-à-vis the public) settings. Tenure would be granted to practitioners who succeeded in reorganizing research agendas and perspectives in profitable ways. Analogous criteria could be developed to evaluate the social epistemologist's attempt to disrupt the institutional inertia of science funding or folk attitudes toward the public impact of science.

Clearly, I envisage social epistemology as the successor subject to philosophy of science in the STS constellation. As it stands, philosophy of science exists only as what may be called a "vulgar sociological

Excavation

formation." In other words, the field exists only in the sense that there are journals which claim to publish work in that area. However, the work to be found within the covers of such journals, while predictable, does not have any obvious integral unity (Fuller 1989b). Some philosophers are essentially doing internalist history of science, others are conceptual underlaborers for the special sciences, and still others are playing the endgame of fifty-year-old debates over confirmation and explanation (postpositivists read: rationality and realism). The Popperians were probably the last to attempt to forge an integral whole out of these ever more disparate parts, but not much has happened in that vein since the Laudan-Bloor debates. The missing philosophical glue is an overarching normative perspective that addresses the ends of science, in addition to its means: a perspective unafraid to suggest how science might change and improve. (Indeed, the best way nowadays for a philosopher to take up these issues is by participating in a funding panel at the National Science Foundation or the National Endowment for the Humanities.) After all, such were the origins of philosophy of science in the nineteenth century, when the likes of Comte, Whewell, Mill, and Mach constructively intervened in the scientific process.

Yet what is also needed is a rhetoric whereby philosophers can see their own interests addressed by social epistemology. For example: Will logicians be allowed to join the club or will they first need to ply an entirely different trade? While it is now a commonplace to say that the social sciences lack their own Newton, it would probably be more correct to say that there are too many pretenders to the Newtonian throne *and not enough Maxwells*—that is, too few outstanding talents who see enough of their own interests represented in someone else's project to devote their energies toward developing that project. Thus, social epistemology needs to attract a few good Maxwells. Perhaps the best way to think about this task is in terms of ways in which the social epistemologist can recontextualize what philosophers of science normally do. In other words, how does one significantly alter the *point* of philosophical activity without having to change its conduct very much? I end here with some possibilities:

> Philosophers skilled in formal theories of rationality and logic can design machines whose handling of scientific tasks is easily confused with that of a competent human. The social epistemological point would be to see whether public reaction to computerizing the task brings the machines closer to being seen as scientists or whether computerization serves only to distance

Two: Interpenetration at Work

the task from the realm of the scientific.

Discussions of the ends of science would at once help break down any artificially maintained science-society distinction as well as the equally artificial distinction between epistemological and ethical concerns. Ironically, an "externalist" perspective on the nature of science—one that keeps "What is science *for*?" an open question—might serve to reunite increasingly disparate branches of philosophy.

Philosophical interest in the history of science need not be "for its own sake" or played out exclusively by the rules of the historians. On the one hand, philosophers can take a cue from Hempel and uncover the hidden social scientific assumptions (i.e., generalizations about human behavior) that historians of science take for granted. On the other, philosophers can take a cue from Ernst Mach's critique of absolute space and time in *The Science of Mechanics* and use the history of science to undermine philosophical legitimation of a currently dominant research program.

Philosophers trained in the conceptual foundations of a special science should not continue to work with members of that science, but rather be placed as catalytic agents in a different science, in order to break down artificial disciplinary divisions between the two fields. Instead of *handmaidens*, philosophers would thus be *matchmakers* of the sciences!

PART THREE

OF POLICY AND POLITICS

7 KNOWLEDGE POLICY:

WHERE'S THE PLAYING FIELD?

As that archdeconstructionist Jacques Derrida (1976) might say, science policy is captive to the "metaphysics of presence." In other words, science policy is treated as something that occurs only when traces of intervention are left (e.g., added funding or regulation) but not when such traces are lacking (e.g., allowing science to continue as is). Yet policy is always being made, even when nothing is changed (cf. Bachrach and Baratz 1962). Indeed, a refusal to steer the course of science policy is itself a very potent form of science policy. One reason why this axiom of policy science is rarely given its due in science policy is that both the public and its policymaking representatives tend to regard science as something that proceeds in a relatively autonomous fashion and to think that science policy is therefore something that intrudes, for better or worse, on this ongoing enterprise. Much of the fire of *Philosophy of Science and Its Discontents* (Fuller 1989a: especially chap. 1, Coda) was directed at this viewpoint, which travels in scholarly circles as the *internal history of science*. Seen in retrospect, my goal was to deconstruct a bad pun that had been masquerading as a sound argument, to wit: if the trajectory of scientific research is subject to *inertial motion*, then the trajectory of science policy should be subject to *institutional inertia*. Even if the antecedent of this conditional were true, which it is *not*, only an inductivist of the naivest sort (or, in political terms, a traditionalist of the most conservative cast) would accept its consequent.

My original deconstruction had two immediate targets that will surface again in this chapter. The first is the tendency of scientists

Three: Of Policy and Politics

(often under the influence of philosophers of science) to calibrate desires to match expectations, so as to appear to be able to get what they want. This usually involves a form of "sour grapes" reasoning, whereby scientists end up defining anything outside their sphere of control as "external" to the scientific enterprise, and hence a drag on the scientific spirit, even though such "external" matters as funding and research prioritization are essential for the conduct of inquiry. In the long term, this strategy serves no one. Scientists start to look like what Marxists have traditionally seen them as, namely, benighted slaves for whom "freedom" is little more than the awareness that their masters can exploit "only" their bodies, not their souls (cf. Freidson 1986: chap. 7, on the difference between managerial and professional control of labor).

My second target is the more general tendency to neglect the material consequences of satisfying intellectual needs. A particularly trenchant way of making this point is to observe that the maintenance of "free inquiry" normally entails the ability to pursue false leads with impunity, which materially involves the freedom to waste resources, which, in an age of increasingly expensive science, means channeling more funds away from other public and private interests. Here, too, we see what Marxists would call alienation of the scientist from both herself and her fellows. After all, what joins scientists to other human beings is the space and time they take in the material world, as expressed in the media of social relations. Moreover, an increasing portion of a scientist's energies is spent on activities that look more like the work of entrepreneurs and managers than that of "scientific professionals" (Etzkowitz 1989). Nevertheless, the scientist continues to believe that she is really in her own element only during the vanishingly small period in which she works with test tubes and formulae.

In what follows, I often use a locution of my own coinage, "knowledge policy," where one would expect to find "science policy." Part of the reason is to remind the reader that, even when the examples are taken from the natural sciences, the range of fields I mean to include for policy scrutiny include all the *Wissenschaften*. Indeed, the fact that we need to evaluate the natural sciences alongside claims to funding and attention made by the social sciences and humanities will figure prominently when we consider in the next chapter the role of science in contemporary democracies. But, in addition, "knowledge policy" is meant to drive home the point that, once cognitive needs are taken in conjunction with their material realizations, the standard policy decisions associated with funding and accounting become de facto epistemological ones.

Science Policy: The Very Idea

The refusal of policymakers to steer science policy is nicely captured in one of the many tacit maxims codified by Harvey Averch (1985), former staff officer at the U.S. National Science Foundation. In contrast to other social programs, scientific research is held not to experience diminishing marginal returns on investment. In other words, *any* research funded for *any* length of time will yield *some* benefit. It is easy to see how this maxim could be regarded as a call to institutional inertia, that is, the tendency to continue a policy, regardless of opportunity costs and rate of return, unless the policy has obviously negative effects–and then these effects must impinge upon a politically sensitive constituency. As an instrument of knowledge policy, social epistemology is designed to address the sorts of issues that would otherwise be decided by institutional inertia.

Right at the outset, however, the social epistemologist needs to persuade policymakers that they do not already know enough about the production and distribution of knowledge to make intelligent decisions. This is easier said than done, especially in the United States, where the bulk of funded research appears as line items on the budgets of agencies that are officially devoted to addressing the public's medical, environmental, energy, or defense needs. Such an occluded accounting procedure both reinforces the idea that scientists are sufficiently self-regulating to be inserted comfortably into any politically sensitive environment *and* impedes the collection of evidence needed to reveal the dysfunctional character of this distribution of scientific effort.

Not surprisingly, then, the American "science policy advisor" is defined as a conduit between science and the government, two institutions that are presumed to work reasonably well by themselves but which can do more for the public at large by extended periods of cooperation. The actual job of the advisor is to communicate the range of public needs to the scientists and the state of scientific research to the politicians. Furthermore, the information required for this two-way exchange is presumed to be fairly accessible if one is an "insider" in the relevant scientific and political circles (the locus classicus is D. K. Price 1965). In short, science policy has been institutionalized to rely almost exclusively on scientists' folk understanding of how knowledge production works–intuitions that are based more on a few anecdotes than on systematic study, let alone sustained criticism or experimentation with a

Three: Of Policy and Politics

course of action that goes beyond a mere extrapolation of "current trends."

The institutional inertia that currently grips science policy reflects the policymaker's relative satisfaction with both our current knowledge of how science works and the policy ends toward which that knowledge is put. This coupling of factual and normative satisfaction is, in turn, indicative of what Daniel Bell (1973) has characterized as our knowledge society (cf. Boehme and Stehr 1986; Stehr and Ericson 1992). Presuming that the workings of science are substantially understood, the knowledge society takes the uses to which science ought to be put as dictated largely by the very nature of science. Thus, among the foremost items on the science policymaker's agenda is the conversion of the amorphous problems that emerge in the public sphere to ones that are tractable by scientific means. Whatever escapes the categories of science is then relegated to a residual irrationalism, pejoratively called (in Bell 1960) "ideology" and euphemistically called "politics."

Yet for all their interest in scientizing the public sphere, policymakers in the knowledge society still operate with what is properly seen as a "folk theory" of how science itself works. And by "folk theory," I mean something like common sense: a set of beliefs that reinforces the "normal" or "natural" character of some phenomenon in the course of explaining it. Thus, strictly speaking, a folk theory is itself "ideological," in that ideas about how science works—well founded or not—are constitutive of science's identity (cf. Fuller 1988a: chap. 2). Policymakers typically despair of identifying any principled (as opposed to "merely political") grounds for shifting science funding priorities because they seem to believe that scientific research never exhibits diminishing marginal returns. As would be expected of a folk theory, this belief is subject to considerable anecdotal support, mainly from cases in which a line of inquiry led to many long-term beneficial products that had little to do with the original conception of the inquiry. But there are no attempts to submit this belief to rigorous tests. Indeed, the "naturalness" of the policymaker's understanding of science is traceable to a metaphysical presupposition of folk theories, namely, that there is no need to explicitly court challenges to one's beliefs because whatever errors those beliefs contain will be revealed in the normal course of events.

Talk of the policymaker's "natural understanding" of science suggests the idea that values are embodied in the very nature of things, which, in turn, harkens back to the Aristotelian notion of "natural ends" (*teloi*; compare MacIntyre 1984: 187ff. on "practices"). However, when the thing in question is science itself, it is easy to envisage that

Knowledge Policy

certain ends have become natural only because the nature of science has not been questioned. Indeed, a useful way of thinking about the occasions that give rise to explicit normative considerations is in terms of what metaphysicians have traditionally called a "sufficient reason" argument: to wit, there would be no need to worry about how science ought to be if science were normally as it ought to be. The great facts-values, descriptive-prescriptive divide was formally fudged by saying—in a way that would please Aristotle—that the indefinitely efficient productivity is an "implicit norm" of science.

But even when science policy practitioners suspect their own folk wisdom, they are nevertheless reluctant to foster research into the exact relation between the production of certain forms of knowledge and social welfare because they ultimately take the sheer pursuit of science to be good in itself, an activity that morally uplifts the society in which it is conducted, regardless of its palpable consequences. Of course, closer scrutiny of this rhetoric reveals that so-called free, nonutilitarian inquiry has traditionally promised some major long-term cultural benefits. Among the short-term indicators that delivery is being made on this promise has been the spread of higher education to larger segments of the population. Even if most college students never directly contribute to the production of scientific knowledge, they are nevertheless exposed in the classroom to exemplary lives in action, scientists, who are regarded as apt replacements for the religious and aesthetic icons of more superstitious and elitist times. (It is a short step from this political sensibility to one, historically tied to the German university system, that uses science as a rallying point for cultural identity and national unity.) For this reason, the increasingly obvious disparity between the value that the academy places on teaching and the value that the public places on it has engendered a public relations crisis in higher education that is unprecedented, even by America's traditionally skeptical lights. By severing teaching from research, academics are now in the process of undercutting the most persuasive case for free inquiry in a democracy.

Of course, to invest cultural significance in the pursuit of what is nowadays called "pure research" or "basic science" is not to say anything about how many people, of which sort, should be doing what, where, or when. Indeed, the arguments surrounding such pursuits make clear that the "freedom" of free inquiry lies largely in its alleged spontaneity or—to anticipate our case study—*unmanageability*. The extent to which this sense of freedom has affected the American conception of scientific inquiry is driven home by David Hollinger's (1990) striking discovery that the expression "scientific community" does not enter

Three: Of Policy and Politics

American English until the early 1960s, with the roughly simultaneous publication of works by Thomas Kuhn, Warren Hagstrom, and Don K. Price. Hollinger's finding was made in the context of trying to explain why there had previously been little discussion of the decision-making process by which science could function as a self-governing–let alone, externally governed–enterprise. It would seem that even the need to establish internal accounting mechanisms is obviated by the (seemingly) spontaneous fair-mindedness of scientists, each of whom, like Rousseau's "noble savage," is free to follow wherever the path of inquiry leads, using up as many resources as she needs, but never so much as to deprive her colleagues of a similar luxury. (Even the four principles that Robert Merton advanced in the 1940s under the rubric of "the normative structure of science" were meant to flow spontaneously from the souls of scientists–at least no institutional mechanisms were specified in case they didn't!) In other words, institutional inertia is supposed to issue in a libertarian's dreamworld of science.

But once we try to materially realize this dream, some peculiar features of science come to light, which make it difficult to imagine that flesh-and-blood "free inquirers" would behave quite like noble savages. If we use as our benchmark capital expansion in the marketplace, progress is measured by the extent to which past products and processes are superseded. And as modern marketing teaches us, such progress can be artificially accelerated by manufacturing goods with "planned obsolescence." This applies no less to transitory fields of inquiry whose sole purpose seems to be to maintain the visibility of researchers until (if ever) something intellectually more substantive comes along. In that case, fully realizing the ideal of free inquiry would produce a system modeled on the convenience foods industry, aptly called *Fast Science*, which would maximize waste by ever quickening cycles of resource use and disposal. Moreover, there is strong evidence that Fast Science has already been with us for some time (cf. De Mey 1982, ch. 9). (The economics of this phenomenon has been analyzed from Marxist [Agger 1989] and neoclassical [McDowell 1982] standpoints.) Once researchers were rewarded for this mentality, it is easy to see how they would start to loathe teaching. Teaching has always seemed attractive to *teachers* insofar as they have regarded the stuff taught as worth *preserving*. Once preservation was no longer valued in the knowledge system, then teaching would seem to offer little more than partial, transitory snapshots from the frontiers of research. Thus, if Fast Science continues uninterrupted, we should expect the continued devaluation of teaching, even by its better practitioners.

Knowledge Policy

The question of values gets to the heart of science policy's inertial character. Perhaps the two most important issues normally resolved by institutional inertia are the relative value of the research produced by academic disciplines, and the means by which a discipline may produce more of value. Does molecular biology "pack more bang for the buck" than high energy physics? And what may be done to address whatever discrepancies exist? Given the vast disparity in the costs and benefits that disciplines have to offer, one would think that this is a major area of systematic knowledge policy research. For the most part, however, the contrary is the case. Indeed, the suggestion that we might be spending too much money on, say, high energy physics is typically treated as exemplifying a "know-nothing" attitude toward science, when in fact the underlying motivation may be a desire to apply science to science itself, specifically, in order to determine the best projects in which to invest, given certain short- or long-term goals (cf. D. de S. Price 1986; Irvine and Martin 1984). In this regard, the STS practitioner is the soulmate of the "philistine" government economist who fails to see why science cannot be subject to cost-benefit analysis, just like every other federally funded social service (cf. Chubin and Hackett 1990: chap. 6). Perhaps some of the philistinism may be removed by looking at science funding through the eyes of a historical counterfactual.

Suppose it were 1870, and I were a knowledge policymaker interested in promoting an atomic view of reality. Up to this point, scientists had been reluctant to think of the quest for "ultimate reality" in terms of getting at the smallest unit of matter because no techniques existed for isolating and analyzing such units beyond a certain level. It was (and still is) common to discuss this impasse as "logico-conceptual" in nature. But why not regard it, instead, as "techno-economic," namely, in terms of a lack of relevant mechanical devices–something on the order of a dynamo or a digital computer–to stimulate the experimental imagination into proposing testable hypotheses about such micro-units? By calling the impasse "logico-conceptual," the would-be knowledge policymaker is left to the whims of scientific creativity with no clear sense of how to focus funding. However, by calling it "techno-economic," she has reason to call for the manufacture of certain gadgets that will enable scientists to hang their abstractions on something concrete, that is, to visualize the analogical implications of mapping properties of a theoretical construct onto those of a material object. If there is such a thing as the "technological determination" of thought, it more likely vindicates McLuhan than Marx, in that technology determines not so much the content as the form that thought takes. Indeed, a striking

Three: Of Policy and Politics

piece of technology may even determine that the thought takes a form at all, and not simply crisscross various levels of analysis. A historical case in point is the focus that the introduction of first the mechanical clock and then the self-regulating steam engine gave to seventeenth- and eighteenth-century discussions of governance in the natural and human worlds (Mayr 1986).

But contrary to these theoretically inspired interventions, science policy research tends to be problem-centered, often to the point of deliberately avoiding recourse to the more systematic cognitive interests fostered by social epistemology. Interestingly, this problem-centeredness has been justified from opposing ideological directions. On the Right, science policy researchers are in the business of solving the problems of their clients in government or industry who are usually interested in manipulating their access to knowledge to serve their own ends. For example, funding for research into the health of factory workers has rarely been done in order to advance the frontiers of medicine–though it sometimes has had this effect. The more immediate goal has been to prevent worker illnesses from slowing down production schedules. On the Left, science policy research has often been prompted by problems that have reached mass media visibility as instances of science "impacting" on the public. Thus, increased attention is now being given to the Superconducting Supercollider project, not because of a newfound fascination with microphysical reality, but rather because the project's costs to the taxpayers are seen as potentially outweighing its benefits (e.g., Lippman 1992).

An Aside on Science Journalism

Be it inspired by the Right or the Left, science policy research plays hardly any role in *discovering* or *constructing* the problems it tries to solve. The same point applies with a vengeance to *journalists*, who rarely track down stories about science with the same investigative zeal that they would a story concerning a politician (Greenberg 1967 is the locus classicus for this complaint, which was partially remedied in Dickson 1984). Except in cases of scientific misbehavior sufficiently grave to worry Congress, journalists will often print watered down or mystified versions of a scientist's own press release, which ends up only increasing the public's confidence in science without increasing its comprehension (Chubin and Chu 1989: chap. 3).

Knowledge Policy

This is an especially curious turn for the "in use" epistemology of journalism to take, since the modern journalistic commitment to "objectivity" has much the same constructivist bent as STS research. Both aim to present as many sides of a story as possible, so as to let the reader decide for herself (Stephens 1988: chap. 13; cf. Mulkay 1985). Just as the public rarely trusts a politician's own words as the last ones on some topic in which the politician has a vested interest, why should not a similar skepticism (politely put: "open-mindedness," "neutrality") be instilled in the public's understanding of scientific pronouncements? Short of generating alternative facts and theories themselves, this is probably the best that journalists have done to raise the public's consciousness about science. I hold the academic practitioners of STS to a higher–or at least more engaged–standard.

However, journalistic objectivity becomes complicated once besieged scientists themselves openly court the press in quest of a "fair hearing." Here journalists have often brought larger political and economic angles into the disputes that start to give science a public face comparable to that of other institutions. Sometimes (e.g., in the sociobiology controversy) this strategy ultimately benefited the besieged scientists, whereas in others (e.g., the "cold fusion" controversy) it did not. To some extent, this is a step in the right direction. However, as Dorothy Nelkin (1987) has noted, in these episodes, the press rarely operates with a sophisticated sense of the methodology of science. In particular, journalists tend to presume that theory choices are winner-take-all contests that turn on some crucial fact or event (i.e., a news item) which will be decided within a limited time frame (i.e., before boredom sets in). Moreover, the more provocative the theory under dispute, the more likely that journalists will champion it, which often serves to shift the burden of proof onto the opponents (typically, the scientific establishment) to design the relevant "crucial experiment." In the case of cold fusion, such experiments were designed and the underdogs lost. But in the case of sociobiology, its distinguished opponents (e.g., Stephen Jay Gould and Richard Lewontin) could offer only more talk to E. O. Wilson's original talk. In that case, boredom soon set in, and the press declared Wilson the winner by default. Nelkin seems to think that, given their role in shaping public opinion, science journalists should be more scientifically literate. On the other hand, I am tempted to advise that sociobiology's opponents take a few lessons in democratic rhetoric.

Independent science journalism also contributes to a subtler phenomenon whose full impact has yet to be gauged, namely, an increased public impatience with the pace of scientific progress. Two images

Three: Of Policy and Politics

worth keeping in mind here are the supermarket tabloids, the public's primary source of information about the latest developments in science (cf. Burnham 1988), and the growing pressure on government agencies from both industry and the public to limit the period of testing on scientific products before making them generally available. Clearly, we are ready consumers of science. But it is part of the folk wisdom of philosophical relativism to believe that each dominant knowledge system excels by the standards set by its own culture. However true this may be of other cultures (and that, too, is an empirically open question), it is certainly *not* true of our own scientific culture. The problem here is that philosophers–not only relativists–fail to take into account the effects that the publicity of scientists' initial expectations have on the standards used to evaluate subsequent scientific achievements. Promises of impending breakthroughs, strategically made to muster funds from Congress, may come back to haunt the scientists concerned once the goods are delivered–and they are either late or somewhat less than promised. Discoveries that would have counted as clear cases of progress by an earlier standard now come to appear as disappointments because they fall short of current expectations. Moreover, one scientist's ill-fated boast may unintentionally set the pace for subsequent researchers, who are then themselves forced to contribute to inflated standards of achievement (Klapp 1991).

Contrary to what may seem warranted, I do not bemoan the fate of science journalism once it decides to pursue an independent course of investigation. As the journalists themselves would have it, their instincts are often good. Readers may have noticed in my account of journalists some telling literal-mindedness at work. Scientists may complain that the newspapers take their arguments and announcements out of context, but are not the scientists themselves the ones who claim universality for their message? What difference *should* it make if the public eavesdrops on promises made only for the ears of Congress? The press simply takes the scientists at their word–and takes it that their word is uttered for all to hear. If the press did not itself so often believe the promises of scientists, it might be able to help scientists realize the situated, and hence rhetorical, character of their utterances. However, as it stands, journalists exude a certain vulgarized positivist sensibility, which sees science as theoretical debate punctuated by crucial experiments. Admittedly, when experiments fail to be forthcoming or crucial, the press simply gets bored, whereas the positivist declares a lack of "cognitive significance" to the proceedings. Yet, as Ezrahi (1990) has astutely noted, both attitudes partake of the mythos of "the

Knowledge Policy

spectacle," that combination of "put-up-or-shut-up" and "seeing-is-believing" that dominates both political *and* scientific imagery in a democracy. And insofar as a free press has historically been democracy's most characteristic medium of expression, its pursuit of the spectacular moment should be seen as an attempt less at debasing scientific thought than at reaffirming democratic values. Here then begins the tension between science and democracy that will figure increasingly in this and the next chapter.

Managing the Unmanageable

To see that I meant no exaggeration when I earlier claimed that the mere fact that science is done is taken to be socially edifying, consider a recent but all too characteristic defense of the purity of scientific inquiry. The very title—"Managing the Unmanageable"—evokes value mystification by appealing to that classic trope of religious rhetoric, paradox. Science *is* the mystery of which the director of Brookhaven, one of America's leading sanctuaries of laboratory life, writes (Crease and Samios 1991). The piece proceeds to argue that the United States is falling behind Japan because too little is being spent on basic scientific research. At least, that is what the author would like the reader to believe. This is a thesis to which apparently many of America's leading scientists subscribe, given the petition that the American Association for the Advancement of Science made to the federal government in January 1991 to double the science budget over the next ten years.

However, from the fine print, it is clear that what Crease and Samios, and their well-wishers, really advocate is that more *unmanaged* money be put into science. Perhaps even the current amount of money will do, but with a smaller portion of it eaten away by such "costly" accounting procedures as grant renewals and program evaluations. Taken at face value, this subtext is much less persuasive: Is the author saying that that most scientific of standards, efficiency, is abhorrent to the conduct of science itself? Tell that to our putative prime competitor, the Japanese! Of course, the argument is not quite so neat, since we are already disposed to believe—without too much show of evidence—that Japanese scientists are less "creative" than American ones, and—again with more presumption than proof—that creativity requires

Three: Of Policy and Politics

spontaneity, the price of which is nonnegotiable. Such biases are rhetorically reinforced by a series of asymmetries that the piece promotes between basic and applied research, the latter being typically portrayed as directed toward achieving practical aims. By calling the portrayal of the two types of research "asymmetrical," I mean to signal that, given an appreciation of the complexities of doing science in the late twentieth century and a little dialectical skill, one could easily render the basic applied or, for that matter, the applied basic.

"Managing the Unmanageable" trades throughout on the false dichotomy of administering science "for its own sake" and administering it as a business venture. Clearly, the intended distinction is between long-term, relatively market-insensitive investment in basic research, on the one hand, and short-term, market-sensitive investment in applied research, on the other. However, the dichotomy is misleading on two grounds.

First, it stereotypes both science and business. As it turns out, a major debating point among philosophers of science, especially Popper (1970) and Lakatos (1979), has been the level of responsiveness that proponents of research programs should have to criticisms lodged by the scientific community. Whereas a consistent Popperian would almost immediately disown a program that was subject to many refutations, the Lakatosian would prefer holding onto a currently unsuccessful research program until either it can attract enough people to capitalize on its strengths or the needs of the scientific marketplace are themselves restructured so as to favor the program over its competitors. By analogy with the Lakatosian approach, entrepreneurial capitalism is based on the strategy of jumping in early and staying the distance with a new product. It explicitly eschews the kind of short-run thinking typically used to sell goods that represent only a slight improvement on those already on the market. In this latter, more Popperian case, once the market for the product dries up, one simply tries to latch onto the next fad. However, if all business enterprises ran on this principle of moderate gains at low risk, no major innovations would ever be made (cf. Brenner 1987).

This brings me to the second stereotype, which pertains to the alleged hardness of the distinction between "basic" and "applied" research. Originally, the distinction was an artifact of government accounting procedures, designed to prevent as much science as possible from being implicated in the manufacture of instruments of destruction (cf. Hollinger 1990). I will say more about this below. Nowadays, of course, the distinction conjures up differences in the motivation for doing science, as well as the content of that science. Indeed, it is typically

Knowledge Policy

thought that we can identify a piece of research as "basic" or "applied" just by looking at the work itself, without actually tracking its consequences as it circulates through society. Thus, no one has ever bothered to show, say, that "basic" research has more impact on the conduct of academic science than "applied" work. It is simply presumed to be the case. Yet a careful reader of "Managing the Unmanageable" will notice the remarkable ability of previously "basic" research to become "applied" under a couple of conditions:

> once the sphere of accountability is extended to include consumers who are themselves not producers of science (e.g., bureaucrats);
>
> once the frequency with which scientists need to give accounts is increased.

This would suggest, conversely, that if the frame of reference for evaluating the outputs of putatively applied research was made solely by other such researchers, and then relatively rarely, the outputs would start to look like basic research. In short, it should not be difficult, at least in principle, to write up any research project as either basic or applied. This point provides an ironic twist to a highlighted quote in "Managing the Unmanageable," to wit: "Basic research flourishes when it is viewed not as a profit-making venture or as an instrument of social change but as an exploration of nature" (Crease and Samios 1991: 83). The irony lies in the quote's implicit admission that "profit-making venture" and "instrument of social change" are, indeed, terms in which the production of "basic research" can be accounted for—albeit not the authors' preferred ones. What the quote suppresses, however, is that these different accounting procedures are relative to the interests of the different groups to whom science might be accountable. In that case, we may ask: In whose interest is it to account for science as the "exploration of nature," the way that basic research does?

This telling question notwithstanding, "Managing the Unmanageable" does a rather deft job of distancing the concerns of science from those of business, thereby keeping the accountants at bay. However, the article will not have completely succeeded at its mission if it did not counteract a verbal symmetry between the two realms that has invaded the public lexicon since the end of World War II, namely, the parallel between "Big Business" and "Big Science." Consider the following counterargument:

Three: Of Policy and Politics

> The suggestion is that large scientific projects unfairly monopolize scientific capital, squeezing out the little guy who might make valuable innovations if given a chance. But the analogy is false. Knowledge generated by large scientific projects, unlike the profits of large corporations, becomes the property of the entire community and restructures the scientific background against which research teams large and small execute new ventures. (Crease and Samios 1991: 83)

Read even on its own terms, this argument trades on what philosophers of language, after Quine (e.g., 1960), call *referential opacity*. In short, the same thing is identified in two different ways. In the first case it is condemned, while in the second it is praised. Big Business monopolizes capital, and is therefore bad. But what is it to monopolize capital, other than to have enough clout in the market to force all potential competitors to orient their activities towards one's own? Yet this very consequence of large scientific research projects is then praised! The very fact that reference to Big Science as Big Business can remain opaque in this day and age testifies to another missed opportunity in the journalistic portrayal of science. Instead of reporting science as if it were the serious side of the entertainment industry, fluctuating between its own kind of dazzle and scandal, the press would do better to accustom people to follow the short- and long-term trends in the public investment of their tax dollars, as they would their private investments in the business section of the newspaper. After all, research is the largest expenditure of federal agencies, and education is premier among local and state authorities. Why not, then, be concerned with performance records? An itemization of projects funded, the proportion of the budget they consume, and their track records at various points would dissipate some of the mystique of unmanageability that Crease and Samios continue to promote. (Of course, this idea would appear more attractive if citizens could reinvest their taxes in other public projects, as they see fit, just as they can with their untaxed moneys.)

Philosophers of language generally cast referential opacity as demonstrating that the same reality can be *described* in multiple terms. However, idle description is hardly the only reason why one might want to identify, or refer to, something. Indeed, from a rhetorical standpoint, it is better to see the multiple identifications of an object as alternative ways of *prescribing* for the future of the object, different ways of treating it, which, depending on the way chosen, could subsequently change the character of the object. Thus, Big Science is untouched if its practices resist the predicate "monopolistic," but they are

Knowledge Policy

likely to change if the predicate sticks, because of the different evaluative standards that are invoked by calling an activity monopolistic. Sociologists will recognize this point as following from W. I. Thomas' concept of *definition of the situation*: "If men [sic] define situations as real, they are real in their consequences" (Thomas and Thomas 1928: 572).

Referential opacity is often compounded by uncertainty about which terms are meant to have positive or negative connotations. Basic research is generally valued above applied research because it more explicitly engages the creative intellect of the scientist, who goes beyond merely enhancing what is already known to discovering something that was previously unknown, and whose ultimate significance may remain unknown for quite some time. However, creativity does not carry a univocal rhetorical advantage in all contexts. Sometimes it pays to seem less creative, especially if the scientist wants to claim credit for something she has done. Patents are a good case in point. The basic researcher may ideally want to distance her concerns from those of applied research by claiming that her equipment enabled the manifestation of a phenomenon that would have existed, albeit undiscovered, even without the introduction of any special equipment. After all, the phenomenon's ontological independence—"realism" in the philosophical sense—is what makes it a genuine discovery, an insight into nature that merits the engagement of the basic researcher's intellect. Unfortunately, if the scientist fails to stress the necessity of her equipment, then it is unlikely that she will be able to acquire the legal rights and economic power that accrue to patents. Indeed, in order to secure a patent, a complete role reversal between the ends and means of research may be in order, whereby the scientist now portrays her discovery as a demonstration of the equipment's ability to work according to set instructions (A. Miller and Davis 1983).

The ease with which scientists switch back and forth between regarding their work as discovering new things and regarding it as extending old ones is indicative of the rhetorical convertibility of the basic-applied distinction to meet specific needs. Much of the dissonance that could be generated by such frequent conversions is mitigated by the recent "ontological turn" in the philosophy of science (e.g., Hacking 1983; Heelan 1983; Ackerman 1985; Rouse 1987), which argues that specific discoveries can be made only from within environments that have already been subject to considerable technological and psychological construction. In most global psychological terms, the convertibility of basic and applied research reflects the two fundamental cognitive orientations identified by Piaget (1971). Basic research isolates the

Three: Of Policy and Politics

logico-mathematical properties of a piece of work by focusing on features that do not rely on such material properties of the work as the particular equipment used. This is in sharp contrast to the *spatio-temporal* orientation of applied research, which does. Insofar as intellectual property law–particularly patent law–requires that research be translated into the concrete terms of the latter orientation, it offers a promising groundwork for charting the material conditions of satisfying intellectual needs (Fuller 1991).

My interest in reducing the mythic elements of scientific creativity to the niceties of legal rhetoric stems from social epistemology's assignment to democratize the intellect. The first step is to acknowledge that all attributions of "creativity" and "genius" are dependent on the reception given to a piece of work, and hence by necessity made in retrospect (Brannigan 1981). As a sheer event in the history of a discipline, such works are anomalies and, as such, may be ultimately diagnosed as the work of either creative genius or foolish effort. It all depends on whether the anomaly manages to change the disciplinary norms or falls victim to them. And so Einstein's 1905 papers marked their author as a revolutionary physicist rather than a harmless crank because of the network of people who came to support, or otherwise rely on, the Special Theory of Relativity. The strength of the network caused the norms of physics to bend to the theory. To claim otherwise is to be faced with the embarrassing question of why it is that a scientist's genius varies directly with the extent of her impact, over which she exerts little control.

I cannot blame the reader for continuing to resist my account, as it flies in the face of certain narrative conventions that structural linguists believe apply to all literary genres, be it folk tales or disciplinary histories. Taking a cue from the linguist Roman Jakobson, I would diagnose my critics' tendency to epitomize a network by a node, or to identify an entire paradigm with its originator, in terms of *metonymic compression* (Culler 1975: chap. 3). It is one of the two mnemonic dimensions that help finite minds comprehend the dispersion of knowledge effects through history. The other dimension, *metaphoric compression*, has been subject to greater philosophical scrutiny (e.g., Meyerson 1930), often under the rubric of the "analogy of nature." The idea here is to manage the multiple levels of reality by envisaging them as governed by laws of the same form, as in the case of Coulomb's law of electrostatic charges microreproducing the mathematical form of Newton's law of universal gravitation. Not surprisingly, the literary genre best suited to these two forms of compression is textbook

Knowledge Policy

mythology—and that is where tales of scientific genius should stay.

Notice that my social analysis of genius does not involve proposing that someone other than Einstein could have come up with Special Relativity. While such a counterfactual may well be true, reliance on it would make it seem as though something about the theory—its "content" perhaps—marked it as a work of genius, no matter who came up with it first. A counterfactual question truer to my analysis is whether it would have been possible at roughly the same time to mobilize some comparably extensive network that would have eventuated in a revolutionary overthrow of Newtonian mechanics. The idea that scientific creativity can be fruitfully subjected to this kind of network analysis is hardly new to social science (e.g., Rogers 1962), though it has been recently revitalized by Latour (1987). The point I mean to raise here may be cast in the vocabulary of evolutionary epistemology (cf. Campbell 1974).

Darwinian evolution requires two sorts of mechanisms, one for genetic variation and one for environmental selection. The traditional epistemological fixation on creativity, genius, and the generation of theories—a focus retained in science policy thinking—stresses variation at the expense of selection. Consequently, policymakers attempt to construct environments that foster creativity before clearly understanding which features of this supposedly self-generative process have been selectively retained in the history of science (cf. Root-Bernstein 1989). And so, while no one can deny that physics was revolutionized as a result of the Special Theory of Relativity, it remains to be shown what it was about the variety of activities associated with the theory's introduction that enabled the revolution to succeed. In order to specify the eminently selectable features of the situation, one would need to consider professional gatekeeping practices, prior expectations and interests of potential allies, and competing research agendas. Thus, my alternative counterfactual is designed to focus thinking on these selection mechanisms, which are relatively independent of the means by which Special Relativity was originally generated, but which nevertheless ultimately determined the theory's survival in the scientific community.

The possibility of socialized creativity raises many interesting and important questions of knowledge policy that are central not only to social epistemology but also to the *modernist-postmodernist debate* that defines contemporary cultural criticism (cf. Harvey 1986). Here is one version of the debate. On the modernist side is Joseph Schumpeter (1942), who proposed that capitalism would devolve into socialism as innovation came to be seen as too risky to a social order that, through

Three: Of Policy and Politics

the diffusion of previous innovations, had attained a level of income and security worth preserving even if by bureaucratic means. Schumpeter has received unlikely support in recent years by the ecology-minded and Habermas-infused "Finalization" movement, which argues that pure inquiry is a luxury that society can ill afford a mature science like physics (Schaefer 1984). On the postmodernist side is Jean-François Lyotard (1983), who appeals to the unsurveyability of the information explosion's effects on today's society as ensuring that we will live in a world of increasing normative tolerance, and hence, wanton creativity. And it is only a short step from the unsurveyability of knowledge effects to the laissez-faire attitude to science promoted in "Managing the Unmanageable." (*The Principle of Epistemic Fungibility*, introduced in the next chapter, is my attempt to raise this debate to a more publicly accessible plane.)

Before leaving the topic of referential opacity, one of its variants is worth noting because of its increasing importance in philosophical contributions to the public understanding of science. Call it *strategic opacity*. The idea is based on the classical trope of *catachresis*, or the misuse of names. If a situation can be described in alternative ways so as to motivate alternative courses of action, then, surely, one of those ways could be a literal misdescription that is nevertheless necessary for the audience to act in a normatively desirable manner. At first glance, this simply sounds like manipulation, but "manipulation" presupposes that the audience is being made to act against its own interests or beliefs, when in this case it has yet to form any clear views on an issue that requires prompt action. If strategic opacity succeeds, then in the long term the world comes to resemble more closely the strategic misdescription.

Consider, by way of example, the role of philosophers of science as expert witnesses on the nature of science. In a trial involving the teaching of Creationism in public school biology courses, philosopher Michael Ruse was asked to demarcate science from nonscience (La Follette 1983). He responded by giving Popper's falsifiability criterion, knowing full well that his answer failed to do justice to the serious objections that philosophers and others have raised to the criterion's plausibility. However, had Ruse attempted to represent to the judge the complex battles that are waged over even the intelligibility of the demarcation problem, he would have probably undermined the credibility of philosophers of science as authorities on the normative character of knowledge production. Thus, Ruse had to see his charge as one not only of representing his personal opinions but also of representing the opinions of others in his field who may be called to testify on

Knowledge Policy

similar matters in the future. And even if these future witnesses would oppose Ruse's theory of science, they would probably object even more to being preempted from offering an opinion. Most important, from a rhetorical standpoint, even if philosophers of science have in fact abandoned Popper, Ruse may still be right that it would be to the advantage of both our knowledge enterprises and the public at large–at least in this case–to act as if falsifiability were true.

The social function of strategically opaque accounts of science is quite familiar to philosophers, ever since John Herschel's *Preliminary Discourse* presented the scientific method to the lay Victorian audience as systematically applied common sense. Herschel was interested in normalizing science's relations with a public that marveled at the spectacle of experimental demonstration but remained skeptical that its results constituted the sort of "humane" knowledge mastered in the British liberal arts curriculum. Herschel's strategy was to transfer the "technical" character of science from the construction of apparatus to the design of nomenclature: a conversion of experiment to rhetoric. He deployed oversharp distinctions in the stages of scientific reasoning–such as the contexts of discovery and justification–that were illustrated by homely examples, which have since become the stock cases from which philosophers of science argue about the nature of science. This last point is important, for in the twentieth century, as philosophy of science came to be practiced more by professionally trained philosophers than by scientists, Herschel's rhetoric was crucial for philosophers to convince *themselves* that they could opine significantly on the nature of science after having mastered some scientific vocabulary and syntax but without laboratory training. And despite the criticisms that it has received, this "shallow," "merely philosophical" view of science has, nevertheless, kept alive a publicly accountable image of science throughout this period of increased disciplinary specialization.

Referential opacity is just one tactic by which public attention is diverted from the more encumbering social consequences of Big Science. The most effective tactic in this vein involves the biggest ruse. It is the presumption that because social science research is more obviously focused on the human and the applied, it is therefore more likely to have socially dislocating consequences than, say, basic research that is exclusively designed to study something as abstract as microphysical reality. (Think of the perceived significance of laboratory aggression research and IQ tests.) The ruse here is to think that the only consequences of research are the officially intended ones. This particular myopia is brought on by overlooking the material character

Three: Of Policy and Politics

of intellectual needs. Moreover, science policymakers exacerbate the problem when they fail to realize that even unintended consequences *need not be unexpected*. In other words, careful empirical study into the social effects of different lines of research could enable the prediction of outcomes that were not intended by the research directors—a valuable conceptual tool for prying open scientists' ex cathedra pronouncements on what their research can or cannot do. I would suggest that policymakers evaluate any practice, including an intentionally scientific one, in terms of the groups that are most likely to be most affected by the consequences of that practice in an appropriate expanse of space and time. Thus, while a series of high-energy physics experiments is intended to affect the community of high-energy physicists, presumably in a positive manner by getting them to rethink their research programs in light of the experiments' findings, it may turn out that, upon examination, the experiments' biggest impact is on another disciplinary community that is more impressed by the results than the high-energy physicists, or, even more to the point, laypeople conceptually unconnected to science who are nevertheless the indirect recipients of subsequent experimental applications.

Let us focus on the typical high-energy physics experiment a little more closely, it being an especially vivid example of the strategic conflation of intention and expectation. What is tested in such an experiment? The intended answer, of course, is some range of hypotheses about the nature of microphysical reality. However, once we think of the material conditions needed for realizing this intention, we should come to expect that other hypotheses will also be tested at the same time—not in physics, though, but in political economy. These social experiments, no less than their "natural" counterparts, involve the enforcement of ceteris paribus clauses. That is, they are designed to exclude all factors from the test site other than the ones that are thought to bear some responsibility for the phenomena under investigation. In this way, scientific research is made to look subject to its own kind of inertial motion. For example, it is now common for high-energy physics experiments to pool the financial and human resources of several countries in terms that are laid out in an international agreement. The wording of that agreement constitutes instructions for converting the physics experiment into a test of a certain theory of international relations. Regarded somewhat more globally, the experiment also tests a certain scheme for redistributing income and personnel. After all, the freedom of physicists to manipulate variables as they see fit rests on the ability of governments, universities, and other scientific support agencies to coordinate labor and capital over vast spaces for long periods that might oth-

Knowledge Policy

erwise move in disparate directions. Indeed, the control exerted over the performance of large-scale natural science experiments makes them both the most powerful testing ground for hypotheses about social interaction and potentially the biggest source of large-scale social dislocation during peacetime.

The conflation of intention and expectation is ultimately a Platonic conceit. Having one's mind in harmony–or in "reflective equilibrium" as students of John Rawls (1972) like to say–is a matter of knowing what one wants and wanting what one knows. The Jesuit moral casuists saw the conflation four centuries ago, and tried to reestablish a distinction between the epistemic (i.e., the expected) and the ethical (i.e., the intended) sides of action with *The Doctrine of Double Effect* (Harman 1983 gives a current formulation and defense). Unfortunately, the Jesuits did this only to play off the former against the latter: that is, you can expect things you didn't intend, and therefore, you can knowingly do something without being culpable–a convenient moral psychology for the religious warrior! By contrast, while I endorse the doctrine as a means of demystifying the idea that, say, a physics experiment is only–or even primarily–about physics, I want the doctrine to empower nonscientists (including policymakers), rather than to excuse scientists. In short, scientists should be held accountable for what can be expected to follow from their hypotheses, regardless of their intentions.

However, the distinction between the expected and the intended has once again been clouded in the contemporary context in which unintended consequences are most often discussed, namely, economic prediction; for economists tend to posit an idealized rational agent, who, though not a Platonist, always seems to intend in proportion to her expectations. When she does not, the consequences are generally beneficial, as in invisible hand accounts of economic order. If the economic agent is not omniscient, she at least remains *blissfully* ignorant. Yet the evaluative asymmetry between basic and applied research creeps into how even this conflation is handled by the defenders of pure inquiry. Only basic research is portrayed as having positive unintended consequences (usually in opening up new lines of inquiry but often in the applied realm as well), while applied research is seen as having primarily negative ones–especially in terms of foreclosing opportunities for pursuing basic research, but also in its unwitting production of instruments of mass destruction. The positive unintended consequences of basic research supposedly flow "serendipitously" from the unconstrained pursuit of inquiry, whereas the negative unintended consequences of applied research appear to be opportunities that ideologically inspired ministers of science are all too eager to exploit.

Three: Of Policy and Politics

Of course, from a strictly scientific viewpoint–the viewpoint from which one might think scientific rhetoric should be judged–all the anecdotes that may be cited as evidence for the beneficial by-products of basic research and the destructive capabilities of applied research are more the stuff of which superstitions are made than rigorous foundations for the conduct of science policy. As the cognitive psychologists say, the privileged anecdotes contribute to a "confirmation bias." Consider the possibilities that would first need to be explored empirically before claiming that basic research unwittingly courts good and avoids evil better than applied research.

> 1. If a large enough expanse of space and time is examined, who is to say that the effects of basic research would not turn out to be just as deleterious as the consequences of applied research?
>
> 2. Or, rather, that the effects of applied research would turn out to be just as beneficial?
>
> 3. Even granting the serendipitous consequences of basic research, who is to say that applied research might not have reached the same conclusions sooner and more efficiently?
>
> 4. And even granting that serendipity is a more efficient way of reaching those conclusions, who is to say that more desirable conclusions would not have been reached by replacing a particular line of basic research with one of applied research?

Whereas (1) and (2) ask the historian to manipulate the parameters within which she examines the actual consequences of applied and basic research, (3) and (4) call for counterfactual historiography, which has established its credentials in economic and social history, but which has yet to take root in intellectual history (cf. Lynch 1989). Without going into the details here, the general strategy would be to go back to the latest point in time when the alternative trajectory in question could have been pursued, and then to estimate the probable consequences of pursuing that trajectory, instead of the one actually pursued, assuming that little else of the actual subsequent history would have changed (cf. Elster 1979; McCloskey 1987). For someone interested in redressing the balance between the consequences of basic and applied research, the goal would be to show that, given the chance, applied research could perform at least as well as basic research, without disturbing too many historical background assumptions.

Knowledge Policy

There is a burgeoning sphere of litigation in which these manipulations of possible pasts and futures make a major practical difference. The cases turn on the liability of scientific research for unwanted environmental change. The battle between Big Science and the Ecologists is often portrayed as a disagreement over matters of fact and levels of risk, but behind it all is a dispute over one's sense of history. Whereas Ecologists typically suppose that the trajectory of scientific research will not veer enough off its current course to preempt or resolve any long-term environmental disasters, Big Scientists presume that most of the potential for disaster will be contained or addressed by research breakthroughs that have yet to be made (J. Simon 1990). Given this contrast in historical vision, it is little wonder that Big Scientists have a fairly short-term conception of liability (since significantly new factors may intervene in the future to confound any current tendencies), while the Ecologists project their legal concern on the long term, wanting to hold scientists responsible for the remote consequences of their actions (Huber 1988). In these cases, the role of the courts is to adopt the standpoint of third parties—the involuntary stakeholders, if you will—who are the potential beneficiaries or victims of whatever the scientists do. In a fully democratized knowledge enterprise, the effects of unsuspecting third parties might serve as the sociological surrogate for the check of an "independent reality" or "external validity." Thus, the judge in an environmental damage case would revert to her etymological origins as the "tester" of alternative causal accounts.

Although the basic-applied distinction is truly clear only in government accounts of science funding, the philosophical history of the distinction has aimed to keep basic research beyond accountability. Consider the pragmatist vision of science, especially as articulated by John Dewey, vis-à-vis the positivist vision articulated by the members of the Vienna Circle. The former saw the epistemic authority of science as resting in its ability to transform nature in the interests of humanity. In Dewey's mind, there was no sharp distinction between basic and applied research, nor any desire to make value neutrality a virtue of science (Proctor 1991: pt. 3). In contrast, the Vienna Circle traced the epistemic authority of science to the intrinsic form of its theories, especially as articulated in logically valid and empirically testable terms. Here we find the relevant distinction and desire. Whereas, for Dewey, "instrumentalism" indifferently referred to a position in epistemology and ethics, for a product of the Vienna Circle like A. J. Ayer, such indifference was the height of philosophical folly. What lay between the pragmatist and the positivist was World War I, in which the German scientific community—generally regarded as equal or superior to

Three: Of Policy and Politics

Britain's on the world scene—openly accepted responsibility for the military hardware that turned out to make the war the most devastating up to that point in history, culminating in Germany's most humiliating defeat. This unleashed an antiscientific irrationalism in the 1920s, to which proscientific intellectuals adapted by promoting a science that the public could see either as itself irrationalistic or as conceptually independent of its destructive technological capability. The indeterminacy thesis in quantum mechanics is an outgrowth of the former tendency (Forman 1971), logical positivism an outgrowth of the latter. The final piece of the puzzle that was needed to put science beyond reproach was the reintroduction of positive value connotations for basic research by emphasizing that some of its best consequences may come in unexpected quarters. The much publicized service of basic physics researchers in the Allied cause in World War II performed the function effectively for the popular imagination (Hollinger 1990), which was celebrated in Vannevar Bush's *Science: The Endless Frontier*, the ideological statement behind the founding of the U.S. National Science Foundation.

But why should scientists, and their favorite epistemologists, resort to these backhanded rhetorical maneuvers in order to avoid accountability? What have scientists to fear from subjecting themselves to greater public scrutiny? Nothing, except a stereotype of what it is to be accountable. That stereotype reaches back to the primal moment of accountability, the academic exam, in which an individual's merit is judged on the basis of an externally driven standard. In this vein, social historians of accounting have recently made much of the fact that the inquisitorial style of courtroom procedure that characterizes Continental European legal systems arose from the practice of university examinations in the Middle Ages (Hoskin and Macve 1986). As Foucault would have it, the society and the individual are "co-produced" in the inquisition, as the latter is made to stand up against the former. But accounting need not be a process for measuring the fit of individual cases to general rules. Rather, it can be a diagnostic procedure that treats cases as symptomatic of the overall state of the rules. Thus, wayward scientists need not fear having their Ph.D.'s revoked if their deeds fail to match up to their words; instead, the scientists' incentive structure may be altered so as to get them to work in a different way or in a different field. Moreover, accounting for science may function compensatorily by awarding damages to affected third parties, especially when the consequences of research stray significantly outside the academy (Huber 1990). For example, I can imagine a polluting laboratory being required to devote a substantial part of its research time to

Knowledge Policy

cleaning up after its messes, or maybe even developing technologies that improve the well-being of the affected parties.

Most significantly, however, science accountancy has policy implications even when knowledge production flows smoothly. Once again, patent law is the prototype for what I have in mind. As the framers of the U. S. Constitution realized, although it is the individual who invents, she will not do so unless the society provides an incentive structure conducive to invention. This means not only rewarding current inventors, but also encouraging others to become inventors in the future. To meet these two somewhat conflicting goals, patent law is designed to allow inventors only a limited monopoly over their inventions. In short, the science accountant treats in systemic terms not only the negative but also the positive deeds done by individual scientists.

However, I would be less than candid if I said that the road to "science accountancy" has been clearly paved. Indeed, if the distinction between "basic" and "applied" research is as constructible–and deconstructible–as the argument in this section has suggested, then one may legitimately wonder how a particular research team or piece of scientific knowledge could ever be held accountable for, say, a technological effect that occurs long after and far away from the site of the original research. Cannot multiple stories be told, some implicating a particular group of scientists and some not? It is difficult to deny the thrust of this postmodern query. If anything, the inevitable open-endedness of all searches for causes should motivate researchers to explicitly associate with and distance themselves from others who take an interest in their work. For if, in the final analysis, the relevant causal trace is determined in the marketplace by competing accounts–and there is no ultimate court of reality that can overturn unjust rulings–then it becomes essential that scientists represent themselves persuasively as having been aligned with "the forces of good," all the while realizing that all such alignments are little more than fallible wagers on the outcome of the case at hand. (Shotter 1984 is probably the best place to proceed from here.)

The Social Construction of Society

Basic-applied, invention-discovery, genius-error–the reversibility of each of these distinctions runs counter to the folk wisdom of science policy. Yet it is the cardinal lesson that STS can teach

Three: Of Policy and Politics

policymakers, one that is central to the doctrine of *the social construction of facts and values*. This doctrine maintains a sharp separation between determining *when a norm applies* and what *follows once the norm applies*. In short form, it distinguishes the *scene* from the *script* of action. Thus, a "norm" is any pattern of social action that is scripted, which is to say that there is a right and wrong way of performing the action. None of this yet seems very exciting, commonplaces that they are of most schools of microsociology. Indeed, there may be little disagreement between a social constructivist and a philosopher of science either about the ways in which scientists justify their research to each other and policymakers or even about the fact that these justifications generally work and hence continue to keep the scientists in business. Both would cite chapter and verse of the hypothetico-deductive method and other positivist scripts.

A difference of opinion does emerge, however, once both the constructivist and the philosopher are asked *why* the script works. Whereas the philosopher will tend to focus on properties of the script (e.g., its logic), the constructivist will turn to the scenes where the script is typically enacted, which allow her to look for features of these situations that enable the verbal performance to elicit the desired effects. And while the constructivist does not presume that these features will be the same from situation to situation, she may believe that the script must be performed somewhere at some point. Indeed, I am such a *script transcendentalist*, someone who believes that arguments and claims concerning the valued form of knowledge, or "science," are necessary for the possibility of society. But I leave open to empirical investigation (of the past and present) and negotiation (in the future) the exact backdrop against which such arguments and claims can be successfully made (cf. Fuller 1988a: chap. 7; Gellner 1989 develops the world-historic implications of this point).

Philosophers of science are themselves no strangers to the study of scenery—except that they shroud it in Latin, the ceteris paribus clause, and shove it still further into the background of their analyses (cf. Fuller 1988a; chap. 4). They probably suffer from the physicist's prejudice of undervaluing in concrete what can be so easily done in abstract, as in the case of deriving the laws of motion from a world of frictionless planes. For their part, constructivists realize that heavy transaction costs are incurred in moving from the abstract to the concrete, especially in terms of the human and material resources that need to be strategically situated (including things that were prevented from getting in the way), in order for "all other things to be equal." The folk wisdom of science policy is symptomatic of a metalevel version of the same preju-

Knowledge Policy

dice, which comes from generalizing too willingly from the anecdotal evidence of respected scientists. Thus, just as the physicist forgets to consider exactly how one would materially construct a frictionless plane on a regular basis, likewise the policymaker forgets to consider what it would take to construct environments to enable future physicists to arrive at their abstractions.

The more locally one considers the construction of scenery needed for enacting a script–say, one laboratory that agrees that certain evidence supports a certain hypothesis–the more social constructivism appears to be a species of dramaturgy. Indeed, the followers of Erving Goffman and Harold Garfinkel who have introduced a microsociological perspective in STS have conveyed this impression. This has led STS practitioners to espouse a bias toward *localism*, or the ontological privileging of the "here-and-now" over the "there-and-then." Sometimes localism is little more than a politically correct way of talking about what positivist philosophers of science have called "the observable"; other times, it is simply a nominalist (i.e., a negative) stance toward the reality of such macrosocial entities as institutions and classes (cf. Harré 1981). In either case, the STS practitioner takes herself to be showing how various localities interlock to produce the dispersal of effects that characterizes today's technoscience (cf. Ophir and Shapin 1991; Shapin 1991). Such a research agenda presumes that the places where scientific work is done are "indexes" for various sorts of knowledge, or, in a more rhetorical vein, reminders of the skills that are called for on particular occasions.

But there is more to indexicality than meets the eye–even the eye of the participant observer in one of these locations. In addition, an index triggers what art historians would call *iconographic* associations, which supply the observable foreground with an affectively charged conceptual background through which the foregrounded experience is understood. For twentieth-century art historians like Erwin Panofsky and Ernst Gombrich, iconography came as close as possible to documenting the *Weltanschauung* or collective memory of a culture, namely, a set of ubiquitously cueable and applicable symbols (cf. De Mey 1982: chaps. 10-11). These associations are verbally elicited when people are asked to explain or excuse their behavior. Thus, engaging in a routine lab technique is more than just the dextrous handling of instruments; it also involves a certain attitude toward the activity that is prompted by its doing and which the participant-observer tries to tap into conversationally. In effect, this attitude toward one's place–one's "station," as it were–is the manner in which the scientist embodies her community's ethos. A good, if perhaps unwitting, example here is

Three: Of Policy and Politics

Michael Polanyi's (1957) ability to go on at (inordinate?) length about science's tacit dimension in *Personal Knowledge*. Polanyi's narrative situated laboratory practices in a world that, in the guise of bureaucratic science policy, increasingly threatened to intrude upon the "natural" performance of those practices. His rhetorical success at conjuring up such an image in the many contexts in which he was read testifies to this more expansive sense of indexicality, one that ultimately calls into question what might be the appropriate contrast term to "local." After all, the very existence of iconographic memory concedes that some of the most important things that happen and matter to locales come from the outside. In short, the nonlocal is already inscribed in the local.

This last point is significant, as it allows us to begin to talk about knowledge policy from a constructivist standpoint. The question that naturally arises in this context is whether interlocking enough locales together could ever produce the sort of "global" picture of knowledge production that would enable a policymaker to set priorities, anticipate outcomes, and adapt to changes in "the system." Both positivists and Marxists, bureaucrats and activists, are skeptical of the constructivist attempt to eliminate such macrostructures as "power" and "objectivity" which seem to slip between the cracks of locales, yet give scientific knowledge its distinct sense of independence from much else that happens in the social world (cf. Fuller 1988a: chaps. 2, 10). Here is a strategy for explaining knowledge production that attempts to respect both local and global sensibilities:

> 1. The translocal uniformity of a piece of knowledge is largely an artifact of the restricted channels in which knowledge must be officially communicated.
>
> 2. Nevertheless, that still leaves open the question of why such a wide range of independently and diversely managed laboratories find themselves communicating roughly similar, if not downright identical, messages.
>
> 3. The answer lies in treating the occurrence of each such message as a predictable outcome of a decision procedure that, while different from the one used in another lab, has precedent as the decision procedure used in other sectors of society, with which lab members would have had some contact.
>
> 4. Thus, the apparent independence of the knowledge produced

Knowledge Policy

in a set of labs from the circumstances of its production is due to a concatenation of individually predictable events that are then rendered uniform by the restricted channels mentioned in (1).

Consider, by way of example, the convergence of the physics community on the existence of neutral currents. Pickering (1984) has shown that the different labs involved behaved in ways that could have been predicted solely on the basis of their particular social arrangements, even if the existence of neutral currents were not at issue. To put the point more provocatively: just because the labs *end up* agreeing on the existence of a particular entity, it does not follow that their agreement is *due to* the existence of that entity. One lab may come to believe in neutral currents because it always follows whatever the research director thinks, whereas another lab may come to the very same belief as a result of a weighted average of what the entire research team thinks. If each lab is operating in its customary fashion, then the convergence in beliefs could have been predicted simply on the basis of knowing the labs' decision-making procedures, without knowing anything about the content of the belief on which they converged. The natural conclusion to this line of thought is that the convergent belief in neutral currents is an epiphenomenon of the diverse social processes that issued in assertions of that belief. This is contrary to such official communications as journal articles, which would easily give one the impression that the various labs reached the same conclusions for largely the same reasons. The decision-making procedures that distinguished the labs above–deference to a superior and the weighted averaging of peers–are ones that can be found in other, nonscientific sectors of society. Indeed, these procedures go to the very heart of how modern society is organized and maintained.

The Constructive Rhetoric of Knowledge Policy

So far, we have been stressing the *social* side of social construction over the *construction* side. However, the construction side brings us into the heart of the rhetoric of knowledge policy. The rhetoric of knowledge policy covers the construction of individual ra-

Three: Of Policy and Politics

tionality out of beliefs and desires, and the construction of collective rationality out of facts and values—the rationality of the *researcher* and of the *research*, as it were. I call these the rhetoric of rationality attributions and of fact-value discriminations. Let us consider each in turn.

THE RHETORIC OF RATIONALITY ATTRIBUTIONS

Rhetorically speaking, rationality is the way we balance the budget of our *beliefs* and *desires*. Whether the context of attribution is ordinary common sense or rational choice theory, one of these two types of mental entity tends to be emphasized at the expense of the other. Metaphorically speaking, desires and beliefs are, respectively, the movers and moved of the mind. Only a small imaginative leap is required from here to David Hume's aspirations for a "mental mechanics" that would parallel Newton's physical mechanics. Not surprisingly, then, a picture of the mind containing mobiles and mobilizers, passive reflectors of nature (beliefs) and active resisters (desires), would support some epistemologically sharp distinctions. For beliefs and desires are usually held to be irreducible not only to anything else but also to each other. Whereas beliefs are typically tempered in light of evidence, desires typically are not, and are in fact often strengthened by evidence to the contrary. But is it all that clear when something ought to count as a belief rather than as a desire?

There is more at stake in this last question than a mere matter of conceptual clarity, namely, the criteria we use to evaluate the rationality of someone's actions. For although we normally explain behavior by appealing to a configuration of beliefs and desires, we tend to lean more heavily on a person's beliefs when we want to evaluate what she has done on the basis of criteria in her immediate vicinity that are not necessarily of her own creation, whereas desires bear more of the burden when the evaluative frame of reference is expanded to cover criteria of her own creation though often not in her immediate vicinity. Thus, in answer to the question why Mary walked out into the rain without an umbrella, we can say either that she did not think it was going to rain or that she wanted to get to the office quickly. In the former, belief-driven account, Mary appears to have simply erred, while in the latter, desire-driven rendition, Mary is portrayed as having taken a calculated risk. Clearly, the two accounts are compatible, yet the first Mary is a victim of misinformation, while the second Mary deliberately suffers short-term losses to achieve longer-term goals. Notice that our evaluation would not change, had Mary gone out into

Knowledge Policy

the rain *with* the umbrella. On the one hand, her correct belief would correspond to a reality (i.e., that it was raining) that did not require that belief for its existence; on the other hand, her risk would have appeared still more calculated, thereby enabling her to minimize even short-term losses.

Generalizing from the above example, we see that a person's rationality can be rhetorically enhanced by giving desires the upper hand over beliefs in the explanation of her actions. Desire-driven accounts have the advantage of mitigating the surface irrationality of an isolated episode in a person's life by subserving it to a more extended life-plan in which the episode figures as merely one of many way stations. This point has recently received some interesting experimental confirmation by behaviorists seeking to reinterpret the many cognitive psychological studies that make people out to be incompetent calculators of expected utility. The behavioral psychologist Howard Rachlin (1989) has observed that these studies typically look at the performance of subjects exclusively in the context of the experimental task. Thus, this research has tended to view the subjects as having false beliefs about some feature of the immediate situation, rather than as relating their response to some long-term goal not directly represented in the test case. It was precisely in order to avoid such "instant rationality" judgments that behaviorists have traditionally assessed the rationality, or "efficiency," of animal response on the basis of a *series* of trials, and usually not until a stable pattern of performance was detected. Ironically, then, in the face of the rigid protocols of cognitive psychology, it has been left to behaviorists to save some of the phenomena associated with the very mentalism whose existence they have traditionally denied. These mental phenomena include foresight, hindsight, and any other form of inference that forces the organism to adopt a historical perspective toward its own behavior.

THE RHETORIC OF FACT-VALUE DISCRIMINATIONS

Whenever knowledge policymakers want to argue for either maintaining or changing a line of research, a strong distinction between *facts* and *values* can play a strategic role in the argument. Specifically, if the policymaker wants to stick to a research trajectory despite resistance from the environment, she can appeal to the "value" of pushing onward, whereas if she is looking for an excuse to abandon the trajectory, an appeal to the countervailing "facts" of experience will typically figure in a winning strategy. The strategy to be outlined is largely an elaboration of this point.

Three: Of Policy and Politics

The practice of replicating experiments is central to science's self-image, especially to its image of having a firm database. Thus, it was to be expected that Harry Collins' multipronged challenge to the feasibility of the norm would prove controversial. According to Collins (1985), not only were there professional disincentives to performing replications (i.e., they were rarely published), but even in cases where replication was crucial for continuing a line of research, important details of the original experiment could be gleaned only by personally contacting the experimenter, since the published text turned out to be singularly uninformative. Here we have the sort of empirical finding that would call enough of the science policymaker's natural understanding of science into question to force her to take a stand on whether replication is part of the "is" or the "ought" of science. To put the point in its most general terms, Collins' recalcitrant cases may be seen either as refuting replication as a fact about science or as violating replication as a norm governing science: one interpreter's falsification may be another interpreter's infraction; it all depends on how one manages the anomalies.

The foregoing line of reasoning builds on David Bloor's (1979) attempt to use Mary Douglas' anthropology of cultural boundary maintenance to make sense of Lakatos' theory of anomaly management. In *Proofs and Refutations* (1978), Imre Lakatos identified four strategies for handling counterexamples to mathematical arguments: monster barring, monster adjustment, exception barring, as well as Popperian falsification. Based on his reading of Douglas, Bloor then argued that a society's preference for one or another of these strategies will depend on its so-called grid-group factors, i.e., the society's internal stratification (i.e., grid) and its external differentiation from other societies (i.e., group). I now propose to give a more explicitly constructivist gloss on what happens in episodes of anomaly management, namely, that they bring into existence the occasions that warrant making the fact-value distinction. To speak more metaphysically, these occasions ground the possibility of there being a fact-value distinction.

I will present a grid-group analysis, first, of possible policy reactions to Collins' counterinstances to replication as a norm of science, and then, of a current science policy issue. It will be instructive to run through the Collins case to show that grid-group analysis is relevant not only to the management of "dangerous objects"–Douglas' own original concern and the direction in which she has subsequently developed the scheme (e.g., Douglas and Wildavsky 1982)–but also to more abstract threats, namely, to one's implicit theory of how the world (or some part of it) works (cf. Thompson et al. 1990). With that in mind,

Knowledge Policy

grid-group analysis suggests that Collins counterinstances can be examined along two dimensions, as epitomized in the following questions:

> (X) Are the counterinstances representative of a more general tendency (X-) or restricted to just those cases (X+)?
>
> (Y) Is replication judged against the counterinstances (i.e., descriptively [Y-]) or is it the standard against which the counterinstances are judged (i.e., prescriptively [Y+])?

The (X)-axis captures the "grid" character of the judgment, in that the policymaker must decide whether there is likely to be a difference between the instances that Collins reports and those that have yet to be observed in the relevant population. "Low grid" suggests no substantial difference between the seen and unseen cases, whereas "high grid" suggests more heterogeneity. The (Y)-axis captures the "group" character of the judgment, in that the policymaker must decide whether the practice of replication will be opened to correction from the counterinstances ("low group") or whether the counterinstances will be banned to uphold the integrity of replication as a norm ("high group"). Thus, combining the possible answers to the above two questions, the following interpretive possibilities emerge:

> (X-,Y-) Replication is an empirical hypothesis about how science works that may be rejected in toto in light of counterinstances. This corresponds to *falsification*.
>
> (X-,Y+) Replication is a normative standard that may be used to discount all counterinstances as cases of scientific malpractice. This corresponds to *monster barring*.
>
> (X+,Y-) Replication is a principle whose empirical breadth may be adjusted in light of the counterinstances so as to render them irrelevant to a proper test of the principle. This corresponds to *exception barring*.
>
> (X+,Y+) Replication is a principle whose normative depth may enable a charitable reinterpretation of the counterinstances so as to render them less contrary than they first seem. This corresponds to *monster adjustment*.

We can easily apply this anomaly management scheme to a major knowledge policy issue that has captured the media's attention

Three: Of Policy and Politics

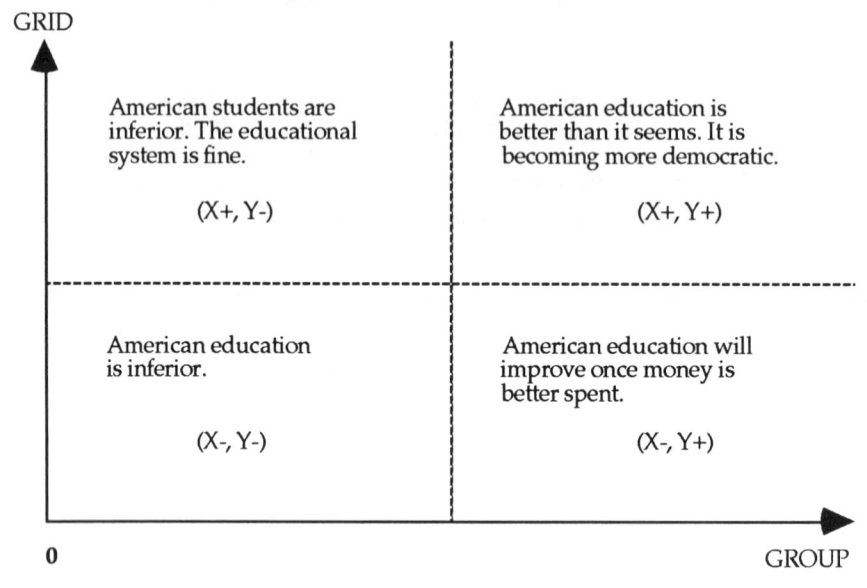

GRID

American students are inferior. The educational system is fine. (X+, Y-)	American education is better than it seems. It is becoming more democratic. (X+, Y+)
American education is inferior. (X-, Y-)	American education will improve once money is better spent. (X-, Y+)

0 → GROUP

Figure 4. A Grid-Group Analysis of Responses to the Crisis in American Education

recently namely, the long-term decline in the academic performance of American students when compared with their counterparts in other countries. (See Figure 4.) This decline comes in spite of the large and increasing amount of funding for education in this country. The decline is such a news item because it prima facie challenges a piece of policy folk wisdom that may be expressed by the following maxim: "Academic performance will improve in proportion to the amount of money spent on education" (cf. Averch 1985: chap. 4).

Commentators on this anomalous state of affairs have occupied every position on Bloor's scheme. Corresponding to (X-,Y-) is a frank admission that American education is inferior, and that clearly current funding patterns are not improving matters. These critics take the trend as symptomatic of a need to radically rethink our educational policy. Being low on both group and grid, these commentators are receptive to the educational initiatives taken in Europe and Japan. Representing (X-,Y+) are those who find the trend relatively superficial, suggesting simply a problem with the accounting procedure used to evaluate education funding. Perhaps moneys are being used to renovate buildings, when they would be better spent on raising the salaries of the best teachers. Tighter scrutiny would presumably remedy such poor managerial judgment. The strategy here is to locate "the enemy within" who can be scapegoated and ultimately exorcised, so as to restore balance to what is essentially a sound educational policy. Position (X+,Y-) is occupied by those who would alter the terms of the argument by pointing out, say, that while American nationals continue to decline academically, a larger number of foreigners are matriculating in the United States, where they form an ever increasing percentage of the excellent students. This suggests that the extent of the problem is more contained than it may initially appear. America is becoming more of a world educational mecca, thereby vindicating the folk wisdom. Unforeseen, however, was that so relatively few Americans would thrive in this competitive environment. Thus, these commentators advise a continuation of the same policy, but with revised expectations about the policy's exact beneficiaries. Finally, the rosiest picture is painted by (X+,Y+), which sees the decline in test scores as symptomatic of the relative democratization of education in this country vis-à-vis other parts of the world. People from all walks of life now go to school in this country, for a variety of reasons, few of which can be satisfactorily evaluated by standardized test scores. Thus, an apparent sign of failure is reinterpreted as a success in disguise.

Someone of a positivist sensibility who has patiently read our grid-group analysis of fact-value discriminations may still want to maintain a "real" fact-value distinction. She would then simply read these categories as suggesting that the "is" and the "ought" pull in opposite directions, since, say, the maintenance of replication as a norm of science rests on marginalizing new information about scientific practice, whereas giving that information its empirical due would undermine replication's normative status. However, if we take seriously the natural understanding that policymakers in the knowledge society have of how science works, then the specifically "empirical" and "normative" features of that understanding will be revealed only in cases where its naturalness is challenged. Thus, as long as replication is regarded outside the context of problematic cases, it is unlikely that the policymaker will feel any need to decide whether replication is descriptive or prescriptive of science. But once the counterinstances are conjured up, the nature of the situation will force the policymaker to take a stand on this issue, as defined by one of the four options outlined above. By the logic of this argument, then, we should expect that the positivist's strong sense of the fact-value distinction would arise in periods when there are severe challenges to a long-standing natural understanding of things, or what the positivist herself would likely describe as genuine

Three: Of Policy and Politics

tests of a set of beliefs. This is, indeed, a major thesis of the first book-length history of value-free research (Proctor 1991).

Armed for Policy: Fact-Laden Values and Hypothetical Imperatives

Concerns about separating "is" from "ought," or distinguishing "fact" from "value," bring to mind that epitome of modernist social science, Max Weber, caricatures of whom continue to haunt the policy potential of STS. Most familiar is the Weber who wanted to keep facts and values separate in order to protect factual inquiries from being tainted by value commitments (i.e., the *value-freedom* thesis). But Weber himself more often thought about this issue in exactly the opposite way: that is, he wanted to protect values from facts (Proctor 1991: chap. 10). Given the fallible and partial nature of our factual inquiries, Weber did not want to make our value aspirations hostage to the latest scientific trends (i.e., the *fact-freedom* thesis). In an interesting sense, Weber's own training in economics reveals the existentialism implied by his position. We are saddled with a surfeit of possibilities for action and a scarcity of the knowledge needed to eliminate all but the best of them. Personal commitments and social conventions must therefore compensate for the uncertainty entailed by this situation. In contrast to all these specters of Weber, mine believes that we value certain social practices and their products—scientific ones, in this case—only because we presume that certain things are true about the role that those practices and products play in society as a whole. But were it shown that these presumptive truths are in fact false, then the value of the practices and products would be thrown into question. In short, I am arguing for the *fact-laden* character of value commitments as more rhetorically revealing than the *value-laden* character of facts, also associated with Weber, that is often said to stalemate rational discourse (cf. Fuller 1988a: chap. 12; Fuller 1989a: chap. 3).

At this point, it may be useful to contrast the view that I am presenting with the pragmatist analysis of the fact-value distinction, classically presented by John Dewey (1958, 1960) and more recently, and specifically in the context of science, by Larry Laudan (1987, 1990b). It is not that I oppose the spirit of the pragmatist analysis, which also emphasizes the fact-ladenness of values, but rather that I take my own analysis to be logically prior to the pragmatist's. In other words, the

Knowledge Policy

pragmatist accounts for the place of facts in value judgments *after* the fact-value distinction has been made. However, my analysis should not be taken as the usual exercise in philosophical one-upmanship. For as we have seen, the confidence that policymakers in the knowledge society have of their understanding of science makes them generally uninterested in distinguishing normative and empirical claims about the nature of science. Therefore, in order to prevent the pragmatist's self-avowed "instrumentalist" analysis of value from turning into a philosophical justification for technocratic reasoning (whereby the policymaker sees herself as merely applying principles that have been shown to work), we need to recall that the fact-value distinction remains an important basis for critique, even if the way it is drawn turns out to be as highly context-dependent as I am suggesting.

The pragmatist argues that norms are really hypothetical imperatives for reaching a certain end by the most efficient means. In effect, the imperatives are experimentally derived regularities for which any ordinary human action is potentially a test case. The primary role of the social sciences is to discover and codify these regularities, evidence for which has been accumulating since the dawn of civilization. But how does one decide on which end to pursue? According to the pragmatist, each end can be regarded as a means to some other end, and so each end may be factually judged by the extent to which it enables the higher end to be achieved. For example, a typical hypothetical imperative would be (assuming that it is true), "If you want to expedite the growth of knowledge, then pick theories that explain the most data by the fewest principles." But why might we want to expedite the growth of knowledge? Is this an end that we must embrace or reject unconditionally? Not so, says the pragmatist. We may regard expediting the growth of knowledge as a means toward improving the quality of human life. However, whether it in fact does so is an empirical question. Indeed, as a matter of empirical fact, distributing currently existing knowledge more widely may turn out to be a more efficient means for improving the quality of human life than simply encouraging the production of new knowledge that only elites will be able to put to direct use. In that case, if we had wanted to expedite the growth of knowledge mainly because we thought that it would best promote the quality of human life, then we had better stop expediting and start redistributing instead.

The pragmatist analysis starts by treating as an open question which of several courses of action one ought to pursue. Thus, the normative inertia that ordinarily engulfs the policymaker has already been interrupted by the time the pragmatist enters the picture. Indeed, given

Three: Of Policy and Politics

Dewey's tendency to define intelligence as one's ability to "react to things as problematic" (Dewey 1960: 224), he is understandably reluctant to admit the robustness of normative inertia among beings as supposedly intelligent as policymakers. By contrast, my own analysis addresses how the policymaker's inertia might be interrupted in the first place, namely, by showing that even an unproblematic course of action presupposes an account of how the world works that makes the course of action a natural one to take. However, once these presumptive facts are challenged, then the policymaker is forced to sort out explicitly facts from values, and consequently to choose from among a variety of means and ends in the manner that the pragmatist suggests. But that is not the end of the story, for my own analysis can be applied to the pragmatist's, leading to the following question: What does the pragmatist's very strategy of constructing hypothetical imperatives presuppose about how the world works, and what if those factual presuppositions turn out to be false? Again, I raise this question not in the spirit of philosophical one-upmanship, but in order to suggest that even avowedly pragmatist accounts of knowledge may themselves contain empirically dubious premises that need to be ferreted out if the accounts are to prove truly practicable. Unfortunately, there is an additional wrinkle in the story, since policymakers often implicitly rely on pragmatist principles to frame their own inquiries. In other words, even the practitioners may have been misled into thinking that pragmatism is more practicable than it really is!

What I have in mind here is the idea that the track record of a hypothetical imperative consists of multiple cases of single individuals or groups (and pragmatists are crucially indifferent between these two possibilities) who have tried to achieve their ends by using a stipulated means. This seemingly innocent assumption is built into the form that a hypothetical imperative typically takes, namely, a statistical correlation between indefinitely many *independent* events of two types, one type covering those who pursue a given end and another type covering those who use a given means. It follows that certain features of human pursuits are not represented in this analysis: How many people are attempting to pursue a given end or use a given means *at the same time?* With what *other ends and means* are these people pursuing the end and means stipulated in a particular hypothetical imperative? These two questions remind us of the commonplace that no one follows a hypothetical imperative in isolation from other people and other imperatives. A very good example of this shortcoming in pragmatist thinking was discussed in the previous chapter, namely, Laudan's attempt to test 300+ philosophical norms of scientific change against a set

Knowledge Policy

of historical case studies.

The pragmatist misses here what has traditionally been regarded as the source of the normative dimension of such imperatives. Following a tradition that extends from the Scottish Enlightenment, I take the feature that distinguishes norms from ordinary statistical regularities to be that norms enable many agents to pursue diverse projects at roughly the same time by drawing on a common pool of resources (cf. Hayek 1973). According to this tradition, norms emerge out of a concern that agents may unwittingly interfere with one another's pursuits, thereby leading to counterproductive results for all involved (e.g., Coleman 1990 is erected on this premise). Thus, a norm is rarely the most efficient means by which any given agent could pursue her ends; rather, the norm offers a relatively efficient means by which a diverse group of agents can pursue their ends with a reasonable chance of success. Therefore, in order to assess the normative range of the pragmatist's hypothetical imperatives, it is essential that we know the social environments in which these imperatives were operative. It is here that the force of pragmatism as an "experimental" approach to knowledge and value may be felt, but only in a way that goes beyond the pragmatist's own analysis.

In a laboratory setting, the experimenter can control the interactive effects of competing subjects or competing ends and means to whatever degree she deems appropriate, and thereby approximate the social conditions presupposed in the construction of the pragmatist's hypothetical imperatives (cf. Fuller 1989a: chaps. 2-3). However, the pragmatist clearly believes that the track records of the various hypothetical imperatives have been established largely outside the laboratory, in the ordinary course of human history. Yet it is methodologically naive–to say the least–to think that the extent to which a given means has succeeded or failed to achieve a given end is unrelated to how many others have tried at about the same time as well as which other ends and means they were pursuing at that time. Recall our original example of a hypothetical imperative: "If you want to expedite the growth of knowledge, then pick theories that explain the most data by the fewest principles." The validity of this injunction may well rest on the fact that in each supporting historical case there were relatively few competitor principles to explain a range of disparate but relatively well-defined data. Therefore, if upon the announcement of this hypothetical imperative too many scientists started trying to follow its advice, then the proliferation of principles and data domains might end up undermining the strategy as an efficient means of expediting the growth of knowledge (Ackermann 1985). Natural science might

Three: Of Policy and Politics

then start to take on the character of sociology, literary criticism, or pre-Socratic philosophy!

The pragmatist's failure to see these consequences of her position is revealed in Dewey's easy recommendation that the hypothetical imperatives be made available to the public at large. Dewey (1946) presumed that human welfare would be best promoted by involving as many informed people as possible in the knowledge enterprise. I have suggested, though, that the success of many, if not most, of the hypothetical imperatives that can be inferred from the history of science has crucially depended on restricting access to the knowledge enterprise in a variety of ways. It is by no means clear that these imperatives would work in environments more democratic than the ones in which science has been normally conducted. I mean this not as an argument against democratizing science, but rather as a cautionary note about the complexities involved in using history as a basis for making science policy. In our own day, feminists are probably the most alive to this point, especially in their deliberations over whether, in the long-term, the influx of women into science will change how and why research is done (Harding 1986: chap. 3; Harding 1991: chap. 3). In this regard, pragmatist intuitions are represented by "liberal feminists" who do not envision that a massive change in personnel will radically alter the character of the enterprise.

On the other hand, while drawing lessons from history is tricky, I do not want to leave the impression that *all* of our problems will be solved by laboratory experiments on groups of scientists working under a variety of conditions. The general concern that ultimately lies behind my critique of pragmatism is its insensitivity to the *frequency* and *distribution* of a given norm across society. In a word, pragmatism seriously lacks a theory of *power*. Seen in this light, the standard methodologies for studying science have some striking shortcomings. Histories (and ethnographies) tend to overestimate the pervasiveness, and hence constancy and even "naturalness," of a readily observable pattern, while experiments (including computer simulations) commit the complementary sin of taking their circumscribed ability to produce alternative results by changing initial conditions as a sign of the malleable and even "artificial" character of the norms that are currently in force outside the lab. If exclusive reliance on the historical method is likely to engender a conservative politics of science, a similar reliance on controlled experimentation should issue in an impracticably radical politics of science: Mannheim's (1936) ideology and utopia revisited!

So is our critique of pragmatism merely propaedeutic to a theory of power? No, there is also a more positive conclusion. To return to Collins'

Knowledge Policy

studies of experimental replication, instead of concluding that replication is either an unfalsifiable norm or a falsified hypothesis about scientific practice, the policymaker may reason that if replication seems to be, in principle, an effective way of ensuring quality control in the scientific enterprise, then the relevant question to ask is not whether individual scientists do it, or even whether they can do it. Rather, the question is *at what level or unit of the scientific enterprise does or can replication occur*. One way of looking at this new question is as a version of the (X+,Y-) interpretation of Collins' cases: even granting Collins that individual scientists do not replicate experiments, that may go to show only that replication is not the sort of thing that *individual scientists* do. Recent work by Augustine Brannigan suggests how the scientific enterprise may be arranged so as to issue in replication as a collective effect without the deliberate replication of an experiment by any individual scientist: Priority concerns typically make scientists quite secretive in their dealings with colleagues. Indeed, Brannigan and Wanner (1983) argue that such lack of communication may be the main source of multiple discoveries. From a policy standpoint, it may appear that selfish considerations are leading to a wasteful duplication of scientific effort. However, another, equally natural reading of this "wasteful duplication" is as unwitting replication of the discovery in question.

This way of looking at the issue recalls the original context in which one of the pragmatists' favorite arguments, "ought implies can," was made. Nowadays, the argument is typically taken to mean that it is unreasonable to require that people do something that it is not within their power to do. On this construal, a would-be norm can be invalidated simply by showing that the norm is not humanly realizable (cf. Goldman 1985). However, Kant first argued "ought implies can" to quite different effect, namely, to show that if we have principled grounds for believing that a certain course of action is the one we ought to pursue, then there must be some faculty (indeed, one we may have yet to discover) that enables us to do it. On this basis, Kant claimed that there must be a special "noumenal" aspect to our being–a "rational will"–that is subject to the moral order, since it is clear that the ordinary physical aspect of our being is swayed amorally by the passions. While Kant's reasoning here may strike the modern reader as perverse, it nevertheless serves to underscore the inherent ambiguity in using "human realizability" as a constraint on the acceptability of norms. Yet the ambiguity is not an unhappy one, given that experimental psychologists have shown that *individuals* are cognitively ill disposed to follow virtually every norm that has been proposed for rational infer-

Three: Of Policy and Politics

ence in economic and scientific matters. In that case, taking a cue from Kant, instead of scrapping all the proposed norms as just so many falsified hypotheses, and thereby concluding that "man [sic] is an irrational animal" (Stich 1985), we may need to turn from the individual to other "units of rationality" where the norms may have bite, namely, as sketches for computer programs or as blueprints for the organization of cognitive labor (Fuller 1989a: chaps. 2-3). Likewise, the policymaker needs to broaden her imagination as to what might count as humanly realizable.

At this point, it is worth noting that some Kant-intoxicated philosophers claim that there are "unconditional" norms, i.e., norms that bind people in all situations, no matter their ends, and even if the immediate consequences are not particularly salutary. These are often called *categorical* imperatives, as opposed to the condition-bound *hypothetical* imperatives we have been discussing so far. At first this may appear to be one of those typically philosophical distinctions in which everything is distinguished from nothing. After all, if norms are supposed to govern the activities of real people, then the norms must be sensitive to the differences in people's situations, which means that all the relevant imperatives will be hypothetical ones. Right? Not exactly—at least if you believe that there is more to a norm than merely a strategy that gets you what you want on a regular basis. As an act of legislative will, a norm is designed to govern an entire community in such a way that one's status in the community does not affect the norm's efficacy. This is the signature modern method for deriving principles of justice, immortalized by John Rawls (1972) as the "veil of ignorance" from which one operates in the "original position" of political theory. It would seem, then, that an important goal of any normative inquiry is to sort out the categorical from the hypothetical imperatives: Which courses of action can be recommended to anyone, no matter what others do? And which can be recommended only after a survey of what others are doing? But wouldn't this exercise involve more than mere sorting, so that the normative inquirer would be compelled to issue norms of her own?

This last question is admittedly a controversial one, especially when applied to science. Most of the hypothetical imperatives that philosophers invoke as "rational criteria for theory choice" emerged quite without legislation, as individual scientists took advantage of situations that they realized would remain unexplored by most of their fellows. Surprisingly, historians of science have said little about this self-selection process whereby, say, a small subset of a scientific discipline at a given time propose explanatory theories that reconfigure and

Knowledge Policy

unify the data in their fields. One possible reason for the self-selection has to do with the nature of theorizing itself—at least the sort of theorizing discussed here, which Popper believes would engulf science in a "permanent revolution" if practiced often enough. This is the sort of revolutionary theorizing defended in this book, which reconfigures entire fields of inquiry by dialectically overcoming existing disciplinary differences. The import of successful theorizing in this sense—as in the cases of Newton, Darwin, Marx, and Freud—is to reorient the research of one's colleagues, and perhaps even to threaten their livelihoods altogether, if they are unable to adapt to the proposed change in milieu.

Philosophers often forget that scientists are generally taught to "theorize" only in the Platonic sense of constructing abstract mathematical models, but *not* in the more Hegelian sense of attempting a dialectical synthesis. Moreover, the typical context in which a scientist encounters a theory is the textbook, where it is presented, not as a challenge to the current disciplinary order, but as a safeguard against posing such a challenge, namely, as a glorified mnemonic device for keeping seemingly disparate notions related in the student's mind. In short, the incentive to theorize—in the synoptic sense that philosophers have traditionally thought to be essential for the growth of knowledge—has never been explicitly built into the normative structure of science. Theorizing is, of course, not prohibited, but it is definitely a risky venture professionally: the payoffs of success are big (for both the science and the scientist), but few succeed, and hence few try. But what would be the benefit of eliminating this risk by elevating the search for explanatory theories to a categorical imperative of science?

Eliminating the professional risk of theorizing addresses the "social justice" of the scientific community. Because theorizing is not a categorical imperative of science, theorists have normally needed "tolerant" institutional settings (i.e., ones permitting long unproductive periods, and much speculation and criticism during the productive ones), access to which has been haphazardly related to a scientist's ability and desire to theorize. But as with any proposal to legislate for purposes of social justice, the question of access is contestable. For example, opponents (i.e., *theory free-enterprisers*) might argue that the determined theorist, if she is really any good, will ultimately find her niche in one of those institutions.

But once we decide to go the categorical route, we have a couple of options, depending on whether we take into account the possible consequences if everyone attempted to adhere to the norm. On the one hand, I may take it to be an inherent good that scientists theorize, perhaps because I believe that theoretical activity embodies the highest intellec-

Three: Of Policy and Politics

tual virtue, and hence I do not care if science starts to look like literary criticism. After all (so goes this view), if scientists aren't ennobled by doing science, then what good is it? I would thus prescribe that scientists theorize to whatever extent, whenever it suits them. On the other hand, I may be interested in science, not as something that is pursued for its own sake, but as something that serves to expedite the growth of knowledge. Too much intellectual virtuosity can get in the way of collective progress, just as too many cooks can spoil the stew. Given this view, then, I would prescribe only as much theorizing as the knowledge production side of science could tolerate if everyone theorized to the same extent. In other words, the normative attitude toward theorizing would not be so permissive that no agreement could be reached on what was important to explain; however, it would not be so restrictive that everyone would conceive of their task as filling in the details of an already existing theory. At this point, our theory free-enterpriser might observe that the result of striking a balance between these two attitudes is none other than the modestly unifying "middle-range" theorizing that has been the typical fare of scientific journal articles *without the need for explicit legislation*.

Machiavelli Redux?

Can all this talk of legislating and experimenting with the normative structure of science ultimately avoid the charge of manipulation? Manipulation typically presupposes a world in which the manipulable have well-defined interests against which the manipulator then imperceptibly acts. If the scientific community is counted among the manipulable, then one can envisage these interests in terms of the direction in which research would naturally develop without nonscientific interference. However, as I indicated at the start of this chapter, I do not believe that science has any such "internally" or "autonomously" defined interests, and so I deny a crucial presupposition underlying the morally repugnant sense of manipulation. Second, this sense of manipulation is possible only as long as knowledge is inequitably distributed across society, so that a set group can always alter the structure of knowledge production, while everyone else is a passive recipient of its products. But one of the major aims of social epistemology is precisely the breakdown of this distinction between production and distribution that enables the morally repugnant sense of ma-

Knowledge Policy

nipulation to take root in society. Lest the reader find my denial of Machiavellianism too glib, let me now take up the charge a little more methodically.

Suppose that each hypothetical imperative associated with the history of science were shown to capture an effect that is emergent on the scientific labor being divided and organized in a certain way. Thus, no particular scientist would be explicitly guided by the imperative, but the imperative offers the best explanation of what implicitly governs their collective behavior. If the policymaker is interested in maintaining the production of such effects, then she will be forced to gauge the advice she gives to individual scientists in terms of the likelihood that their subsequent actions will contribute to producing the desired effects. In Machiavellian short form, the ends will justify the means that the policymaker selects. Is this line of reasoning objectionably manipulative?

In rough-and-ready terms, manipulation occurs when one person knowingly gets another person to do something unknowingly that goes against her own interest but benefits the first person's interests (cf. Goodin 1980). Clearly, the knowledge policy strategy advanced in this chapter satisfies some of these criteria: the policymaker's bird's-eye view of the scientific process gives her an advantage over the average scientist in determining the overall significance of that scientist's work, and, given the broader scope of societal aims within which science policy must be made, the policymaker's interest is arguably somewhat different from that of the average scientist, and, indeed, the scientist is being made to serve the policymaker's interests. Conspicuously absent, however, is the idea that the policymaker wants the scientist to do something not merely different from, but demonstrably against, the scientist's own interests.

Now, admittedly, a virulent antipolicy tradition exists within the scientific community (largely associated with Michael Polanyi 1957) that would blame all the deleterious consequences of scientific research on meddlesome policymakers who force scientists to act against their better judgment by making funding hostage to the production of ideologically sanctioned knowledge. But while such cases of deleterious consequences are all too familiar (e.g., Lysenkoism, Nazi genetics, the atomic bomb project), it has yet to be shown that, in these cases, there was a clear alternative research trajectory that the scientific community would have pursued had they been left to their own devices. What is more likely is that some other policy imperative would have given direction to scientific research, which, if socially beneficial or neutral, would be credited to the "autonomy" of science, but which, if equally

Three: Of Policy and Politics

deleterious, would once again be laid at the doorstep of meddling policymakers. Obviously, however, this point cannot be settled by speculation alone (for some evidence, cf. Aronowitz 1988). Still, it makes one wonder whether the scientist has any specifically *scientific* interests that are different from those of the policymaker, or whether the differences in interests are entirely non-scientific in nature (e.g., a scientist's interest in receiving a bigger cut of the available funds). The potential for policymakers to be objectionably manipulative diminishes as it becomes more difficult to identify a natural trajectory to scientific research independent of policy considerations of one sort or another.

But now let's reverse the burden of proof: Why would anyone have thought that scientists had a distinct set of interests that could be disentangled from those of the rest of society, and especially those of the policymaker? After all, the idea of "interest" itself is an anthropomorphism that implies that events in the world neatly correspond to outcomes having different values for rival groups. The world, of course, is not so willing to oblige our efforts at totemism. As the Doctrine of Double Effect would have it, every self-interested course of action will be received by others who do not share our interests. Yet, I submit, science policy enables scientists to ignore this point by indulging a deep-seated psychological bias that would not normally be tolerated in other less esteemed groups. The most striking case of this bias is what social psychologists call *the fundamental attribution error*, which explains the asymmetry in the stories we tell about ourselves in relation to those we tell about others (cf. Hewstone 1989: chap. 3). Roughly speaking, when thinking about ourselves we tend to explain the good things that happen in terms of enduring ("internal") personality traits and the bad things in terms of ("external") situational accidents, whereas we tend to explain what happens to others in reverse (i.e., people fail because of fatal flaws in their character and succeed out of sheer luck). Thus, internalism seems to be integral to the construction of self-identity, in which case we might speculate that a good way to identify the dominant perspective–the "hegemonic authority"–in a society would be in terms of whose self-identity story is presumed by all. Clearly, if all classes of people are susceptible to the fundamental attribution error, then social coherence can be attained only by privileging some of the asymmetrical accounts of self-versus-others at the expense of other such accounts. In that case, much of contemporary science policy can be readily seen as privileging the scientific community's commission of the error.

The fundamental attribution error is such an error, not simply because it leads people to draw inaccurate conclusions about the causes of their own and others' behavior. More important, it fosters the illusion that one's self-interest is something that can be discovered by examining oneself and finding, say, a stable personality trace. On the contrary, insofar as "interests" denotes anything at all, it makes essential reference to utilities that exist outside of oneself, cognitive access to which is likely to be no better than to any other external object (Goodin 1990). Indeed, even the interest groups that one identifies with at the beginning of a course of action may not be the interest groups with which one identifies later on, once some consequences of that action have been revealed. Given the apparent psychological irrevocability of self-interest, this point can then be used as leverage in getting scientists to realize that as policy is projected into the indefinite future, one's own interests become less distinct from those of others. As a nuclear physicist, I may want unlimited funding for my field, but that is only if, after the funding period is over, I still plan to be in nuclear physics and, hence, identify with the community that will be receiving the funds at that time. As the quickening pace of scientific change forces a perpetual turnover of specialties, the odds that this will be the case—even without any government directive—diminishes, and hence it becomes more rational, from a purely "self-interested" standpoint, to act in a way that is not likely to harm others in the course of benefiting those who are hypothesized as one's own successors (cf. Parfit 1984).

With this in mind, policymakers can be interpreted as manipulative in a way that benefits the scientific enterprise, namely, by counteracting two sorts of nonscientific interests which scientists *themselves* possess:

> 1. the tendency to see one's own research as the very center of all that is worthwhile in science;
>
> 2. the tendency to satisfy the norms of science with as little effort as possible.

Hobbes would have recognized these two interests in the inhabitants of the state of nature, and would have rightly diagnosed them as being born just as much from ignorance as from desire. In the case of (1), the policymaker arranges funding patterns so as to force scientists to think of their research not as ends in themselves but as parts of larger projects, the realization of which may exceed the cognitive grasp of any of the participating scientists. This policy scenario is most prevalent in attempts to bring multidisciplinary perspectives to bear on pressing but ill-defined social problems. More modestly, the strategy may also be

Three: Of Policy and Politics

applied to consolidate the knowledge base of a discipline whose research has become highly fragmented through specialization (cf. Fuller 1988a: chap. 12). In the case of (2), the policymaker designs accountability procedures that force scientists to endure various probative burdens before being licensed to claim a cognitive achievement as their own. Depending on the extent to which others are expected to rely on the putative achievement, the probative burdens may be as light as simply reporting that one has carried out the appropriate procedures or as heavy as sustaining the scrutiny of other researchers with a vested interest in debunking the achievement or claiming it for themselves (cf. Fuller 1988a: chap. 4). If policy intervention ensures, so to speak, the productivity levels of science in the case of (1), in the case of (2) it ensures quality control.

So far I passed over the trickiest area in which STS may have policy relevance, namely, in terms of the image that scientists have of themselves and their pursuits. Most STS research has earned its scandalous reputation by revealing discrepancies between scientists' words and deeds. Indeed, I would say that STS researchers nowadays take it for granted that scientists are laboring under false consciousness of one sort or another. While this belief could well motivate the STS researcher to intervene in the practices of scientists, explicitly imputing "false consciousness" to scientists is unlikely to motivate them to change their ways. Indeed, if it is agreed that science has met with a fair degree of success, then it would seem that at least some forms of false consciousness have much to recommend them. Consider two hypothetical imperatives in this vein:

> If science is to enable prediction and control over larger portions of the environment, then scientists had best think of what they are doing as probing ever deeper levels of reality (and not simply as applying craftier techniques).

> If science is to produce ever more policy relevant consequences, then scientists had best think of themselves as autonomous inquirers (and not as high-paid civil servants).

Suppose that these two imperatives were shown to work. Would that license a redoubled effort to educate fledgling scientists—and maybe even the public—in the myths of the profession? Here the issue of science policy as *ideological* manipulation is raised with a vengeance. The social psychology of creativity offers some clues as to how to treat this matter. Amabile (1983) has observed that creativity is tied to a

Knowledge Policy

strong sense of one's work as "intrinsically motivated." Given the constructivist approach to knowledge policy pursued in this chapter, what is interesting about this finding is that Amabile presents intrinsic motivation, not as an independently identifiable fact about the subjects' psychology, but rather as a feature of the accounts they give for their activities. Thus, the content of the expression "intrinsic motivation" will vary from person to person, depending on how one demarcates internal factors from those external to one's work. Presumably, the lines of demarcation are up for renegotiation, as scientists come to "internalize" political economy and social accountability as part of their motivational structure. However, this will not be easy, especially as long as scientific language remains autonomous from the greater society.

A Recap on Values as a Prelude to Politics

How does the social epistemologist propose to mobilize STS research in order to alert policymakers to alternative strategies for funding and evaluating research? In a nutshell, the answer is to shake them from their unreflective stance of presuming that the "is" and the "ought," facts and values, are fused together in some "implicit norms" or "natural trajectory" of knowledge production. The desired state of mind would be one in which policymakers do not automatically equate statistically normal behavior with normatively desirable action because they realize that the social construction of facts and that of values pull in opposing directions, and that any perceived sense of "normalcy" is really only a temporary resolution of this tension.

No doubt the image that my answer initially evokes is that of each policymaker negotiating in her own mind (or for her own jurisdiction) which claims will be treated empirically and which normatively, thereby leading to a multiplicity of independent decisions resulting in incommensurable sensibilities about where the fact-value distinction should be drawn. Admittedly, the localistic bias of much constructivist STS literature could easily suggest such a conclusion. It is true that, possibly in the long term, and probably in the abstract, any piece of research can figure in any sort of value scheme, as disparately interested parties find use for the research. And so, in that regard, research, while not value-neutral, may be inherently *value-indiscriminate*.

Three: Of Policy and Politics

But perhaps the best way for the social epistemologist to dispel this lingering image of value-neutrality is to observe that if values are generally defined as "the sphere of freedom" and facts as "the sphere of resistance," then one person's treatment of a claim as normative may turn out to cause another person to treat that same claim as empirical. Indeed, this sort of asymmetry would be definitive of a power relation between the two parties. A good case in point is provided by sociologist Chandra Mukerji (1990), who observes that most "basic research" in the United States is funded not by the National Science Foundation (the only federal agency officially devoted to such funding) but by task-driven agencies like the National Institutes of Health and, especially, the Defense Department. The reason she offers as decisive is *not* any potential technological spinoffs from the research (though there may be some), but rather the role that the research may someday perform as background information to policy decisions. (For example, oceanographic mapping does not contribute to the production of weapons, but it may come in handy to justify where they are placed.) The potential for research to serve this "intelligence" function is enhanced by the fact that the government has privileged, if not exclusive, access to research findings. In that case, the ends that policymakers project as realizable in light of such findings appear as brute facts to a public that has no way of evaluating the government's course of action because it cannot survey the range of options available, which would be a prerequisite to trying to hold the government accountable for the consequences of its actions.

8 KNOWLEDGE POLITICS:

WHAT POSITION SHALL I PLAY?

Philosophy as Protopolitics

The specifically anti-inertial, interventionist cast of social epistemology's policy orientation is closely aligned with the normative mission of philosophy qua philosophy, as opposed to its other historical roles, such as qua inchoate special science (cf. Rorty 1979). Philosophy qua philosophy is protopolitics. At its best, philosophy is at once partisan and nonpartisan. It always takes a stand, but one that is defined oppositionally to the unreflective modes of understanding at a given time and place–no matter how normal, acceptable, or even exemplary–insofar as they prevent people from seeing issues of mutual concern. Indeed, half of a philosopher's problem is always that her interlocutor doesn't already see the problem. Consequently, I have made a point of nodding favorably toward the Popperian and Marxist traditions, both of which–in their ideologically quite opposed ways–have carried this core philosophical attitude into our own day (cf. Adorno 1976). Neither tradition has been especially moved by the "argument from repair," i.e., *if it ain't broke, don't fix it*. Whereas the garden variety Sophist intervened on behalf of someone who had already perceived a problem (much as a lawyer would today), Socrates went out of his way to make people see problems in aspects of their thinking that they would normally treat as unproblematic. He typically managed to get his inquiry off the ground by persuading his interlocutor that seemingly isolated problems of judgment and action, the existence of which the interlocutor would easily admit, were really symptomatic of the same deep conceptual disorder.

The expression "seemingly isolated" in the above sketch of the Socratic strategy signals that what the normative inquirer specifically

Three: Of Policy and Politics

challenges is a *frame of reference*, a *perspective*. The problem, then, is how to get people to see things from a "better" perspective. The scare quotes around "better" immediately signal that we have a rhetorical problem on our hands. The sort of perspective that a philosopher is likely to consider "better" is one that her interlocutor would probably see as "better" only once she adopted it (hence, my discussion of normatively corrective presumptions in Chapter 10). Prior to that point, it seems like an arbitrary imposition of philosophical will, a challenge to the special scientist's disciplinary turf. This challenge is simultaneously a normative and a rhetorical problem, since we need to be clear as to *whose* frame of reference we are trying to change for the better. Consider a feature of this problem that has perennially led philosophers to monger a particular class of norms known as "methods."

In the cases where our behavior is hardest to change, yet the incentive for doing so is great, we often already know the right thing to do. That is, we have *merely intellectual* knowledge of the right way of seeing things. For example, we can nod sagely and discourse volubly about the importance of class, status, and power in determining what scientists do, but we still intuitively evaluate science as if there were nothing more to it than a bunch of individuals running around in laboratories. (The acuteness of this problem can be seen whenever someone says the social character of science is "true but trivial.") The standard solution to this problem is to somehow raise the underlying reality to the level of appearances, or, which amounts to the same thing, to give some perceptual embodiment to our intellectual understanding, or better still, to enable us to intuit what we can now only infer with great difficulty. A method is precisely a way of discounting and reinterpreting our natural forms of experience so as to arrive at the prescribed way of seeing things. For Descartes, a method was essentially a verbal recipe for changing your mind. But for those less sanguine about the mind's native capacity to set itself straight, experimental intervention has been the preferred methodological route. The idea here is to restructure the environment, usually by introducing controls and eliminating distractions, so that one can see the effects of the hypothesized causes.

However, philosophers often forget that specifically Socratic intervention requires conversation, which is to say, the verbal collaboration of those whose minds we would change. Philosophers are typically quite adept at persuading their colleagues that some benighted group of social or natural scientists need to change their ways, as stipulated by the "canons of rationality" or what other authoritative name the norms are given. As for the unsuspecting target population of scientists, however, they are hardly approached at

Knowledge Politics

all. These rhetorical misfirings are no mere tactical blunders, but grounds for concluding that the proposed norms are themselves "invalid." The standard of validity that I am invoking here is the one typically used in psychoanalysis, in which the patient's acceptance of the analyst's account is a necessary (though not sufficient) condition for the adequacy of that account. Once adopted as the patient's own, the account can then serve as the touchstone for her recovery. By analogy, the philosopher of science should not propose norms that would improve the conduct of science *if* scientists were to follow them, but, rather, she should propose norms that would likely gain the consent of scientists and *thereby* improve the conduct of science—even if in ways that the scientists themselves had not anticipated.

In a sense, then, I am proposing a cousin of the formula for expected utility maximization in rational choice theory. Whereas rational choice theorists would have the agent, for each option available, multiply the probability of a possible outcome by the amount of utility she would derive from the outcome (and then elect to pursue the highest product as her course of action), I am suggesting, instead, that a rhetorically savvy normative theorist would multiply the probability that her intended audience can be persuaded of a course of action by the product of the amount of improvement that would result from being so persuaded and the probability that such an improvement would indeed result.

But, to repeat, what I have been characterizing is philosophy *at its best*—not necessarily as it is normally practiced today. The second half of the twentieth century has been marked by philosophy's steady withdrawal from the sensitive business of advising people on what they ought to do. This retreat from prescription has taken two forms.

One line of normative retreat can be detected in the work of some latter-day pragmatists, followers of the later Wittgenstein, Heidegger, and even Habermas' theory of communicative competence: Since the relevant norms are already implicit in what we normally do, we should simply alter the way we understand our practices so that their normative structure becomes more apparent, which will then enable us to correct the few isolated infractions that remain. In this context, norms are often said to be "immanent," which means that we could not get rid of them even if we wanted to. I take this born-again stoicism to be an "adaptive preference formation," the social psychologist's way of diagnosing the attitude of "sour grapes" (Elster 1983).

The second line of normative retreat is associated with analytic philosophy's turn to "meta" issues in ethics and epistemology, in the wake of legal and logical positivism. Since the concern, in this case, is

Three: Of Policy and Politics

with identifying the distinctly "normlike" feature of norms, norms end up being severed from every other aspect of human practice (MacIntyre 1984). This results in a conception of norms that can evaluate from afar but offer little by way of guidance for local improvement. Politically speaking, philosophers afflicted with the second strain are like the utopian socialists whom Marx condemned for espousing an idly "transcendent" radicalism. In short, normative retreat of the first sort undercuts the possibility for a normative inquiry with radical import, while that of the second sort subverts the practical thrust of such an inquiry—the Scylla of Mannheim's (1936) *ideology* and the Charybdis of his *utopia*.

All of this makes the prognosis for a politics of knowledge look very dim. But maybe in our own day, to recall Hegel's image of philosophy, the Owl of Minerva no longer takes flight at dusk because its soul has transmigrated to a more evolved species. For, nowadays, questions of the scope and urgency associated with Socratic inquiry are identified in what prima facie appear to be very *un*philosophical terms, specifically as ones of resource allocation and management, often attached to a strong "ecological" or "global" orientation. However, the sort of practice in which a political activist is engaged when she tries to alert people to the precarious ecostructure that supports their everyday lives is no different in kind from the practice of the philosopher who tries to get people to see deep problems in how they ground their knowledge claims, problems that go beyond the authorizing conventions of particular fields. (The typical range of responses from interlocutors is also similar in both cases.) I do not make this comparison to reassert the continuing importance of academic disputation over the problem of skepticism; nor is my point, *pace* Rorty, that the Big Philosophical Questions have been domesticated by disciplinary conventions. On the contrary, these questions have been reincarnated as a form of political practice, one whose sense of urgency is defined in terms of the threat of environmental despoliation rather than of the inscrutability of the Cartesian Demon's web of deception (cf. Sassower et al. 1990, for a discussion of the role of economics in this reconceptualization.) Does this mean that ecology is epistemology continued by other means? Almost.

Unfortunately, along with philosophy's soul, politically inspired, empirically informed calls to global consciousness have also inherited philosophy's rhetorical incapacities. These open the door to the question introduced in the next section, which sets the pace for this chapter: *Have science and democracy outgrown each other?* A vivid case that lends support to an affirmative answer has been recently offered by Craig Waddell (in press), who has examined the rhetoric of the most

Knowledge Politics

distinguished spokesperson for population control, Paul Ehrlich (1978), author of the best-selling book *The Population Bomb*. Waddell calls him "a modern Cassandra." In brief, Ehrlich engages in what I dub *preemptive contempt* of his audience by confessing at the outset just how difficult it will be to convert them to his position. Ehrlich opens by reminding the reader of the biases toward shortsightedness that evolution has built into the human hardware. The effect is to convey, "You are probably prejudiced against me, but if it so happens that you can rise above your prejudices, here is what I have to say . . ." Thus, in the course of trying to prescribe for our future survival (the utopian vision), Ehrlich casts his gaze above the heads of his audience (the metalevel condescension). A couple of things seem to be going on here. The arrogance of privileged insight is combined with an expectation of failure and maybe even humiliation. Perhaps this self-defeating strategy arises when one knows too much for one's own good, and hence feels confident in second-guessing the audience's negative response. The audience is, indeed, provoked, but polarized as well. Those who came already sympathetic to Ehrlich's case will leave more sanctimonious than ever, as they have clearly overcome their biased biological hardware, whereas those who came opposed or indifferent will leave feeling dismissed and downtrodden, feelings that are hardly the best preludes to constructive action. In the phenomeon of preemptive contempt, we thus get a glimpse of the ambivalent relations between science and democracy, experts and the public. Let us now take a longer look.

Have Science and Democracy Outgrown Each Other?

Is science compatible with democracy? The classical, modernist, Enlightenment answer is yes, because "science" and "democracy" are simply alternative ways of identifying what Karl Popper suggestively called "the open society." The more abstractly we conceive of science and democracy, the more plausible Popper's line seems. In effect, the idea of the open society invites us to think of epistemically and politically desirable states as involving no tradeoffs: as one increases, so too does the other. However, once we try to make this point explicit, doubt begins to set in: Is *advanced* science really compat-

Three: Of Policy and Politics

ible with *maximum* democracy? The distinction that I drew in the Introduction to this book, between plebiscience and prolescience, implies a negative answer to this question—that once we try to operationalize science and democracy as decision-making processes, we find that the two states vary inversely with one another. What accounts for this reversal, and can it be remedied? In what follows, I will address this issue dialectically. After briefly presenting Popper's "straight" conception of the open society, I will show its conceptual instability, and diagnose that instability. The next stage of the dialectic will be a reformulation of the relation between science and democracy that takes this diagnosis into account, but also attempts to show that there is, indeed, an ironic sense in which science is compatible with democracy. It is a sense suited for a "postmodernist" understanding of science and democracy. However, upon closer examination, this postmodernist rapprochement will also be shown wanting, which will, in turn, reveal that the idea of "liberalism" is no less contestable within democratic theory than "knowledge" is within theory of science.

In *The Open Society and Its Enemies* (1950), Popper attempted to show the roots of totalitarian thinking in idealist philosophy, with Plato and Hegel highlighted for especially critical scrutiny. These roots confer legitimacy on "closed societies," so-called because they operate by the principle of institutional inertia, or the presumption that societies strive toward order, that such order is not natural but must be imposed, and that whatever order has been imposed ought to be maintained. This is all in contrast to "open societies," which accept human fallibility as the perennially legitimate grounds for challenging any standing set of beliefs. Such fallibility is the basis of human equality, which, in turn, empowers democratic forms of government. Ironically befitting a Hegelian scenario, Popper follows John Stuart Mill and other liberals in suggesting that the open society begins from within the closed society, as science emerges as an island of free inquiry, which is destined to transform the entire closed society into an open one. This image has led the American sociologist Robert Merton (1973: chap. 13) to identify the normative structure of science with the regulative principles toward which modern democratic societies more generally strive: "communalism," "universalism," "organized skepticism," and "disinterestedness." The vehicle by which society is rendered more scientific is something called "education," but the question remains of how much education does the average citizen need to have before science can safely open its doors to the scrutiny of the rest of society. The instability of the open society ideal is revealed in the two polar directions in which philosophical thinking has gone in response to this question.

Perhaps it would not be unfair to identify a *Left Popperian* and a *Right Popperian* interpretation of the stage of development at which contemporary society finds itself with respect to the goal of a truly open society. Paul Feyerabend (1975, 1979) represents the Left Popperian response in believing that the time is long overdue for science to be made the subject of complete public accountability. He clearly has in mind the forum of classical Athens, in which any citizen could raise any objection to any proposition on the floor for debate. Feyerabend sees the democratization of science as simply the reflexive application of the scientific ethos of free inquiry to science itself. STS researchers, who have observed the ability of scientists to account for their activities when pressed by nonscientists, reach the same conclusions by empirical means. It may be inconvenient for scientists to make sense of their activities to a larger audience, but they are not precluded from doing so merely because of the work they do. By contrast, Michael Polanyi (1957, 1969) would stand for the Right Popperian construal of the open society, arguing that we are still in the early stages of scientific enlightenment, and that science budgets still need to be protected from public scrutiny because, in its rage for quick fixes, the public is likely to pervert the spontaneous course of scientific development. The point that I now wish to raise is that the feasibility of all of these visions of the open society—Popper's, Feyerabend's, Polanyi's—is relative to the *scale* in which science is done. Specifically, the larger the society, and the more extensive the scientific networks in that society, the less plausible any of the arguments for the open society will seem.

Popper's vaunted method of conjectures and refutations works only in small intimate groups, such as research teams, whose members have earned the mutual respect that enables the free flow of making and taking criticism. In order to ensure the prompt feedback that is necessary for one to benefit from criticism, the mode of interaction ought to be face-to-face. Brainstorming comes to mind as an activity that exemplifies the qualities of a Popperian open society, but this is a social structure that is quite uncharacteristic of the signature features of modern Big Science. Indeed, in examining the types of scientific and technical systems that populate society today, sociologists Shrum and Morris (1990) have concluded that while advances in electronic and print media have increased the opportunities for exchanging ideas, such that one can now more than ever incorporate the work of others for one's own purposes, the intellectual contagion spawned by these networks makes the identification, diagnosis, and correction of error more difficult than ever. In short, there may be actually a *tradeoff* between the free flow of information and the feasibility of rational criticism. (On a smaller

Three: Of Policy and Politics

scale, something similar happened during the initial introduction of the printing press to Europe in the fifteenth century: what enabled the cumulative growth of knowledge *also* permitted an unprecedented diffusion of rumor and superstition [Febvre 1982].)

By contrast, Feyerabend's vehement opposition to Big Science presupposes an awareness of scale in a way that Popper's conception does not. Feyerabend follows Rousseau, and the anarchistic-libertarian tradition, in arguing that participatory democracy can flourish only in societies whose homogeneity enables agreement on fundamental value issues, so that any other differences will be freely tolerated. Societies of this sort must inevitably be small, as those who fail to agree on the fundamentals will be encouraged to form their own society. Each society will sponsor its own science, which, because of the size of the societies, will be unlikely to interfere with the well-being of people uninvolved in the sponsorship of the science. The dangers posed by Big Science to modern society are a direct result of its being the beneficiary (as well as protector, cf. Mukerji 1990) of an artificially inflated state apparatus that is able to extract revenue from vast numbers of people for projects about which they are never consulted. An invitingly radical way of reading Feyerabend is that science as the quest for knowledge *disappeared* once it outgrew the dimensions capable of sustaining participatory democracy. Unfortunately, our models for thinking about what has taken its place have yet to catch up with this reality. Regardless of what one makes of the overall Feyerabendian position, there is much to think about in this point.

No one could be more opposed to this sentiment than Polanyi. But once we factor the dimensions of Big Science, the prognosis for Polanyi's picture is not particularly pretty. If science is an open society only insofar as both the means and the ends of inquiry are "free," then the long-range forecast is that increasingly technical equipment will be focused on increasingly specialized issues, whose overall relevance to society will remain obscure for increasingly long periods of time. Assuming that we do not live in a world of infinitely taxable wealth, this will mean that, in the long term, a vanishingly small number of people can be permitted to participate in the scientific enterprise. The start-up costs for a lab alone will price virtually everyone out of the market. To be sure, science will have engulfed all of society in its maintenance operations, as everyone slaves away just to enable a diminishing few to enjoy the sustained luxury of conducting inquiry. This sustained luxury is essentially *the freedom to waste resources* (cf. Hardin 1959, especially chap. 13).

Knowledge Politics

The tenor of this discussion of the scale sensitivity of the open society has been to suggest that science may have expanded to the point of being unrecognizable. However, democracy seems to have been left intact—or has it? The argumentation theorist Charles Willard (e.g., 1990) has done much in recent years to consider the implications of a large and diversified scientific community for the future of participatory democracy. His arguments serve, albeit indirectly, to undermine the corporatist conclusion that we must rely on a rule by experts. Willard (1991) challenges the idea that knowledge claims ought to be evaluated in terms internal to the knowledge production process itself. If the nature of science itself is scale-sensitive, why too shouldn't the nature of *justification*? Here is Willard's line of reasoning, which explicitly takes the form of an argument for the obsolescence of epistemology as the basis for justifying knowledge claims.

> 1. Epistemology offers no guidance for the evaluation of knowledge claims in the modern world, because given the number of knowledge claims that are nowadays made and the differences in background knowledge that the claims presuppose, it is simply no longer rational to try to do as the epistemologist tells us to do, namely, to evaluate each claim on its own merits.

> 2. Under these circumstances, the much-maligned appeal to expert authority starts to make some sense. However, ultimately, we need to get beyond this appeal too, because it is not clear that the "experts" are competent in the area where we tend to want their advice, namely, in the "public sphere," an undisciplined sort of place populated by ill-bounded social problems. It is hardly the sort of place in which one could ever have expertise. *The impossibility of expertise in the public sphere is the best epistemic argument for democracy.*

> 3. But then, how does the public decide on which expert to believe—and to what extent? At this point, Willard proposes a new discipline (or profession, perhaps), *epistemics*, which is a special kind of meta-epistemology—a social epistemology, if you will. In particular, epistemics uncovers the sociopolitical networks that enable someone to command "expert" status on a given issue at a given moment. Since, by the reasoning in (2), there are no true experts in matters concerning the public sphere, the public must judge avowed experts by the sort of net-

Three: Of Policy and Politics

> works that sustain them: Who benefits? Who loses by buying what this expert says?
>
> 4. In other words, the impossibility of making rational judgments about individual knowledge claims implies *not* paralyzing skepticism, but rather that the objects of our judgments must be something other than we originally thought. After all, a public crisis does not disappear just because we are unable to find secure epistemological foundations for our judgments. Epistemics is the field that identifies the pragmatic successor to an epistemology that has outlived its usefulness—assuming that it ever had any—in the modern world.

Willard's reasoning presupposes that there are clearly distinguishable networks of interests corresponding to the epistemic options from among which society must decide. If, however, networks intertwine, so that following the advice of one expert rather than another does not guarantee that the outcomes will be significantly different, then the very point of having a choice—the essence of participatory democracy—gets called into question. And this is a real possibility. After all, even if reality is a social construction, it does not follow that a society's members can determine which of many possible worlds turns out to be the one that obtains in a particular time and place. The error here resembles the one involved in saying that because all officeholders in a democracy are elected by the people, it follows that the people always elect the candidate they want. The inference is faulty, of course, because it confuses the fact that each individual's vote contributes to the election's outcome with the fact that the outcome need not conform to any given individual's expectations—or preferences for that matter (especially if individuals try to vote "strategically" based on a false understanding of what their fellows think). In Sartrean terms, Willard speaks the language of *praxis*, the "socially real," as it were, in which each possible course of action carries a distinct social meaning. By contrast, I am drawing attention to the *practico-inert*, the "really social," i.e., the material residue of social practices that often unwittingly undermines the differences in meaning that those practices try to establish (cf. Sartre 1976). An excellent example of the practico-inert getting the better of praxis is afforded by the history of intelligence testing.

Alfred Binet developed the IQ test as a method uniquely suited for French educational reform, since once it was possible to specify the extent to which a child's cognitive achievement fell below the norm for

Knowledge Politics

the child's age, the teacher could then design an appropriate course of study that would enable the child to catch up. No longer (so thought Binet) would slow learners be written off as intractably stupid. Of course, over the last hundred years, the IQ test has been used by many educational interest groups, most of which have been diametrically opposed to Binet's. Indeed, as the recent work of Stephen Jay Gould (1981) and others vividly shows, an unintended consequence of the IQ test's having passed through so many hands is that it is no longer the passive tool of any particular interest, but rather has become an object of study in its own right (a branch of psychometrics) with standards of interpretation that can be used to decide between the knowledge claims of competing educational interests. For example, arguments concerning the racial component in intelligence are typically adjudicated by seeing whether the difference in means for, say, Blacks and Whites in an IQ test is statistically significant. These developments in the "objectification" of intelligence testing take us far from Binet's original intentions, perhaps even to a point that would make him regret ever having introduced the concept of IQ. Thus, "objectivity" in the scientist's sense triumphs in spite of (indeed, *because of*) the efforts of locally oriented agents who have been characterized in terms acceptable to the postmodernist. The tale is one familiar to sociologists (from Simmel 1964, on "third parties," including money, as mediating agencies) of a means designed for a particular pursuit–in this case, the IQ test–unwittingly turning into the standard by which the success of all pursuits are judged.

How should our normative sensibilities respond to this story? Outrage or acceptance? Take the latter response first, as it evinces the doubly ironic sense in which postmodernist philosophers of science like Richard Rorty (1989) and Jean-François Lyotard (1983) see us as already living in the open society. Yes, our conception of science is fragmented, but that enables the products of science to circulate freely in the marketplace. And yes, our conception of democracy is equally fragmented, but that prevents any one faction from monopolizing the marketplace. The politics of knowledge so ironized captures a certain sort of image of the open society, the "laissez-faire" side of the postmodernist's liberalism, whose noninterventionism is at the same time much more in line with the spirit of Rorty and Lyotard's wanting to, as Wittgenstein would say, "leave the world alone."

> *Equal-in-Principle Liberalism*: Since all viewpoints are "created equal," in the sense that none has any a priori advantage over the rest, it follows that whatever success particular

Three: Of Policy and Politics

viewpoints turn out to have historically will be solely the result of their having adapted to contingencies in the marketplace of ideas.

Like the more strictly political versions of laissez-faire liberalism, this one extends a false hope, which is suggested in the ambiguity of the term "adapted." There is one sense of "adapt," Lamarckian, that implies that individuals can intentionally adapt to their circumstances. This sense is implicit in postmodernist descriptions of agents as trying to capture the spirit of the time and thereby maximize their own advantage. If the agents are properly attuned to the sorts of arguments that will persuade their intended audiences, then they are likely to succeed. However, the second, more Darwinian, sense of "adapt" is not nearly so sanguine, insisting that what turns out to have made various individuals either adaptive or maladaptive to their intellectual environments is an unpredictable and emergent feature of their collective activities in those environments. Thus, while the IQ test began by enhancing Binet's instrumentalist views on education, it turned out to be an even more potent weapon in the hands of his foes, the racialists. And in the process, the statistical trappings of the IQ test have become the part of the standard by which any theory of intelligence is judged.

But there is also a "welfare state" side to the postmodernist's liberalism, a side that believes that a rich culture is one rich in viewpoints. Consider these two variants:

> *Equal-Time Liberalism*: The doctrine that efforts should be taken to ensure that all parties to the conversation of mankind are *always* on an equal footing, no matter what actually transpires in the course of the conversation, including a radical change in the attitudes that the parties have to one another's views.
>
> *Separate-but-Equal Liberalism*: The doctrine that since a viewpoint is valid for the culture from which it arose but invalid (or at least inappropriate) for any other culture, it follows that efforts should be taken to protect the viewpoint of a culture from outside interference.

These two versions of liberalism are basically two ways of translating epistemic relativism into political terms. They are expressions of Left Popperianism. They share a commitment to egalitarianism that is sufficiently strong to justify the application of force to prevent equality

Knowledge Politics

from disintegrating under either (in the first case) the emergence of a dominant voice or (in the second case) cultural imperialism. The implication, of course, is that such disintegration would *naturally* occur without the liberal's intervention. Thus, insofar as the postmodernist is a liberal in either of the above two senses, her social policy is given to a certain amount of artifice. An example of the equal-time case supported by someone like Feyerabend would be to compensate for the meager attention that astrology has received in the recent past by allocating it a disproportionally larger share of attention in the future. An example of the separate-but-equal case would be to ghettoize knowledge production into mutually exclusive domains of inquiry by the institution of technical terminology and departmental bureaucracies. This is the way academic disciplines coexist in the university on a day-to-day basis.

These two species of liberalism propose to enforce some sort of cognitive egalitarianism. As such they suffer from the ancient conundrum of trying to maintain the equality of things (in this case, forms of knowledge) that embody seemingly incommensurable values. Seen through Aristotelian spectacles, the *equal-time* and the *separate-but-equal* case reproduce, respectively, the quandaries of *commutative* and *distributive* justice. In the equal-time case, where we are interested in redressing an earlier injustice between, say, astrology and astronomy, it might at first seem that the desired end-state is the *literal* availability of air time for the two disciplines, as measured by courses given, journals published, and the like. However, if the audience for this air time includes the general public, then arguably one or the other discipline may have an advantage in terms of receptiveness and background knowledge. In that case, it would make sense to grant more exposure to the less advantaged discipline in order for the two to have, so to speak, "comparable epistemic effects." But what would such effects look like? *How plausible* would *how much* of each discipline need to appear? A similar problem arises in the separate-but-equal case, once we consider that astronomy and astrology do not utilize air time in the same way. Contrast the two disciplines with regard to the nature and need for communication among practitioners, as well as the nature and need for special research environments. In Aristotelian terms, distributive justice here would demand that unequals be treated unequally. And so, assuming that astronomy is a much more technology-intensive field than astrology (e.g., advances in telescopes make a difference to the former's knowledge base that they do not to the latter's), astronomy could receive five times the funding of astrology and still argue that its activities are "externally constrained," because astronomers are not being

Three: Of Policy and Politics

funded in proportion to their needs as astrologers are to theirs. The diagnosis: a failure to consider the material conditions for meeting intellectual needs.

Back from Postmodernism and into the Public Sphere

A postmodernist reflecting on the varieties of liberalism under discussion might well make the following response to the so-called deep problems of equality and justice I have managed to elicit:

> Aren't these problems simply artifacts of the rather modernist way in which you have framed the state of contemporary society? After all, you seem to presume that there is a concentrated and finite amount of "concern," "attention," "resources"—call it what you will—that focuses the efforts of the various disciplinary perspectives. In large measure, that seems to be what you (and others, such as Habermas) mean by "the public sphere." But, as a matter of fact, there is no such carefully circumscribed epistemic field of play. The only feature of today's liberal societies that bears any resemblance to this picture of the public sphere is voting, which is increasingly seen as a mindlessly performed ritual, when performed at all. If voting is defined as the focal activity of democratic liberalism, then most of the action would appear to be happening offstage.

Although there is some merit to this critique (which Joseph Rouse originally raised in private conversation), it unfortunately carries with it the suggestion that the problems traditionally associated with the public sphere, such as the existence of asymmetrical power relations, have likewise disappeared. That suggestion, however, is unwarranted. One way of looking at the fate of the public sphere is that the "forums" that used to focus the life of a *polis* are now commercially licensed to the media for purposes of mass consumption. Consequently, reformers—be they politicians or intellectuals—who begin by addressing the public to make a particular long-term policy commitment often end up as disposable media icons, whose widespread exposure serves to outwear their welcome before they have had a chance to make a lasting impression.

Knowledge Politics

Not surprisingly, it is common today to confuse *entertainment* with *influence*, one's ability to command people's time and money on an occurrent basis with one's ability to transform people's underlying dispositions. Yet none of this puts an end to power. It just means that power can be effectively exercised only by people whose projects are insulated from, and perhaps even camouflaged by, the endless circulation of limelight. Thus, the diffuse marketplace atmosphere of postmodernism's "antipublic" is ultimately a playground for Machiavellis (cf. Elster 1989, on market vs. *polis*).

Political naiveté aside, the postmodernist herself trades on a certain epistemic asymmetry, allowing a privileged perspective to her own position that she explicitly denies to her opponent's. If the postmodernist were correct that contemporary democracy lacks a public sphere to center its activities, and hence is no longer the sort of thing that one can centrally (or philosophically) plan, then how could she have come to know such a thing? What sort of global understanding would she have had to attain–what type of surveillance operations would she have had to perform–to reach this conclusion? What would be the implicit "center" of her own conception of contemporary democracy? My point is that the postmodernist needs just as sure and comprehensive a grasp of the knowledge system as the modernist to conclude that any major change in our epistemic institutions would be misguided.

In a brilliant critique of the "reasonableness" of laissez-faire skepticism of public policy, the distinguished political economist Albert Hirschman (1989, 1991) has noted that liberal and radical politics–the politics of change–have typically been bolstered by claims to knowledge that entitle their possessors to construct a new order: "Enlightenment is Empowerment!" or 'The truth will set you free!" (cf. Fay 1987; Sowell 1987). Not surprisingly, conservatives and reactionaries have been able to occlude their own politics by simply disputing the epistemic grounds of such calls for change: Do we really know more now than the accumulated wisdom of the past has taught us? Are utopian promises worth risking a world to which most of us have grown accustomed? While I do not wish to begrudge postmodernism any affinities with right-wing politics (cf. Rosenau 1992: chap. 8), Hirschman's point is still worth making in this context. If our understanding of human affairs is as partial and indirect as both postmodernists and conservatives think, then their criticism should apply doubly to metalevel claims about our own knowledge of this partiality and indirectness: If history allegedly teaches us that central planning always fails, then maybe we should equally distrust the central planning of the historical record that was required to draw that conclusion–in which case, we are

Three: Of Policy and Politics

entitled to at least a modest optimism about the prospects for social experimentation.

But what kind of social experiment should we make to circumnavigate the three liberalisms that today dominate the thinking in contemporary democracies? The question calls for the construction of a new model of the public sphere, or "the forum," to put it most vividly. I will answer this question from the standpoint of social epistemology, and especially the rhetoric of interpenetrability fostered in this book. Postmodernist qualms have not convinced me that the forum has become a bankrupt image. To be sure, postmodernists are right to reject a nostalgic view of the public sphere that finds no counterpart in recorded history. Habermas, Dewey, and even Popper sometimes write as if there were a time when discourses were commensurable and the variety of ends of concern to significant sectors of the population were openly disputed and ultimately resolved. These are clearly pleasant mystifications, which have led postmodernists to jettison the forum as just so much excess normative baggage. On the contrary, I argue, the rise of incommensurability is precisely what has motivated the mass translation and communication projects associated with "reductionism" in epistemology and "the forum" in politics. For an ironic consequence of the ever-increasing division of cognitive labor in society is that more of us, for more of the time, share the role of *nonexpert*. This universal sense of nonexpertise, I maintain, is the epistemic basis for constructing the public sphere today.

One reason why nonexpertise has not been given its due is that it is frequently occluded in both the philosophical (e.g., Stich and Nisbett 1984) and sociological (e.g., Giddens 1989) literatures by discussions of the rational grounds for "deferring to authority" or "trusting the relevant experts." These rather panglossian discussions tend to rest on an unanalyzed conception of *trust*, one that places too much positive value on the fact that we are rarely in a position to scrutinize the activities of our fellows. Consider, for starters, two quite different conclusions that one may draw from the fact that in a highly complex society like ours, we are forced to trust others for things that we are not able to do ourselves:

1. Everyone is as competent in their field as I am in mine.

2. Everyone is as incompetent in their field as they are in mine (or I am in theirs).

Both (1) and (2) are, prima facie, inductions that I could equally make from my own experience. The difference between the two inductions would simply seem to be the amount of interpretive charity that I am willing to bestow on the actions of others. But is that the entire epistemological story? Put the matter this way: What is my evidence supporting each of the following claims?

>1a. I am competent in my field.
>
>2a. Other people are incompetent in my field.
>
>2b. I am incompetent in other people's fields.
>
>3a. Fields are sufficiently similar to each other that I can infer what "competence" would mean in other fields from what it means in my field.

If the reader is like the author, then she will conclude that most of the evidence at her disposal is of a "default" nature, namely, it is driven by the relative absence of evidence that contradicts the above claims. This is hardly the stuff of which robust epistemic commitments should be made. And so, now, consider the following two interpretations of what "trust in action" amounts to:

>A. I have a live option to check up on someone, but don't do so in deference to the presumed character and ability of the person.
>
>B. I have no such option because I lack the time and skill to do so, and so I am forced to rely on that person's judgment.

(B) is more commensurate with our real epistemic situation than (A), but the point is easily masked, as the cognitive dissonance created by our near-universal inability to scrutinize the actions of others is remedied by an implicit lowering of standards for what we expect from those who purport to act on our behalf. Thus, "competence" dissolves into a measure of the number of irreversible errors (the fewer the better). The term need not imply performance significantly better than would be expected of a nonexpert. Thus, given the diminished expectations that have accompanied our "society of trust," it should come as no surprise that experts and nonexperts may perform equally well at so-called expert tasks, as measured by "real world" standards (cf. Arkes

Three: Of Policy and Politics

and Hammond 1986). In short, the egalitarianism required of the public sphere has reentered through the back door!

After (rightly) stressing the pervasiveness of incommensurable discourses in contemporary democracies, postmodernists (wrongly) shift the motivation that one might rationally have for making a knowledge claim from *communication* to *self-expression* (cf. O'Neill 1990). However, if one retains a rhetorical interest in communication–in spite of this admitted incommensurability–then an image of the forum is necessary as a reminder that an expressive environment is self-defeating if its design inhibits responses to whatever claims happens to be expressed. Not only can I not countenance the feasibility of an indefinite number of voices, but I also doubt that any of those voices would want there to be indefinitely many others clamoring for attention. The scarcity of air time presupposed by the forum is, therefore, the mark that communication is of ultimate concern. Thus, it will not be enough to simply argue–as Habermas (1985, 1987) tends to do–that a normative conception of a public sphere is *already presupposed* as an ideal limit to our everyday talk. On the contrary, this is to put matters exactly backward. The need for norms emerges from the material exigencies of the speech situation–the need in real space and time to discipline expression in order for communication to elicit action in a timely manner. In a world without exigence or scarcity, there would be no need for the sort of norms that Habermas so rightly seeks. With all that in mind, I will proceed by, first, sorting out the wheat from the chaff in the three liberal models of the forum.

The strong suit of the *equal-in-principle* model is that it assigns a central role to contingency in the relative standing of knowledge claims, thereby instilling an "adaptationist" mentality in claimants interested in surviving the vicissitudes of the marketplace of ideas. However, the major disadvantage is that there is nothing more to the selection and survival of claimants than the outcomes of their contingent interactions. Here I would say that there must be a normative dimension that interestingly complements this natural state of contingency. In the case of *equal-time* liberalism, the wheat is the sustained concern that all major positions have equal access to the means of knowledge production. Unfortunately, the chaff appears in the form of the tendency to inhibit people from changing their minds in light of the sort of vicissitudes that the first model stresses. That something akin to a monopoly might spontaneously emerge from the marketplace leads the equal-time liberal toward wanting to restrict free trade. Clearly, what is needed here are notions of equality and contingency that do not pull in opposing directions. Finally, the *separate-but-equal* model is to be applauded for

Knowledge Politics

its attempt to preserve differences in positions. But the liberal here seems to be willing to pay the cost of reifying those differences as "cultures," which limits the possibility of redefining differences, especially as new claimants to knowledge enter the marketplace.

From this selection procedure, I conclude that the desired liberal forum is a communicative environment that simultaneously sustains epistemic discourses that are mutually adaptive, indefinitely alterable, and equally available to the public. The norm that I now propose to stabilize this environment is taken from the economist's notion of *fungibility*, that is, the extent to which a good is interchangeable with some other good in a consumer's preference structure. A highly fungible good is one that the consumer is willing to trade for another, under appropriate circumstances. For example, I may possess a lot of food, more than I can eat right now. For the right price, I would be willing to exchange a large amount of that food for a good that would be of more use to me now. Consider, by contrast, the case of the car that happens to be my only possession that is not directly implicated in my survival. I may wish to buy something that costs considerably less than the car is worth. However, it would make little sense for me to take the car apart and present a part of comparable worth–say, the carburetor–to use in trade for the desired good. The reason, of course, is that my car would no longer work, and my trader would probably have no use for a stray carburetor. Thus, in these examples, the car is less fungible than the food. My earlier example of astronomy and astrology suggested how the analogy applies to knowledge production, which I now propose as a principle:

> *The Principle of Epistemic Fungibility*: In a democratic forum, an epistemic discourse must be aligned with practices whose fungibility increases with an increase in the demand that the discourse places on the cognitive and material resources of society.

Epistemic discourses may be more or less fungible, depending on the ease with which their knowledge claims can be translated in nonnative idioms without causing the natives to claim a loss of epistemic value. And so, as high energy physics is typically understood today, its claims cannot be so translated. There are no cheap substitutes for particle accelerator experiments, advanced mathematical calculations, and the like that would enable more people either to participate in physics or to partake of its current budget. Any attempt to find more economical and less discursively formidable means of testing physics claims will be

Three: Of Policy and Politics

met by cries of "vulgarization" on the part of the physics community. Thus, one must sequester funds and expertise specifically *and exclusively* for the conduct of high energy physical inquiry—or not at all. There is no middle position, no room for negotiation. Philosophers of science, Polanyi most notably, have traditionally portrayed this nonnegotiability (or "autonomy," to use the euphemism) as an appropriate aspiration for the special sciences. Even philosophers such as Nicholas Rescher (1984), who fully realize that this goal is bound to be economically unfeasible, nevertheless continue to endorse it on principle. I ask: *What principle?* To underscore the intuition behind the Principle of Epistemic Fungibility, one should stop thinking of Big Science as a self-sustaining and progressive enterprise whose trajectory we may someday be forced to curtail for "merely practical reasons." Rather, one should imagine contemporary physics as a large fossil-fueled industry that became overadapted to an environment that no longer exists and now resists converting to ecologically sounder energy sources. The age and size of physics would thus mark the discipline as a dinosaur whose continued existence—in its current form—threatens the livelihood of other discourses also in need of resources.

The crucial phrase here is "in its current form," for it *is* possible to make physics safe for democracy. This fungible physics would recognize the deeply conventional character of expressing its theoretical claims in certain sorts of terms that are then tested on certain kinds of machines. While it would be easy to show the academic, military, and industrial interests that found it to their mutual advantage to configure physics in this way in the aftermath of World War II (cf. Galison 1987; Galison and Hevly 1992), those interests have moved elsewhere since that time, yet the configuration survives as an atavism in today's world. Fungibility would require a new configuration of verbal and other material practices that either would enable others currently in the forum to pursue their interests (should physics wish to retain its current large scale) or would downsize physics to a point at which its own esoteric pursuits no longer threaten the viability of other epistemic discourses (should physics wish to retain its current autonomy). Along the lines of the first option would be for physicists to encourage social scientists to do the sorts of inquiries that would make it easy to understand and evaluate high-energy physics research as large-scale political, economic, and cultural phenomena. (At the moment, social scientists often professionally suffer when they attempt such intensive scrutiny, e.g., Traweek 1988). Along the lines of the second option would be for physicists to agree to decide their high-level theoretical disputes by using only advanced mathematics or relatively inexpensive

Knowledge Politics

computer simulations.

As a normative model of communication in the public sphere, fungibility, on account of its economic origins, may leave something to be desired. Do I really want to inject "the market experience" (cf. Lane 1990) into the forum, especially its tendency to collapse heterogeneous value dimensions into a money-based standard of utility? However, the market need not be rendered in the image and likeness of neoclassical economics, that is, as a field of utility maximizers in an overall state of equilibrium. A more attractive image of the market has emerged in recent years, an "economic sociology" designed to articulate what Max Weber and Joseph Schumpeter had seen earlier in this century as the source of capitalism's cultural dynamism (Swedberg 1989; Block 1990 provide contemporary statements). According to this picture, the producers and consumers of a good are oriented differently, which makes their interaction in the marketplace always somewhat adventitious. Under normal circumstances, producers are primarily oriented toward each other as they internally differentiate a niche for goods that they suppose consumers will regard as competing for their attention (Harrison White 1981). Consumers, however, are primarily oriented toward types of functionally equivalent goods that place conflicting demands on their appetites. A market is present insofar as producers and consumers orient their activities in terms of each other's projected array of options.

To put the last point crudely, whereas knowledge producers today would like consumers to think in terms of disciplinary alternatives, say, "microphysics versus microbiology" (and hence support the research trajectories projected by one or more of these fields), the knowledge-consuming public (which includes not only government, taxable laypeople, and industry, but also other professional knowledge producers in search of ways of deploying their labor) really think in terms of such problem areas as "nuclear power versus cancer research." This difference is of considerable importance in understanding the role of innovation in reconfiguring consumer needs in a market economy. A "consumer need" cannot be identified independently of the relevant set of goods that consumers take to be functionally equivalent, and hence interchangeable in a given transaction–the original meaning of "fungible." Thus, as the array of rival products changes, so too does the nature of the need. The most successful innovations are not ones that presume the objectivity of consumer needs, and hence some metric of efficiency by which such a need can be better satisfied. Rather, successful innovations reconfigure a market niche by causing consumers to make choices between products that they previously did not take to be func-

Three: Of Policy and Politics

tionally equivalent, which is to say, in competition with each other (Brenner 1987).

Thus, innovation can go in one of three general directions:

> 1. Producers can compete to satisfy an already existing consumer need more efficiently. This strategy is the principal source of competition in economic markets (cf. Layton 1977: 200 on "demand pull" as an explanation of technological progress). But given the control that professional associations exert on the manufacture, sale, and assessment of knowledge products (cf. R. Collins 1979, on "the credential society"), major external (e.g., state-induced) incentives typically need to be in place before this "seller's market" is broken, which would then enable the formation of interdisciplinary coalitions needed for addressing consumer concerns. An obvious case in point is the establishment of well-endowed National Institutes of X (where X is some pressing social problem) that entice researchers away from their pet projects. As it stands now, it is entirely possible for a knowledge producer to make a comfortable living by simply addressing discipline-specific problems.

> 2. Of course, producers may try to reconfigure consumer needs so as to bring them into optimal accord with producer capabilities ("demand management" in Galbraith 1974). Advertising often functions as a precipitant of wants for things which people previously had no desire for. Universities typically operate as de facto advertising agencies for knowledge producers, as they inform students that if they are interested in, say, solving the mysteries of cancer, instead of dealing with that problem directly and comprehensively, they should accredit themselves in some subfield of biology by contributing to one of its standing research programs, which may, with some luck, eventually solve part of the mystery. In this epistemic bait-and-switch, a yearning for civic relevance is all too often satisfied by academic filler.

> 3. Then, there is the sort of radical innovation that so impressed Weber and Schumpeter as the lifeblood of capitalism. The innovator succeeds in reconfiguring consumer need by restructuring the relationships in which producers stand to each other. As STS would have her do, the innovator regards the current organization of personnel and equipment in the scientific

Knowledge Politics

community as a conventionally divided pool that may be redivided to strategic effect. For example, a new research program that promises much without demanding major retooling from interested personnel might incline a wide variety of scientists to invest their efforts, and ultimately turn into a disciplinary arrangement that generates needs and products quite unlike anything previously seen. The strategy worked, for example, to Wilhelm Wundt's advantage in showing that philosophers, physicists, and medical researchers could contribute their expertise to a new science of the mind, "psychology" (R. Collins and Ben-David 1966).

Finally, my formulation of the Principle of Epistemic Fungibility is influenced by the history of mass media law in the United States (Lichtenberg 1990), especially the debates surrounding the so-called fairness doctrine, whereby a public medium is required to offer free response time to someone criticized in the medium. This doctrine has been expanded over the years to include the presumption that a medium will include the major sides of a controversial issue that it plans to air. Interestingly, the expanded fairness doctrine has been criticized from two quite opposite political quarters, but for what amounts to largely the same reason.

The fairness doctrine, which was developed with television and radio broadcasts in mind, has been questioned by those who see broadcasts as sufficiently continuous with print media to be worthy of the same legal coverage. Newspapers are typically not obligated to print the responses of people who are criticized on its pages because readers already understand that a newspaper is a partisan medium. The appropriate course of action for the criticized person is to find a paper symapthetic with her views and to write for it. To obligate newspapers to publish responses would thus serve only to dilute expression and confound public debate.

On the other hand, the fairness doctrine has been attacked from those who believe that certain positions on an issue—especially if they are morally repugnant—do not deserve any air time whatsoever. Like the protectors of a free press, these defenders of censorship also worry about the resulting dilution and confusion, only in this case it happens to be of "right-minded" opinions that may not appear that way unless they receive a clear and exclusive public hearing. Indeed, the main fear of sophisticated censors (e.g., Bloom 1987) is not conversion to a "wrong-minded" opinion, but rather the tolerance that equal access breeds, which leads the public to question the significance of having to make a

Three: Of Policy and Politics

choice between opinions (cf. Jansen 1988).

Both the independent newspaper and the moral censor take for granted a certain lack of natural fungibility in opinions, so that what can be said clearly in thirty minutes can be only said confusedly or dilutedly in fifteen minutes, especially alongside a competing opinion. I have tried to argue here that knowledge production in contemporary democracies cannot afford to make this assumption. And while I do not wish to endorse the "sound-bite politics" that the fairness doctrine seems to promote in our day (i.e., where everyone is limited to thirty seconds so as to enable all six candidates to speak), I believe that we leisure-ridden academics can never be reminded too often that the need for norms governing our epistemic pursuits arises from the sorts of "real world" constraints that the mass media have directly addressed–even if not to everyone's satisfaction.

Beyond Academic Indifference

Academic students of knowledge production are reluctant to recommend courses of action on the basis of their research. One classic humanist argument in this vein asserts simply that if the past is to be understood "as it actually happened," then historical inquiry cannot be subserved to contemporary interests. After all, denizens of the past were addressing each other, not us. To presume otherwise would constitute epistemological malfeasance. Humanists impressed with this argument often recommend an "antiquarian" research strategy that makes the identification of dissimilarities between past and present a scholarly desideratum. The only response that need be made to this argument is that, as a matter of fact, antiquarianism's rigorous pursuit of historical incommensurability has been rhetorically *very effective* in delegitimating contemporary practices by showing that societies have functioned quite well without what is now taken to be necessary for the continuation of our own society. Indeed, the more self-contained the past is made to appear, the more today's trenchant "necessities" look like dispensable "contingencies." Much of Michel Foucault's account of the role of "madness" in European society prior to the emergence of psychiatric and penal institutions is presented in this spirit, which is endemic to the French historiographical tradition from Fustel de Coulanges to the *Annales* School (cf. Fuller 1988a: chap. 6). As with ethnographies that stress the salutary divergences of native lifestyles from our own, an exaggerated, not a mitigated, sense of antiquarianism is likely to jar any ethno-presentist complacency. Thus, the potential corruption of scholarly methods is not the real issue here, but rather how explicit the humanist needs to be in drawing out the intercultural differences implicated in her own line of inquiry. For it is hard to take note of those differences in a completely neutral manner. Much will depend on the standards by which–and the audiences to whom–humanistic scholarship accounts for itself. What sorts of things should one expect to learn from such inquiries?

From the more immediate ranks of STS researchers, social constructivists particularly have made abstinence from policy look fashionably radical, even though they end up playing a familiar "value-free" policy role when they lay out the variety of implications that follow from reading the evidence in different ways. Even granting that, in constructivist hands, the "facts" are multiply interpretable texts that no longer speak in one voice, the studied neutrality of constructivists on policy issues makes one wonder whether they are high-minded ("beyond politics"), opportunistic (available to the highest bidder), paralyzed (anxiety-ridden, in the manner of Pontius Pilate), serenely cynical, or simply oblivious to the fact that neutral acts are still acts, and hence no less interventionist than committed ones. Actually, a rigorous self-consistency is the stance most characteristically adopted by the constructivists, as evidenced by their logically relentless pursuit of *reflexivity*, which will be given a detailed consideration in Chapter nine. But for the time being, let us call the enigma posed not only by constructivism, but by all officially uncommitted scholarship, *the inscrutability of indifference*.

Yet, such inscrutability does not deter policymakers from making use of STS scholarship to suit their own purposes–especially once they have been persuaded to fund some of it. For, like most scholars, those of us in STS are willing to play enough politics to get funded, but rarely enough to take responsibility for the extramural consequences of that funding. Since the case studies produced by social constructivists are designed to show that seemingly ironclad instances of scientific reasoning or technological application can be called into question under sufficiently close scrutiny, it should come as no surprise if these studies are used to slash the budgets of both military and medical research, projects in both artificial intelligence and social work. Thus, the intellectually radical metamorphoses into the politically capricious, as many, if not most, of these cuts would probably be condemned by the constructivists as "private citizens."

Three: Of Policy and Politics

Even to grant that the scholar can live the Weberian dream of neutrally presenting the courses of action available in a given situation is to say nothing about the number of possibilities that should be presented. Here I am talking about the introduction of values, not as something that distorts choice by clouding judgment, but as something that enables choice by focusing judgment. In the fashionable terms of cognitive science, a scholar's value judgment functions as a "heuristic" for her audience. If the scholar (or teacher, more generally) lays out more possibilities than a policymaker (or student) can reasonably be expected to weigh in her mind, she is effectively subverting her charge to motivate action. The move is tantamount to saying that incapacitation, confusion, and arbitrariness are preferable to the following of specific advice if the advice is anything less than foolproof or unbiased. Faced with this scholarly obstruction of action, the audience will be able to act intelligently only if they decide to ignore outright–rather than discount upon reflection–what the scholar has told them.

And even if the scholar does succeed in presenting a cognitively manageable range of options to the policymaker, the scholar's satisfaction with *that* state of affairs–that she has no further obligation to resolve the issue–reveals not so much the suspension of moral commitment as a positive commitment to *moral vagueness*, by which I mean the academic's latent preference for keeping ideas in a state of perpetual play over advocating one such idea for the purpose of changing the audience's mind. It is as if intellectual assent–the judicious nod–were assent enough. And it certainly is, if one is interested in reinforcing the idea of an internal history of science that is subject to its own inertial motion until arrested by an external force. This point would be more apparent if the usual academic presumption about ideas were reversed, so that ideas were regarded as *normally* motivating action, unless willfully prevented from doing so by, say, the sort of moral suspension of practice required to "entertain" ideas. As it stands, however, academics are professionally disabled from distinguishing serious from playful utterance. John Dewey, where are you when we need you!

Admittedly, the alternative to gaining mere intellectual assent, namely, changing minds, is no easy matter, and academics these days are not particularly up to the task. One reason is the fear that open advocacy will be perceived as "dogmatic," the ultimate academic breach of tact. In the public sphere, we normally regard the taking of stands as a sign of intelligent engagement with the world. Not so in the academy, which overestimates "the power of ideas," and so advises a policy of self-restraint as a courtesy to a world unprepared to properly assimilate those ideas (cf. Bloom 1987). As an antidote to this line of reasoning, I suggest thinking of dogmatism as something that is rhetorically accomplished. In other words, dogmatism should not be seen as the inherent property of an opinion or even of the person holding it. Rather, it is a social fact that is constructed whenever a speaker takes a position *and the audience refrains from resisting it*. Lack of resistance inhibits the emergence of a standard by which the position can be held accountable. Such standards typically arise from the rhetorical obstacles that interlocutors pose in the way of their acceptance of the speaker's position. By somehow trying to remove the obstacle, the speaker implicitly acknowledges the presence of a standard of accountability. Advocacy compromises objectivity *only* if the advocate is addressing a captive audience. To the extent that people worry that education is slipping imperceptibly into indoctrination, to the same extent classroom conventions have yet to emerge which redress the asymmetrical power relations enjoyed by the lecturer.

Thus, where charges of dogmatism are lodged one can expect to find a rhetorical vacuum, in which hardly anyone is uttering opinions, and probably little communication is taking place between those who dare utter. Unless public encounter is actively encouraged in the academy, the official policy of bland tolerance is likely only to exacerbate this tendency. Indeed, a very unfortunate by-product of placing the tolerance of alternative viewpoints above all other intellectual virtues is that it leads to what may be called *reverse dogmatism*, or the tendency of all opinion to gravitate toward the epistemology of existentialist theology, namely, the essential irrationality, and hence uncriticizability, of any beliefs that matter to their holder (Bartley 1984). Under such circumstances, any attempt to assert that one's viewpoint is superior in some way to another's is perceived as a charge of "bad faith" against the people holding the allegedly inferior opinion. Hurt feelings, not counterarguments, are the likely result. In many ways, the existentialist scenario is not the worst possible one because I can further imagine an academic culture that becomes so accustomed to intellectual self-restraint that members of the culture unwittingly turn themselves into classical skeptics filled with the spirit of *ataraxia*, the peaceful indifference that comes from not feeling compelled to take a stand on anything, either publicly or privately.

But even more than lacking a taste for public encounter, academics are simply ill practiced in the art of changing minds. They are cursed with captive, docile audiences–students and colleagues–who are largely forced (or paid) to listen to them, usually by an institutional mechanism that is only tangentially related to the promotion of the interests of either speaker or audience. As a result, in this rather ironic

Three: Of Policy and Politics

way, the academic audience listens for its own sake *because the exercise serves no other useful function!* Simple facts sum up the case here. Since the livelihood of academics is dependent not on the size of their audiences but on the bare inclusion of their courses in the curriculum or their papers in conferences, only in a very weak sense do they need to compete for "air time." (This may be more true in the United States than in, say, France, where the boundary between the academy and the general culture is more permeable: cf. Debray 1981.) Moreover, air time itself is implicitly devalued as oral presentations are often heralded and judged as surrogate writing events; hence, lectures are "read." Given such insensitivity to media, one's rhetorical skills can easily grow fallow. An academic is typically under no professional obligation to render her viewpoint as a natural extension of one that the audience already holds, and hence is typically unused to treating the audience's intellectual and material resources as necessary means for achieving her own ends. Indeed, the academic audience is typically made to feel guilty for *its* failure to grasp what the speaker has said.

Policymakers frequently remark on the inability of academics to express themselves in the manner demanded by our harried, postmodern times. The masters of this desired form of communication run management training seminars, which are quickly becoming standard weekend events at hotels throughout the United States. These seminars are intensive, but they are modularized into clearly delineated "mind-bites," so that the audience is capable of chunking the information presented at a manageable rate. The oral presentation is animated, memorable, flexible, and interactive–the management trainers being unafraid to tailor the relevant principles to the needs of the audience. Indeed, they even draw attention to the value of such tailoring, which academics would be inclined to see as instances of equivocation best kept hidden if unavoidable. What the academics miss, however, is that 'The Top Ten Tips to Talk Turkey" are not meant to be vulgarized empirical generalizations about successful negotiation strategies, but rather mnemonics that the negotiator can call to mind to stimulate lateral thinking about her current situation, where the hidden puns and equivocations in the principles serve as the source of opportune associations. In *Social Epistemology*, I spoke about this rather unacademic use of language as characteristic of maxims and aphorisms found in the law and literature. These principles are worded so as to offer the most economical expression of an idea that is intended to have the widest possible application (Fuller 1988a: 204-5). The need to design principles that are both parsimonious and inclusive ensures an interpretive flexibility that prevents the principles from stereotyping the cases to which they are applied, and decreases the likelihood that the principles will be renounced in light of a conclusive test case.

In short, then, academics often fail to impress policymakers–even when they try–because they misunderstand the social function that their words are being asked to perform. Policymakers want to be served up language that can be used as a tool, as part of a course of action, with a clear aim in sight. What that aim is, however, turns out to be more negotiable than academics are usually willing to credit. Indeed, ideally, policymakers would like the academic to offer advice *as if it mattered to the academic herself,* that is, to assume a stake in the outcome of the policy issue under consideration. The academic's failure to take up the challenge is probably a greater source of disappointment and resistance than any hidden agendas or ideological preconceptions on the part of the policymaker (cf. Weiss 1977). Craig Waddell (1990) is right that the devaluation of *pathos* in academic rhetoric is principally to blame here. Consequently, not even self-avowed "socially responsible" academics interested in bringing about policy changes seem to know how to convey commitment in their speech and writings so as to motivate the appropriate sorts of actions. (Ecologists seem especially vulnerable to this charge: cf. Killingsworth and Palmer 1991).

An academic is taught to speak and write as if her audience were going to evaluate her utterance "on its own terms." Thus, the academic is expected to mobilize facts and reasoning that are sufficient for her intended audience to license the conclusion that she wants to draw. Taken at face value, this is quite a weak rhetorical charge. All that is demanded of the academic speaker is that she provide "good reasons" for her claim. At most, such reasons will convince her audience that there is nothing irrational about holding the view expressed, but the audience will still be left wondering why the speaker would *want* to hold such a view, and, more important, why anyone else *should*. After all, a defensible view is not necessarily worth defending. As we will see in the next section, a rhetorically adept speaker typically enables an audience to see her viewpoint as an extension of theirs. This strategy certainly smoothes the passage between intellectual assent and motivated action. Unfortunately, academic communication often seems to move in the opposite direction, namely, toward forcing audiences to accommodate their own agendas to what the academic speaker says. Indeed, academics have been known to wear their rhetorical intransigence as a badge of integrity–or at least disciplinary purity. However, the academic desirous that others recognize her integrity must hope that her audience does *not* follow her own example! Kant and Habermas would be very disappointed by the lack of symmetry in the expectations of

Three: Of Policy and Politics

speakers and audiences in the typical academic speech situation.

A truly democratic rhetoric, one comprehensive enough to cover academic discourse, requires that change of mind not be the product of what may be called the *belligerent syllogism*:

> One of us must move.
> I won't.
> Therefore, you will.

Instead, change of mind must result from the *facilitative syllogism*:

> We're already trying to move in the same direction.
> There is an obstacle in your way.
> Therefore, let me help you remove it.

The form of the facilitative syllogism reminds us that the task of changing minds begins only once the speaker already detects a common core of intellectual agreement with her audience, but the audience has yet to see that agreement as a basis for action. There are two rather opposing strategies that I recommend in this context.

The first strategy appeals to an aspect of everyday cognition for which academics have a trained incapacity. It involves seeing that conceptually unrelated items may be materially inseparable. The use I made of referential opacity in Chapter seven, when critiquing "Managing the Unmanageable," would be an example. Following the human geography literature, Anthony Giddens (1984) has called the process *space-time binding*. The idea is for the academic to show that by acting on her seemingly rarefied point, the policymaker will *also* be in a position to do what she has wanted all along. In the pedagogical appendix to this book, I outline a course designed to make students more proficient in urging what rhetoricians have traditionally called the "timeliness" of action.

The second strategy, however, requires less special training, as it caters to the academic's taste for discriminating essential from non-essential features of an object. To wax Aristotelian, the idea here is to argue that by concentrating too much on the immediate "matter" of the object, policymakers have failed to do what is best to realize the object's underlying "form." Less metaphysically speaking, policymakers tend to fetishize a particular means, while forgetting the end that it is supposed to serve. This distancing of ultimate interests from current courses of action is a strategy whose power is matched only by its versatility. To make the point, consider two arguments that might be made to persuade policymakers that it would be in science's *own best interest* to be downsized.

What is essential to the maintenance of the scientific enterprise, and what are mere accretions on that essence? One argument says that if science does, indeed, aim to increase the storehouse of knowledge, then the quality of communication between inquirers will matter more than the sheer quantity of inquirers communicating. In that case, it may be shown that, beyond a certain number, each additional scientist *reduces* the likelihood that *any* of them will make substantial contributions to knowledge. Conclusion: both established and novice scientists have an interest in restricting their own numbers.

However, what is euphemistically called "personnel redeployment" is rarely a popular cause. Thus, the academic may have to reverse her tactics. This second argument identifies science essentially with its practitioners, and only inessentially with what they do. For example, if science is portrayed as a democratic process that works better as more people's opinions are incorporated, then what is needed is a strategy to maximize everyone's involvement, within resource constraints. Given that high-tech equipment has become so expensive that only a privileged elite can participate in cutting-edge science, policymakers would then need to be persuaded that only an unreflective pursuit of convention leads them to think that advanced theories in physics, say, must be tested by such means. Indeed, reducing the cost of apparatus and training needed for testing theories would improve the quality of scientific judgment by expanding the pool of potential testers.

Postscript: The Social Epistemologist at the Bargaining Table

In what frame of mind should the social epistemologist approach the epistemic bargaining table? How may she ply her interpenetrative trade to maximum effect? At the outset, the social epistemologist must face a "coordination problem": *How does she get her foot in the door without putting it in her mouth as well?* Levity aside, the social epistemologist has to be prepared to justify her existence to audiences jealously guarding their autonomy from unwanted normative incursions. In this context, it is important to bear in mind that she who raises a problem is not necessarily in a privileged position to solve it. If the social epistemologist can persuade her audience to confront a prob-

Three: Of Policy and Politics

lem, then the fact that she was the one who first articulated it is immaterial. What we have, then, is a shared problem, one which all parties identify as their own and hence one requiring a collective judgment.

Next, the social epistemologist needs to appreciate just how much the knowledge production process has changed in the 350 years since the first politically sanctioned scientific societies. The social epistemologist cannot simply assume the mantle of earlier philosophers of science, given the change of scale in the scientific enterprise. A change of scale typically implies a change in causal structure, which in turn implies new pressure points for intervening in the system. An illuminating analogy can be drawn between the three stages in the history of industrial management proposed by geographer David Harvey (1986) and the stages undergone by knowledge policy during the same period:

A. *Traditional*:

Free labor dictated its own terms, as both the conception and execution of work remained in the hands of the same individuals, who together constituted a "guild" with exclusive rights over a "craft." The normative structure of work would be initially passed on through apprenticeship under a master of the guild. Eventually, the individual would be allowed discretionary power over the conduct of her work. Although rules of thumb may be proposed for the performance of labor, these rules will offer little guidance to someone not already a part of the guild. In terms of knowledge policy, this corresponds to the period from the seventeenth to the nineteenth century, when philosophy of science was done primarily by scientists reflecting on their experience as "natural philosophers."

B. *Modern*:

The terms of labor are dictated by management. That work is conceptualized and executed by two mutually exclusive groups of people amounts to a class difference, as labor and management receive different training, and, indeed, are in contact with each other only in the formal work setting, during the evaluation of labor's performance. Thus, management may have little more than a cursory, often stereotypical, understanding of the actual practices of the labor they supervise. Given this pattern of interaction, it should come as no surprise that management tends to think that the execution of any particular task can always be streamlined along dimensions that the task shares with other seemingly unrelated tasks. Also unsurprising is labor's response, which is to resist management's strictures by asserting the heterogeneity of tasks.

The introduction of "external" managerial standards for labor parallels the rise of philosophy of science as a field of inquiry distinct from science in the late nineteenth century. As the positivist movement illustrated most clearly, all sciences were subject to the same principles of evaluation, regardless of content, method, or stage of development. The philosophers were themselves sometimes trained in the formal aspects of the special sciences, but more likely in logic and epistemology. Perhaps the clearest example of the alienation of philosophical conception from scientific execution was the introduction of a strong distinction between the contexts of justification and discovery. In addition, the philosophers explicitly raised the issue of "the ends of knowledge"–often under the rubric of principles of scientific progress–the presumption being that scientists produced knowledge for some larger purpose of which they might be only dimly aware. This axiological discussion corresponds to Frederick Winslow Taylor's original conception of the industrial manager as someone who steers the course of labor in the direction of the "public good," something that was not necessarily served by labor left to its own devices.

C. *Postmodern*:

Today's transnational corporations have highly diversified financial interests spread throughout the globe. Effective management in this context demands a flexible power structure that can exploit new markets as opportunities arise. A flexible power structure implies not only a proliferation of decision makers, each of whom can act in relative independence from the rest, but also a mobile labor force. Today's manager fully realizes that highly skilled labor is difficult to monitor and replace, and so the modernist tactic of dividing the conception from the execution of work will no longer prove effective. However, management can ensure that labor remains loyal to corporate ends by preventing the emergence of any local power

Three: Of Policy and Politics

base. The idea here is to keep labor circulating around different work settings, encouraging temporary collaborations that are designed to develop new products, but at the same time inhibiting the formation of group attachments that could jeopardize the corporation's adaptability to future changes in the market.

The corresponding tendency in knowledge policy is the one promoted by social epistemology. The academic division of labor has rendered absurd the idea of philosophers telling scientists how to run the day-to-day activities of their laboratories—especially by giving all scientists the same advice, as the logical positivists tried to do under the rubric of methodology. Moreover, it is far from clear that this "competence gap" is appropriately addressed by the social epistemologist's acquiring a smattering of formal training in a "hard science," since that would be to reinforce the science-society boundary that STS claims has nothing more than convention in its favor. A fortiori, this point applies to more educationally extended attempts at meeting scientists on their own turf. I question whether the social epistemologist could obtain the degrees and skills that would mark her as a science "insider" without becoming coopted in the process. For the longer one spends in professional training, the more psychologically primed one is to find something worthwhile in it. It is very difficult to learn only negative lessons from one's experience. A subtle site for cooptation of this sort is the science criticism that scientists, such as Stephen Jay Gould (1981), practice.

While scientifically trained science critics are fully aware of the error and deceit that have traveled under the name of "science," they nevertheless are prepared to find fault only with particular individuals. Rarely do they extend their critique to the institutional structure of science itself. Indeed, the institution's tendency toward epistemic equilibrium is usually credited with ultimately uncovering the individuals at fault (Chubin and Hackett 1990 demystifies this ideology as it affects science policy). Thus, in the hands of scientists, science criticism often turns out to be just another opportunity to celebrate science's capacity for self-governance. Readers of such works are able to vent their indignation at the outlaws of science, without any spillover effects that might lead to a reconstruction of the scientific enterprise itself. An apt analogy for capturing the difference in the normative sensibility between this form of science criticism and social epistemology is the

Knowledge Politics

relation in which the Protestant Reformation historically stood to fully secularized European culture.

Yet, strange as it may sound, none of this should encourage social epistemologists to avoid the company of scientists. In fact, following the current fashion in STS training, the social epistemologist should engage in what ethnomethodologists call "participant observation" of scientific practices. In other words, she should learn to ply her trade in the presence of those whose company she is most likely to loathe. That is really the only way to avoid the trap of all Enlightenment projects, namely, *preaching to the converted*.

A gambit true to the character of social epistemology is to try persuading scientists that it is in *their own* interest to become social epistemologists. In particular, scientists need to realize that competence is context-dependent. A scientist is not competent per se, but competent relative to standards of performance, and especially to the control that the scientist has over the circumstances under which she is expected to perform. The patina of expertise enjoyed by physicists and economists stems, in large measure, from their ability to dictate the terms in which they display their knowledge. They always seem to get to play in their own court. But this patina would fade by making interdisciplinary projects unavoidable. For if scientists are required to pool their resources with those in other fields, then the terms of epistemic exchange will need to be continually renegotiated, which means that no group of scientists will be able to gain the sort of power that accrues to workers who routinely have discretionary control over the use of their labor. This, in turn, will force scientific discourse to be intelligible to a larger constituency, and, in so doing, indirectly open it up to greater public scrutiny. The relevant normative instruments will no longer be methodologies, but *incentive structures*–strategies that enable disparate scientists to see that it is in their own best interest to work together on projects that will have generally beneficial consequences for society at large. None of this should be taken as antiscience in the least. It is only to underscore the priorities of social epistemology's brand of Enlightenment politics: Before society can be *scientized, science must first be socialized.*

So, let us say that the social epistemologist has arrived at the bargaining table, trying to mediate between conflicting or non-

Three: Of Policy and Politics

communicating groups of researchers, typically representatives of different disciplines, in order get them to collaborate on some pressing need, be it broadly "political" or narrowly "cognitive." What might she do under the circumstances? To appreciate the types and levels of intervention, let me start by paraphrasing a question originally posed by Georg Simmel (1964) in defining the sociology of conflict: If you see two groups of researchers in conflict, what do you do:

>   a. ignore the conflict (isolationism);
>   b. engage both sides in conflict (jingoism);
>   c. take a side (ideological alliance);
>   d. make yourself essential to *any* resolution?

Alternatives (a), (b), (c) represent familiar philosophical roles. Professionalism over the last fifty years has made (a) an increasingly common response, as philosophers regard their task as the production, correction, and maintenance of philosophical texts–full stop. The difference between (b) and (c) captures, in rough-and-ready form, the legendary antagonism between positivists and metaphysicians, respectively. Metaphysicians would try to build their favorite sciences into the groundwork of reality, while positivists opened up the sciences to as much logical and empirical contestation as possible. The social epistemologist aims to be both more opportunistic and more useful by adopting role (d), which Simmel called the strategy of the *tertius gaudens*.

Strategies for following (d) can be ordered from least to most involvement with the conflicting parties. (This ordering is drawn from a standard model for conceptualizing the role of the legal system in resolving interpersonal disputes. Cf. Golding 1974):

> *Facilitator*: you simply provide a neutral forum for the combatants to work out their differences (philosophical precedent: Habermas' ideal speech situation)

> *Negotiator*: you present each side to the other divested of unnecessarily polemical trappings (philosophical precedent: logical positivism's reduction of claims to their "cognitive content")

> *Arbitrator*: you design the mechanism that resolves the dispute for them (philosophical precedent: Popper's crucial experiment).

Knowledge Politics

Now, what might result from this mediation? Consider four possible outcomes of the current border war being waged between philosophy and psychology which was partially aired in Chapter three. At the risk of sounding cynical, I imagine that movement down this list will accelerate as university budgets tighten.

> 1. Psychology and philosophy recognize that they are engaged in completely different activities that do not yield to direct comparison.
>
> 2. Psychology and philosophy complement each other's activities, thereby enabling an integration of disciplines.
>
> 3. Psychology asserts its authority over philosophy by placing empirical constraints on any adequate philosophy.
>
> 4. Psychology replaces philosophy as its successor discipline.

Moving from general strategies to particular tactics of interdisciplinary mediation, one of the most important obstacles to success is the belief that certain things–such as the truths espoused by a discipline or the character of the knowledge that it produces–are, for all intents and purposes, unchangeable, and hence *nonnegotiable*. One example would be arguments to the effect that philosophy can be only one way because it is in the "nature" of philosophy to be that way. The fallacy here–to confuse what is innate or original with what is fixed forever–can be diagnosed as a violation of STS's Conventionality Presumption.

The rhetorical solution to such nonnegotiability is what may be broadly called *compensation tactics*, which, in turn, may be *prosthetic* or *corrective*, depending on whether one takes the interlocutor's nonnegotiability at face value and thus proposes a course of action to mediate or transform its inevitable effects, or one takes the nonnegotiability as itself negotiable under the right circumstances, say, at the right price. After deploying either compensation tactic, one may then proceed to more explicit forms of persuasion. One can, thus, envisage a *continuum of rhetoric* ranging from the nonnegotiable (and hence coercive), through the compensatory (and hence manipulative), to the negotiable (and hence truly persuasive):

> a. mechanically move the interlocutor to do what you want;
> b. threaten the interlocutor to do what you want;

Three: Of Policy and Politics

    c. pay the interlocutor to do what you want;
    d. subtly change the interlocutor's situation so as to cause the interlocutor to do what you want;
    e. persuade the interlocutor to do what you want by appealing to the interlocutor's interests, while keeping yours hidden;
    f. persuade the interlocutor to do what you want by appealing to the prospect that both your and the interlocutor's interests would be served;
    g. the two of you agree (probably as a result of the interlocutor's changing your mind somewhat) that your mutual interests would be served by pursuing a common course of action.

If the social epistemologist were interested in changing minds as unobtrusively as possible, without opening herself to a possible change of mind, then a compensatory tactic, such as (c) or (d), would be preferred to either (a), (b) or (e), (f), (g). However, cognitive dissonance research suggests that corrective compensations may eventually backfire as the presence of payment or otherwise artificial conditions for judgment may continue to remind the interlocutor of the distance between her "real" position and the one to which the social epistemologist has managed to get her assent. After all, if the interlocutor naturally held the position, why would she need to be compensated for holding it?

For reasons similar to these, the Machiavellian school of political sociology (Vilfredo Pareto, Gaetano Mosca, Roberto Michels, and their followers) has tended to see the power struggle between "lions" who appeal to brute force and "foxes" who appeal to negotiation as ultimately won by the former. Whereas the lions simply eliminate potential opponents, the foxes treat them as potential coalition members, and thus appease them by various compensatory tactics. However, such activities eventually absorb all the foxes' energies, as payments need to be increased in line with the coalition's increasing awareness of its centrality to the foxes' remaining in power. The Machiavellian moral to this story is that the social epistemologist should avoid rhetorical tactics that are *nonreusable*, that is, likely to wear thin over time. It is a rhetorical lesson as well, in that even if the boundary between "natural" and "artificial" (or "internal" and "external") is continually renegotiated, nevertheless it remains essential that an interest which the interlocutor originally held to be "artificial" is subsequently interpreted as part of the interests that the interlocutor considers "natural"; otherwise, the appeal to artificial will eventually wear itself out. Thus, if the original artifice is financial, as in the cognitive dissonance case, then the social epistemologist may have even more of an incentive

Knowledge Politics

than usual to get the interlocutor to see political economy as integral to her activity, so as to divest financial interests of their "artificiality".

Aside from reusability, the social epistemologist needs to be reminded of the *humility* of her own position. That is, the person whose mind you are trying to change may have good reasons to resist your efforts, which, if you gave her half a chance, she would tell you about and which would perhaps even change *your* mind. Within science, the issue of humility has become especially relevant to the notorious inability of psychologists to convince their subjects of the "errors" of their ways in postexperimental debriefing sessions (cf. B. Harris 1988). In a fit of perverse consistency, psychologists have traditionally believed that subjects whose behavior deviates so strikingly from their own folk theories of themselves would probably resist any attempt to acknowledge such deviations. However, the scope of the psychologist's own inquiry comes into question at this point. For if the psychologist believes that she has indeed detected deep inconsistencies in a subject who considers consistency one of her great virtues, then the psychologist would seem to be obliged to have the subject come to see this. Yet I would also guess that an important reason why the psychologist might not care to put in the effort is that she sees the inconsistency as being of little significance for the subject's everyday life–an observation which need not be made to the advantage of everyday life! Not only is this a self-fulfilling hypothesis, but also it is a meta-appeal to humility in order to preempt a more genuine application of the humility principle, namely, one in which both the subject and the psychologist negotiate the exact the relevance of the detected inconsistencies for both science *and* everyday life.

If the social epistemologist is not sensitive to the power relations in which her position is embedded, she too may unwittingly be met with the passive resistance of those she is supposedly trying to help, as they conceal vital information or perspectives. Imagine representatives of two disciplines politely engaged in dialogue with an interdisciplinary mediator, and then going about their business as usual after the meeting. It will thus be important for the mediator to uncover latent disagreements, hostilities, and misunderstandings. The critical literature on *colonialism* is an excellent source for thinking about this entire problem, as colonized peoples tend to use the explicitly cognitive appeals that colonizers make for their authority ("We know what is best for you") as the basis for subtle, usually negative judgments of the colonizers' moral worth (cf. Forrester 1985: pt. II). In sum, then, to avoid the colonizer's fate, the social epistemologist needs to abide by two principles of *epistemic justice* (cf. Rawls 1972):

Three: Of Policy and Politics

> *The Principle of Reusability*: When trying to get someone to change her ways, avoid tactics that are nonreusable, or are likely to wear thin over time. (This captures the pragmatic punch of more ethereal appeals to the "universalizability" of the means of persuasion, namely, that the tactics must work not only here and now but at any place and any time; hence coercion and less than seamless forms of manipulation will not work in the long term.)
>
> *The Principle of Humility*: The person whose ways you are trying to change may have good reasons to resist your efforts, which, given the opportunity, she would tell you and which would perhaps even change *your* mind. (This safeguards against the high-handed tendencies of demystification and debriefing, in which the zeal for remaking others in the image and likeness of one's own theories can prevent the reformer from catching potential refutations of her own theory.)

Finally, it is worth considering that, contrary to the received wisdom of STS, the ethnographer may not be the purest exemplar of humility. As an alternative, consider the *student*. Unlike the ethnographer, who is ultimately interested in the natives in order to have something to bring back to her own tribe, the student training in a particular discipline wants to "go native" largely for its own sake—and not for the sake of some other enterprise, such as the enhancement of anthropology. Moreover, this fact is incorporated in how the natives (i.e., the professors) treat the newcomer (i.e., the student). The tolerance that natives often show the ethnographer disappears once the interlocutor is recognized as no mere visitor but, for better or worse, as a collaborator and perhaps ultimately a successor in the continuation of native culture. Since the stakes are indeed this high in the case of the student, the standards imposed on behavior and its interpretation are stricter. This is an important point, for it is all too easy to confuse the polite tolerance of one's colleagues in other fields with genuine interdisciplinary negotiation (cf. McGee and Lyne 1987). Students are treated more harshly than ethnographers because, in an important sense, the students are taken more seriously by the natives.

PART FOUR

SOME WORTHY OPPONENTS

9 OPPOSING THE RELATIVIST

Ever since Socrates first confronted the Sophists, philosophers have tried to defeat relativism on conceptual grounds as "self-refuting." However, most self-avowed relativists, from the ancient Greek Sophists to present-day sociologists of knowledge, have been drawn to their position on *empirical* grounds, and thus have failed to be moved by Socratic charges of conceptual incoherence. In this respect, relativists are the original naturalized epistemologists (Quine 1985). But, if anything, this makes their position *more* vulnerable, as well as more interesting, to the various empirical disciplines whose research can bear on the relativist's claims. In what follows, I will argue that relativism is, on empirical grounds, an obsolete position for studying science in society, *especially* if one wishes to derive a point of normative intervention on the basis of such a study. In the course of making the argument, I will elucidate the sorts of sociology that social epistemology countenances, and will also settle the score with the problem of "reflexivity" that has traditionally dogged both relativist and normative projects, and which has recently taken hold of the STS imagination.

The Socratic Legacy to Relativism

That Socrates was the most artful Sophist of them all is a recurrent theme in the history of Western philosophy. The idea is that Socrates outwitted his sophistic interlocutors by using their own rhetorical skills (Billig 1987). One trick in particular deserves mention in this context. With only a hint of hindsight, we may say that Socrates managed to persuade his audience to treat *relativism* and *antirealism* as one and the same position. In other words, he got them to confuse the thesis that (epistemic or moral) standards are relative to a

319

Four: Some Worthy Opponents

given locale with the thesis that standards are none other than what one says they are at a given moment. It is a confusion that we continue to suffer from today. Call it the *Socratic Conflation*. Evidence for the Socratic Conflation may be found in the way philosophy students are most often taught to interpret the Protagorean maxim, "Man is the measure of all things." Whereas today the "man" in the expression is taken to mean the solipsistic individual, who is a standard unto himself ("true for me" truth), *anthropos* in its original Sophistic use referred to the "average man" in a community, in terms of whose standards one could tell whether one was in the right or the wrong.

As a dialectical strategy, Socratic Conflation converts relativism from a positive to a negative thesis. And so Protagoras advises that when in Athens do as the Athenians do; however, Socrates interprets him to mean that when not in Athens one need not do as the Athenians do. Protagoras thought he was respecting local customs, but Socrates managed to portray him as cynically trying to appease the yokels. We would now say that Socrates obscured for future generations the possibility that relativism might be aligned with *realism*–that there may be spatiotemporally indexed "facts of the matter." For in successfully reframing Protagorean deference as cynicism, Socrates made it seem as though if a fact is *determinate*, it must also be *universal*. Moreover, Socrates managed to suppress the deep cynicism implicit in his own position. For as soon as Socrates granted the universality of standards, he denied that any particular native understanding of those standards was adequate. Indeed, it was up to philosophy to relieve the natives of their confusions by informing them of the principles that have all along implicitly underwritten their sense of right and wrong.

The fact that Socrates was able to make the Sophists look bad suggests a couple of interesting points about people's psychological reaction to relativism. First, relativism is not the attitude that people normally have toward their beliefs. In fact, it is an attitude that needs to be explicitly cultivated, as when one engages in "disinterested" research into people's beliefs. For example, anthropologists typically have a clearer sense of the differences between their own culture and the cultures that they study than the natives of those cultures would. (Indeed, would it be so farfetched to say that the discipline of anthropology could only have arisen in the West, which since the time of the Greeks has been fascinated by its own cultural identity?) In a similar vein, David Bloor (1976) and Harry Collins (1981) are quite right in seeing sociologists of knowledge as "professional relativists." Second, people would prefer to think that there were universally shared beliefs or standards–even if they have only imperfect access to them–

Opposing the Relativist

than to think that such beliefs or standards had merely local purchase on people's actions. Another way of making the point is to say that what makes norms "normative" is not knowledge of their specific content but the fact that everyone abides by them, whatever their content.

The Sociology of Knowledge Debates: Will the Real Relativist Please Stand Up?

Philosophers have continued to reenact Socrates' original ruse in today's encounters with relativists. A good case in point is Larry Laudan's portrayal of the relativist–designated as a sociologist of scientific knowledge–in his own set of dialogues, *Science and Relativism*. Laudan (1990a: especially 74) is principally concerned with what determines theory choice in science. His version of the Socratic Conflation occurs by having the relativist slide from saying that *nature* does not determine theory choice, to her saying that *evidence* does not determine it, to her concluding that *reason* fails to settle matters. The relativist, then, is made to look like a skeptic and an irrationalist. Laudan makes his job easy by taking advantage of the rhetorical appeals that Harry Collins and other radical sociologists have made to Quine's thesis that data always underdetermine theory choice. By endorsing this thesis, the sociologists unwittingly buy into Laudan's (1977) arationality assumption, which provides a place for social accounts of science only once accounts based on "rational methodology" have been exhausted. The sociologists think that Quine supports their case because Quine seems to believe that the methodological accounts are *always* exhausted. However, this sense of exhaustion leads critics like Laudan to infer that relativists believe that the grounds for theory choice are never more than makeshift.

Now, in Laudan's defense, it must be said that the more radical "reflexivists" among the social constructivists *do* assimilate their relativism to a form of antirealism that opens them to the above charge. The bluntest form of the charge comes as a tu quoque: If it is always left to happenstance which theory should be selected, then doesn't this point also apply to the relativist's own account of science? To their credit, reflexivists such as Steve Woolgar (1988b) readily concede the point, but then try–in classic Pyrrhonian fashion–to convert their dialectical ambivalence into an instrument for destabilizing any presumptions the reader might have about how scientific knowledge is

Four: Some Worthy Opponents

constructed. The "New Literary Forms" that Woolgar (1988a) and his colleagues in discourse analysis have pursued in recent years are Borges-inspired attempts to ensure that the reader's ruminations never reach a resting point. Thus, the reflexivists forsake the "cognitive," or "representational," function of language, in favor of exploiting language's ability to provoke and interrupt thought processes. Whatever they may privately think of the efficacy of this project, Laudan and other logically trained philosophers of science can respect it for its self-consistency: at last, relativists gladly eating their own words!

Unfortunately for Laudan, however, the relativists that he explicitly attacks—Bloor and Collins—are not antirealists, and hence have felt no need to exchange empirical assertion for more exotic forms of verbal expression. Given Laudan's Socratic view of relativism as antirealism, it is perhaps not surprising that he argues with thinly veiled contempt against Bloor and Collins. And although, there is a sense (to be explained below) in which these relativists *do* deny that nature can determine theory choice, they most certainly do not deny that reasons can. Rather, Bloor and Collins restrict the scope in which any set of reasons applies. In other words, they hold that there are no *unconditionally* good reasons for selecting a particular theory. This is normally called the *instrumental theory of rationality*: the justifiability of beliefs is relative to the epistemic constraints under which one operates—in particular, the methods available and the ends toward which inquiry is directed. But this explication puts us dangerously close to Laudan's (1987) own "normative naturalism," whose attendant theory of rationality consists of a set of historically verified hypothetical imperatives.

Truth be told, it may even be argued that Laudan's instrumental rationalist is *more* of a Protagorean relativist than is the image of the scientist who emerges from Barnes and Bloor's (1982) Strong Programme in the Sociology of Knowledge. After all, Barnes and Bloor hold that instrumental rationality is fundamental to the human condition, science simply being a particular set of situations and utilities that frames instrumentally rational action at certain times and places. The Strong Programme's four methodological tenets—impartiality, causality, symmetry, reflexivity—ensure that instrumental rationality can figure in the explanation of any human action, if it could, in principle, figure in the explanation of *every* action (regardless of, say, our approval of the action's consequences). Laudan hardly aspires to such universality. However, this point is often obscured because Laudan samples from the entire history of science for instances of instrumental rationality. But only a few figures and episodes are eligible to be drawn from each period. They include people who, in retrospect, can be seen as having been driven by epistemically appropriate ends–in short, the progenitors we would have chosen as our own. Although a "culture" that encompasses both Newton and today's best scientists is more spatiotemporally diffuse than the paradigm cases of culture familiar from anthropology, Laudan's relativism here is unmistakable. Because he sets stricter conditions than the Strong Programme for the presence of rationality in science, Laudan outdoes his sociological foes in contributing to the image of science as a rather idiosyncratic human practice–the very image that one would expect from a relativist!

Yet it is generally agreed that Laudan scored a major rhetorical coup by avoiding all association with relativism. He succeeded by highlighting certain claims by Bloor and Collins that suggested the irrelevance of nature to the selection of scientific theories. Perhaps the most notorious of these claims is this often quoted one by Collins (1981: 54): " The natural world in no way constrains what is believed to be." Laudan would like us to infer from this quote that Collins holds that we are so embedded in our social constructions that nature can never have any purchase on our beliefs. Now, even if this is what Collins was trying to say, such a belief would not necessarily commit him to a social idealism or solipsism. On the contrary, it would be possible to relate this interpretation to a widely held view among ethologists, namely, that in comparison with other members of the animal kingdom, human beings are sheltered from any direct contact with the forces of natural selection, largely because we are encased in a socially constructed environment within which our behaviors are selectively reinforced. In fact, according to Byrne and Whiten (1987), the perceived complexity of the natural world may be little more than a function of the complex social relations in which one must engage in order to have access to nature. This is true whether one is talking about getting a bite to eat or getting a publishable scientific finding. Byrne and Whiten thus claim to be able to correlate primate intelligence with sociological complexity.

But, as I said, we need to appeal to such a thesis, only if Laudan has got his intended sociological targets right. However, the following quote from Barnes and Bloor (1982: 34) would suggest, however, that this is not the case:

> The general conclusion is that reality is, after all, a common factor in all the vastly different cognitive responses that men

Four: Some Worthy Opponents

produce to it. Being a common factor, it is not a promising candidate to field as an explanation of that variation.

This, I would argue, puts an entirely different slant on things. Nature cannot determine our theory choices because it is *always already* a component of those choices. Barnes and Bloor make this point in the course of arguing against a view often supposed by rationalists, namely, that the scientists whose theories have stood the test of time were somehow in closer contact with nature than the scientists whose theories have not. In other words, Barnes and Bloor want to *oppose*, not support, the idea that epistemic differences reflect ontological ones, which implies that their relativism presupposes, not antirealism, but realism.

Interlude I: An Inventory of Relativisms

The careful reader will notice that I have countenanced at least three different positions that are legitimately called "relativism." For the sake of analytic clarity, I present the following inventory, designed to show three different contexts in which relativism figures in opposition to some other position in science studies debates. And so, what might "relative" mean?

R1: *Local (vs. Universal)*: This is the relativism of Protagoras, Mannheim, and the Strong Programme. It presupposes realism in two senses: (a) there is a fact of the matter as to what is true and false, right and wrong, but this fact is spatiotemporally indexed, often specifically to cultures; (b) all of our thoughts and actions—not just the ones we deem true or right—are grounded in a reality independent of our conceptions, which serves, in Kantian fashion, to convert all questions of metaphysics to ones of epistemic access.

R2: *Indeterminate (vs. Determinate)*: This is the relativism of the later Wittgenstein and more moderate social constructivists of science. It is antirealist in the sense that there is no fact of the matter as to what is true and false, right and wrong, until closure is brought to an interpretively open situation. These episodes of closure constrain the justification—though not necessarily the commission–of future action. They establish *conventions*. There are two general reasons why interpretively open situations might call for conventions: (a) a *surfeit* of competing interpretations, as in the variety of tradeoffs that can be made when no single theory maximizes all the relevant cognitive criteria or no course of action harmonizes the interests of all the relevant parties; (b) a *dearth* of competing interpretations, as when certain conceptual (i.e., theoretical) distinctions fail to make any empirical (i.e., practical) difference, until practices are instituted–such as alternative experimental outcomes–that operationalize the distinction.

At this point, notice that it is possible to be both an (R1) and an (R2) relativist. For example, most moderate social constructivists, such as Collins and Knorr-Cetina (1981), are (R1) relativists with regard to social scientific discourse (and hence are, after a fashion, "local social realists"), but (R2) relativists with regard to natural scientific discourse (and hence are "antirealists," in the sense that philosophers of science normally use the term). What this means, in practice, is that these constructivists respect the integrity of science as a culture, but they refuse to privilege the scientists' own understanding of their culture. As Woolgar and other more radical constructivists have observed, this view suffers from a lack of reflexive consistency, since clearly (R1) privileges the sociologists' scientific understanding of *any* culture.

R3: *Irrational (vs. Rational)*: This is the original relativism of Edward Westermarck (1912), Max Weber, and the logical positivists. It involves a *de gustibus non est disputandum* attitude toward values. It also captures the Pyrrhonian side of the reflexive social constructivists of science. In a backhanded way, this form of relativism presupposes a *deep* ontological distinction between what is real–and hence representable and cognitively accessible–and what is not. Values fall in the latter category because they allegedly rest on subjective choices and emotional commitments for which no independent rational grounding can be given. Verbal reinforcement (i.e., "ethics") and ritual then serve to routinize these commitments, which–from a more objective standpoint–may no better contribute to a society's survival than would some other combination of behavioral and verbal conditioning. However, the ultimate test of a morality is not what some outside observer thinks, but whether the insiders can "live" with its strictures.

Four: Some Worthy Opponents

As a point of reference, the history of anthropology has exhibited all three forms of relativism. (R1) reflects the "idiographic" commitments of orthodox ethnographic method pioneered by Franz Boas and still dominant among symbolic and cultural anthropologists. (R2) captures the reflexive ethnography that "inscribes the ethnographer in her own text" (cf. Clifford and Marcus 1986), and in that way removes the last epistemic vestiges of imperialism. However, in the process, this move may also eliminate anthropology's traditional object of inquiry, the self-contained alien culture. Finally, (R3) may be observed is structural-functionalist social anthropology (Malinowski, Radcliffe-Brown), especially in versions that stress discrepancies between the anthropologist's and the native's perspectives, as in the "latent functions" performed by seemingly irrational social practices. The skeptical side of constructivism results from a reflexive application of (R3), as will become clear in my critique of Malcolm Ashmore's work.

Interlude II:
Mannheim's Realistic Relativism

Two German-Canadian sociologists, Volker Meja and Nico Stehr, have recently provided some assistance by translating the debates surrounding the initial reception of Karl Mannheim's (1936) sociology of knowledge in Germany (Meja and Stehr 1990). Those who have participated in the latest round of the sociology of knowledge dispute–involving Laudan, Bloor, Collins, et al.–would be struck by several turns that the dialectic has taken since Mannheim first met his critics. For whereas today's sociologists of knowledge tend to define themselves as *opposing* philosophy, Mannheim usually tried to blur the difference between the two disciplines. In fact, he displayed his sympathy with the classical philosophical aspiration to universal truth by explicitly opposing antirealist forms of relativism, and proposing instead the doctrine of *relationism*, which states that social conditions determine which truths are epistemically accessible. This doctrine was elaborated in a discussion of the social significance of the sort of synthetic thinking championed by Hegel. According to Mannheim, Hegel was part of a generation that was in a position to pull together strands of thought that were left unraveled by earlier generations. Although Mannheim certainly did not consider the Hegelian synthesis as final, he seemed to think that it marked genuine progress in thought that would not have been possible had Hegel not had specific precursors, and had he not lived in the time and place that he did. The idea, then, seems to be Hegel's very own, namely, that universal truths may be glimpsed only at certain moments in history. Or, to put it as a question: *If there are, indeed, universal truths, then why have we not always known them?* An interesting way to read Mannheim, which some of his critics picked up on, is as claiming that if one takes *very seriously* the idea that certain things are true for all times and places, then sociology of knowledge simply takes up the traditional tasks of epistemology by explaining the differential access that people living in different times and places have had to those truths.

Mannheim's critics tended to raise doubts about whether the sociology of knowledge was equipped to subsume the philosophical enterprise of epistemology. In retrospect, Mannheim's strategy seemed very much like Quine's (1985) "naturalization" of epistemology. Both held that the relevant special science–be it sociology of knowledge or behavioral psychology–can subsume epistemology by showing that the sorts of positions which traditionally distanced epistemology from the sciences (i.e., absolutism, foundationalism) are empirically untenable. Perhaps more than Quine, Mannheim took this to be not a capitulation of philosophy to the special sciences, but rather a consistent application of philosophical reasoning to the point of transcending the disciplinary boundary separating philosophy from the special sciences. (After all, is it not only the institution of philosophy in the twentieth century–and not philosophical thought itself–that clearly demarcates philosophy from the sciences?) Indeed, Mannheim periodically cast his own interest in the "existential connectedness of thought" as continuous with Heidegger's search for existential structures in *Being and Time*. In this way, Mannheim managed to answer most of his critics' charges of relativism.

However, Mannheim failed to stave off the concerns raised by his Frankfurt School critics, Herbert Marcuse and Max Horkheimer (Meja and Stehr 1990: 129-57). They located Mannheim's latent relativism in the sociology of knowledge's failure to specify the sense in which a form of thought "reflects" its social conditions. After all, a body of thought, such as Marxism, may be very much a product of its time, yet it may serve, not to reproduce the existing social order, but to radically transform that order so as to enable a completely different sort of thought to be generated in the future. In other words, Mannheim's implicit sociological functionalism dampened the prospect that substantially different consequences might follow from the political options

Four: Some Worthy Opponents

available in a given time and place. Not surprisingly, then, Marxists have tended to distrust the surface radicalism of the sociology of knowledge as masking a politically quiescent worldview.

Is Relativism Obsolete?

The Frankfurt School's political dissatisfaction with Mannheim's sociology of knowledge can be analyzed in more strictly epistemological terms, and generalized to other forms of relativism. To claim that a knowledge system is adapted, or "existentially connected," to its social context is to suggest that people exert considerable control over their thought processes–probably more than is warranted by the evidence concerning cognitive biases and limitations (cf. Elster 1983). If we set aside cultural differences that are marked primarily on racial grounds, what is striking about the phenomenon of cultural diversity is just how *invisible* it is to most people most of the time. Consequently, when anthropologists try to get the natives to reveal their local customs, the natives often find themselves attending to their behavior in ways that they never did before. Indeed, when anthropologists "go reflexive," they begin to wonder whether they might be subtly coercing the natives to draw distinctions where none exist. This is not to deny that laying claim to cultural identity and difference is a pervasive social practice. What I question is whether the practice amounts to anything more than a mobile rhetoric that is deployed on various occasions to achieve various ends. In short, while the average anthropologist knows enough to put the native's distinction between "good magic" and "bad magic" in scare quotes, she has yet to learn that the same policy should apply to the more seemingly fundamental line dividing "them" from "us."

If what I have said about the rhetorical character of cultural differences is correct, then Mannheim's question should be turned on its head. Instead of explaining what appears, from the inquirer's standpoint, as the *real diversity* of beliefs, the deeper concern ought to be with explaining the *apparent uniformity* that the different believers themselves experience (or, rather, presume). Recall the realist epistemology that motivates Mannheim's enterprise: If there is indeed one reality, or nature, with which we are always in contact, what explains, then, the difference in access to that reality that is implied by the existence of alternative knowledge systems? Now, let us turn the tables on Mannheim's realist presumption by subjecting it to the same test of epistemic access: if there are indeed deeply diverse knowledge systems, which nevertheless affirm a belief in a common reality, why then should we think that instances of such a belief imply the existence of such a reality? For if the mere existence of one world were sufficient to cause different people to experience a world that they presume others also to experience, then there should be no diversity at all. However, the fact that diversity exists suggests that people unwittingly presume different worlds of one another, differences that can be best seen at the group level in the form of spatiotemporally grounded "cultures." The mechanism at work here may be a generalization of the argument that I first made in *Social Epistemology* and will elaborate in the next section: The illusion of epistemic agreement is maintained by a failure to detect real differences that emerge in the process of knowledge transmission.

From an epistemological standpoint, Mannheim's all too easy "adaptationist" approach to the role of knowledge in society is the product of two distinct conflations: (a) between a culture's system of beliefs and its beliefs about those beliefs; (b) between the consequences of one's beliefs regarded abstractly as a system of thought and their consequences regarded concretely as the product of linguistic transmission and other forms of social interaction. In the case of (a), the inquirer's "clarity" about a culture's system of beliefs may give a highly misleading picture of what members of the culture make of those beliefs, if their "metabeliefs" are sufficiently different from the inquirer's. Thus, Mannheim and other methodological relativists fail to consider why they alone (and not the cultures they study) enjoy the privilege of being relativists. The case of (b) points to Mannheim's tendency to ignore the material, unintentional (sometimes counterintentional) character of knowledge-based action. This point will become increasingly important in our critique, as it highlights the empirical ambiguities involved in trying to demarcate a region of space-time "relative" to which a certain knowledge system is "legitimate" or simply just "operative."

Both (a) and (b) appear most noticeably as a blind spot about the critical role of intellectuals in society, one which prevented Mannheim from appreciating the normative project of the Frankfurt School. In particular, because the relativist thinks of culture as a historically and geographically well-bounded unit, every epistemic standpoint must be either "inside" or "outside" the culture under study. The former is said to be "naive," the latter "critical." Taking the metaphor of standing "outside" a culture to its most literal extreme, Mannheim (1940) ultimately characterized the intelligentsia as "free floating." Although

Four: Some Worthy Opponents

the Frankfurt School is not normally regarded as the most realist or materialist of Marx-inspired intellectual movements, its Hegelian reliance on a reflexive, or embedded, conception of critique makes it just the right antidote to Mannheimian relativism.

Here is what I take the Frankfurt critique of relativism to be: Once it is realized that knowledge is embodied in action (or, more precisely, in the disposition of people to act), and that action has consequences that transcend the intentional horizon of the original agents, then it is possible to gain critical leverage over the members of one's own culture —namely, by having come *after* them in history. Of course, this is not to preclude the possibility that today's critic will be surpassed by one in the future who can comprehend the first critic's blind spots. The point is, rather, that one cannot underestimate the epistemic advantage that accrues to someone who stands at the end of a sequence of events. Sometimes, in a Popperian vein, this state is said to enable one to "learn from mistakes," but this way of putting matters is too strong, as it suggests standards of performance that do not vary over time, completely accurate recall, and other historically and psychologically implausible assumptions. More modestly, the critic need only say that she sees things her predecessors did not. In any case, the burden of proof is squarely on the relativist to explain how history is incorporated into societies that have existed for any length of time. That is to say, relativists typically forget to include a notion of *institutional memory* (cf. Douglas 1986) in their conception of culture. As a result, they end up treating all moments in the history of a culture as epistemic equals: as far as relativists are concerned, one may have come at any point in the sequence.

One conclusion that emerges from this argument is that there is something empirically misbegotten about ongoing epistemological disputes between relativists and realists or rationalists. Do particular communities devise standards for evaluating knowledge claims? The answer is, of course, yes. But, *pace* relativists, it does not follow that those standards are used primarily to judge current members of that community. In other words, the context of *evaluation* and the context of *conduct* are quite different. If one is already a member of good standing in the community, then charity is more likely to operate in interpreting any disparity in the person's behavior. Thus, outrageous sounding hypotheses may be entertained by a scientific community a little longer when a Ph.D. utters them than when a mere B.A. does. However, if one has yet to prove oneself, then stricter, more "official" standards of evaluation apply. Under those circumstances, accidents and innovations are more likely to be seen as the products of ignorance and error. And as our critique of Mannheim suggested, such official standards also figure in judgments made about one's predecessors. In any case, these standards may well be quite different from the norms that implicitly govern the behavior of the community's own members when they are not under especially tight scrutiny.

Because a community's official standards tend to be used to judge various sorts of people who had nothing to do with their design or ratification, the standards achieve an aura of "independence" that gives heart to the realist–especially if a very wide array of people are so evaluated. Here the relativist rejoinder is on target: "Independence" in this sense mainly reflects an absence of resistance to the evaluation made of the groups in question. It remains to be shown whether there is anything else going on (cf. Latour 1987). Of course, there are many possible reasons for this lack of resistance, including the relative powerlessness of the groups in question and the indifference of those who are in power. (Who speaks for the past but zealous exegetes?) But such powerlessness and indifference should never be confused with outright acceptance of an evaluation (cf. Fuller 1988a: 207-32). In short, from an empirical standpoint, the battle between relativists and realists is most fruitfully seen as being about how people come to speak for other people (not necessarily themselves).

A crucial antirelativist assumption in the foregoing analysis is that the principles governing a society need not coincide with actors' construals of what those principles are. This would seem to commit me to an especially virulent form of sociological realism–"eliminative sociologism," as patterned after Paul Churchland's (1979) antipsychologistic "eliminative materialism." In other words, there is a fact of the matter about a society's epistemic practices that may elude that society's members. Indeed, members of the society may normally act on the basis of an empirically false "folk sociology" that functions as a kind of "false consciousness" (cf. Fuller 1988a: App. B). With the exception of cultural anthropology, the social sciences have typically justified their existence with a claim of this sort. In any case, epistemologists need to explain how it is that knowledge producers continually do things with which other such producers find fault, whether it be an error, a failure to persuade, or simply a failure to communicate.

One plausible way of casting this situation is to say that the identity of epistemic practices is very much like the identity of stock market trends: they are constituted in the course of being anticipated, or "guessed at," where the guesses pertain to what other relevant people

Four: Some Worthy Opponents

will guess. Because feedback from the guesses is often delayed and imperfect, the market is prone to display considerable volatility, which would lead to complete financial collapse if government did not insure the legitimacy of the transactions. This "Keynesian" perspective provides some justification for the office of epistemologist as someone who does something useful that individual knowledge producers or knowledge-producing communities could not themselves do. Moreover, we need a Keynesian–rather than a strictly socialist–approach to knowledge production because the fact that all the knowledge producers do not have the same sense of what the epistemic practices are, and indeed none may have a particularly good grasp, does not prevent the emergent result of their activities from turning out much of the time to good epistemic effect. Yet to say that the knowledge enterprise often works by means of an "invisible hand" is not to downplay its social character. If anything, it is to reinforce it, for if everyone had the same epistemic practices, then it would be possible to study a randomly selected individual to understand how the entire knowledge production process works (cf. Wrong 1961).

While the last point may seem obvious, it nevertheless cuts against the desirability of a political stance traditionally associated with relativism, one whose most articulate proponents have included Jean-Jacques Rousseau and Paul Feyerabend. The stance goes by a number of equally misleading names, including "libertarianism," "anarchism," and even "democratic communism." However, the outlines of the view are clear enough. Communities are portrayed as voluntary associations sufficiently wellbounded–and perhaps even spatiotemporally isolated from other communities–that both the possibilities and the outcomes of actions are surveyable by the members of that community. In that case, action is readily treated as a projection of the collective beliefs and desires of the community. If the community's actions have unforeseen negative consequences for other communities, then it follows, on this view, that the community in question is too large, or at least is having impact on those from whom consent has not been secured. The proposed remedy is for the community to restrain itself in some way, perhaps by splitting up into smaller, more homogeneous units that can survive without unwittingly involving the lives of others. The flaw in this political vision is twofold. On the one hand, it overlooks the point I earlier raised against Mannheim, namely, that apparent uniformity in beliefs can mask real diversity that, when finally articulated in the political arena, turns out to be a major source of "betrayal" and "disappointment" (cf. Hirschman 1982). Ostracism would become a routine activity, as it was in the Greek city-states (cf. Winant and Ross 1991). On the other hand, the politics of relativism neglects the fact that people with beliefs radically different from one's own can do things in remote times and places that end up limiting, if not jeopardizing, one's ability to act.

As sociologists–though not yet sociologists of knowledge–have turned increasing attention to the "globalization" of the human condition, some interesting diagnoses have been offered for the persistent popularity of relativism. Most have pointed to the "reactive" character of relativist epistemology and politics, partly born of resentment and partly of nostalgia. In particular, these diagnoses point to a sense on the part of relativists that they are losing control of their own fates to forces that they do not fully understand. Thus, Wallerstein (1990) interprets nineteenth-century nationalism, with its emphasis on a historically segregated, geographically bounded "homeland" or "society," as a backlash to the homogenization processes of the capitalist world-system. Moreover, as Sztompka (1990) has suggested, relativists try to foster the illusion of distinct peoples with distinct causal lineages by artificially maintaining local modes of understanding, long after contact with other cultures have rendered them obsolete. Indeed, the evolution of trade languages ("pidgins") into more generally applicable forms of communication might prove a useful source of models of how people from different communities come to understand, accept, and express the fact that they are bound together in a common fate. It is this ecologically minded ethic, rather than respect for local sovereignty, that is likely to foster the mutual calibration of interests and standards that characterizes the global consciousness appropriate for our times.

In conclusion, I would hate to leave the impression that I see no use whatsoever for relativism in today's world. On the contrary, I believe that a certain form of relativism is in fact quite necessary for "the pursuit of truth," in the way a realist might understand that expression. First, there is the point recently elaborated by Stich (1990), that given the infinitude of truths for any possible domain of inquiry, to urge that one simply "maximize the truth" is to offer no guidance for action, since virtually anything one might do is compatible with the injunction (including covering up errors in the short term in the hope that they will cancel each other out as one approaches the truth). Indeed, most of the truths that philosophers have regarded as epistemically most valuable have emerged as unintended consequences of attempts at satisfying local interests, specifically as the resistance that comes from the mismatching of means to ends (Popper 1972). One might call this the *counterpragmatic* or *disutilitarian* theory of truth. As a social

Four: Some Worthy Opponents

phenomenon, truth first appears as individual disutility, but ultimately it contributes to the maximization of group utility. In other words, *if everyone benefits from one person's error, then a truth has been produced*. Thus, should the naturalized epistemologist remain interested in prescribing methods for maximizing truth acquisition, then she should focus her energies on designing *knowledge maintenance systems* that can disseminate relevant information about an individual's error to those members of her group who are likely to be in a similar situation in the future, which would then improve what management theorists call the community's "living organizational memory" (cf. Argote and Epple 1990; Engestrom et al. 1990). The sphere of computer software engineering known as "knowledge acquisition" already designs systems of this sort for such undervalued epistemic communities as the telephone company (e.g., Terveen 1992).

Thus, while it is clear that I believe the pursuit of truth is best understood as a social practice, I do not draw from that the relativist conclusion that any social practice has to be accepted as it is. In fact, the sorts of practices that advertise themselves as pursuing "truth for its own sake" may be the very ones whose social organization is most epistemically suspect, precisely because they do not receive enough external resistance. One need not impute conspiratorial thinking to the forces of Big Science to observe a couple of contexts in which the rhetoric of autonomous inquiry has transformed even *de jure* realists and rationalists into *de facto* relativists. One is the widespread belief among science policy advisors that if an expensive scientific project does not actually harm the citizenry and offers the vague hope of beneficial technologies, then it deserves, ceteris paribus, to be maintained at current levels of funding. Another is the somewhat subtler phenomenon in which scientists acquire "adaptive preference formations" as they come to identify the epistemically relevant aspects of their craft with those over which they have relatively direct control, which are, of course, manipulable by the political environment in which scientists find themselves (Fuller 1989a: 161-62). Thus, it would be interesting to see if scientists would continue to see such a sharp difference between the "intellectual" and the "economic" value of research if they were solely responsible for raising and distributing their own capital. Both rhetorical contexts impede the pursuit of truth by encouraging inquirers to turn a blind eye to their social setting. The remedy requires social practices that counteract these rhetorics in the strategic manner of someone who truly believed that knowledge is a product of its social organization. Perhaps a good name for this remedy would be *counterrelativism*.

Counterrelativist Models of Knowledge Production

GENERAL WAYS OF THINKING ABOUT THE
INTERPENETRATION OF SCIENCE AND SOCIETY

In *Philosophy of Science and Its Discontents*, I advanced some proposals for overcoming the idea that the "cognitive content" of science is something other than its "social context." I tried to make good on Bruno Latour's (1987) insight that science has incorporated all of society into its networks–so much so that to claim that science is done only by technicians in laboratories is just as misleading as to claim that finances are transacted exclusively by tellers in banks. In order to capture the totalizing character of science, its sociocognitive identity, I suggested that the variety of solutions offered to the mind-body problem provides the appropriate model for understanding the possible interrelations of "cognitive" and "social" factors in science. In terms of that debate, most philosophers and sociologists remain "dualists" of some sort, in that they presume cognitive and social factors to be separable entities. Thus, philosophers imagine that knowledge and reason subsist independently of any social embodiment, whereas sociologists still tend to see knowledge and reason as the epiphenomenal projections of social factors. By contrast, there are few confessed dualists these days in the philosophy of mind. Instead, one finds functionalists, reductionists, and eliminativists, all of whom believe that mind is, in some sense, a property of certain arrangements of matter. Likewise, social epistemologists hold the correlative view that knowledge and reason are ways of embodying certain kinds of social relations. Consider these possibilities that follow from pursuing the mind-body analogy (a good overview is Churchland 1984):

> *Functionalism*: Any of a variety of social structures–though probably not all–can instantiate a given cognitive relation. For example, we are psychologically ill disposed to falsifying our own theories; hence, it is unlikely that Popper's falsification principle would be instantiated in individuals. It is also equally unlikely that criticism will be effective if individuals

Four: Some Worthy Opponents

do not receive prompt unequivocal feedback from people whose judgment they respect. In other words, cognitive relations are sensitive to the spatio-temporal dimensions, or *scale*, of the social enterprise in which they are embedded.

Reductionism: The categories of social and cognitive accounts of science diverge, not because they refer to ontologically different features of science, but simply because they have not been developed in conjunction with one another. Just as psychological states can be more closely monitored and refined if we attend to their physiological correlates, so too the cognitive character of science may lose its seeming disembodiment if we attend to the social circumstances in which cognitive claims are made.

Eliminativism: Although cognitive categories are the vehicles by which scientific claims and practices are officially justified, in fact, those claims and practices can be best explained and predicted solely on the basis of social categories. For example, one explains the widespread acceptance of a scientific theory, not by a common perception of an underlying reality, but rather by examining the social mechanisms of belief acceptance. These may be quite heterogeneous across cases, leading ultimately to a denial that there was some common cognitive content upon which all sides agreed.

The first social epistemologist to break away from the dualist mindset was Karl Popper, who clearly envisaged scientific rationality, not as a detachable abstract logic, but as an embodied community of "conjecturers and refuters." One interesting consequence of adopting this more monistic mindset is that it enables one to think of science either as transpiring throughout society (e.g., each purchase you make helps decide between rival economic theories) or as a site for reproducing all the major institutions in society (e.g., science contributes to both family breakdown and capital development). These options are in addition to the two that will be raised in the next subsection: to wit, the complementary character of *standardization* and *incommensurability* in knowledge transmission. That is, on the one hand, society is reconstituted in the image of science through the introduction of standards for speaking and acting correctly, while, on the other, science itself acquires the marks of modern society's own diffuseness as science is communicated from context to context. All four options, then, are distinctive contributions of STS. If we take them all equally seriously,

Opposing the Relativist

	SCIENCE SOCIALIZED	SOCIETY SCIENTIZED
SCIENCE/SOCIETY AS PLACE	Science is the site for reproducing social institutions	Society is the testing ground for scientific theories
SCIENCE/SOCIETY AS PROCESS	Science becomes diffuse as it crosses social contexts	Society becomes standardized as it comes under the rule of science

Figure 5. How To Make The Science-Society Distinction Disapper

then any residual "ontological" difference between something called "science" and something called "society" should disappear.

A MODEL OF KNOWLEDGE PRODUCTION SPECIFIC TO
SOCIAL EPISTEMOLOGY

Social epistemology requires an appropriate conception of the "social." In *Social Epistemology*, I pursued this issue in terms of an expanded reinterpretation of the incommensurability thesis. Unlike most recent philosophers of science I believe that incommensurability is a real and unavoidable feature of modern knowledge systems. Indeed, I regard the incommensurability thesis as Kuhn's most important contribution to the understanding of science. I hold that the main source of conceptual change is the emergence of undetected differences in the way words are used, which is, in turn, a natural consequence of the expansion and proliferation of epistemic communities. In short, as the knowledge system grows, normative control becomes diffuse, as it is exerted less through face-to-face interaction, and more through the evaluation of official accounts. This ascendance of the written over the oral display of knowledge enables the standardization of scientific discourse at the cost of

Four: Some Worthy Opponents

permitting an enormously wide range of activities to travel under a common rhetoric. This explains, for example, the appearance of "consensus" that characterizes a scientific community during its "normal" phases. A good way of seeing my point here is through its reinterpretation of a cardinal tenet of scientific realism, namely, that if a theory is true (or, true to some extent), its truth (or, the extent to which it is true) is the best explanation for its acceptance by the scientific community.

Why do realists take a theory's truth–insofar as the theory is true– as the best explanation for its acceptance? The concept of *acceptance*, like that of justification, presupposes a broader range of scientific contexts than the ones involved in the theory's "discovery." Since these contexts are generally embedded in rather diverse social circumstances, realists suppose that the theory must have some "content" that remains true across these circumstances. Moreover, invariance of this sort is often taken as necessary for the transmission and growth of knowledge. However, as I see it, this line of thinking confuses the uncontroversial claim that the truth itself does not change with the more controversial claim that the truth is transmitted intact by reliable linguistic means. Whereas the latter claim is about the *medium* of communication, the former is, so to speak, about the *message* it conveys. For, if it can be shown that the linguistic means at our disposal to transmit truths over time and space is less than reliable, then whatever invariance we seem to find in scientific theories accepted across socio-historical contexts is likely to be due not to the invariant nature of the truth transmitted, but rather to cognitive mechanisms that mask the differences in interpretation that would have naturally resulted from the theory being *un*reliably transmitted to different times and places.

A good way of seeing what I am doing here is as reflexively questioning the fundamental assumption of Karl Mannheim's (1936) sociology of knowledge. We have seen that although Mannheim is often mistaken for one of those self-defeating relativists that philosophers since Socrates have loved to criticize, he in fact saw the need for a sociology of knowledge arising from the following mystery: If there is indeed one reality, or nature, with which we are always in contact, what accounts, then, for the differences in access to that reality implied by the existence of alternative knowledge systems? As I indicated earlier, I want to turn the tables on Mannheim's realist presumption by subjecting it to the same test of epistemic access: If there are indeed deeply diverse knowledge systems, which nevertheless affirm a belief in a common reality, why then should we think that instances of such a belief imply the existence of such a reality.

Opposing the Relativist

My model of the above process—one in which the transmission of knowledge content is not invariant, but diffuse yet undetected—involves what cognitive psychologists now see as an interaction of *hot* and *cold* mechanisms in the knowledge process (Elster 1983; Elster 1985: chap. 8, discusses this in the context of Marx's theory of ideology). Hot mechanisms are interests and passions that externally drive, or "bias," rational cognition (sometimes to self-destruction), whereas cold mechanisms are such internal cognitive liabilities as fallacious reasoning skills and limited memory capacity. By analogy, one can envisage these mechanisms operating on entire scientific communities.

Start with a scientific theory like Newtonian mechanics. What is primarily transmitted to various research communities is a common language for the transaction of knowledge claims. In each community, however, there will be differences in what counts as appropriate and inappropriate applications of the language, reflecting the local interests of the research communities, which may be quite far afield from the intended applications of the original Newtonians. These differences would be the result of the "hot" social mechanisms and would be conveyed in teaching and other face-to-face, oral transmissions within the theoretical language. However, each community will also have an interest in linking up its research with that of other communities and perhaps even with that of other periods (especially if there is a need to establish a "tradition"). Scientists will therefore jump at the opportunity to draw comparisons and analogies, which should be relatively easy to do, given that Newtonian mechanics is the theoretical language common to all the communities. However, the facility with which these similarities can be found, coupled with the need to budget one's efforts between doing one's own research and finding such similarities, will lead the scientists to neglect important local variation in how the Newtonian language has been applied. This will have the effect of covering over real differences in these scientists' thinking, which is probably fine for legitimation purposes but not for keeping an accurate record of the growth of knowledge; hence, the "cold" social mechanism, which is most likely to be found in written transmissions within the theoretical language.

RELATIVISM REVIVED:
CAN SOCIAL EPISTEMOLOGY SURVIVE THE REFLEXIVE TURN?

The status of "reflexivity" in STS today is similar to that of "political correctness" in scholarship more generally in the 1960s. Both notions

Four: Some Worthy Opponents

aim to situate the inquirer as an integral part of the world in which inquiry takes place. They are also both ethical stances, reminiscent of the Golden Rule or Kant's categorical imperative. The common intuition is that we should not apply analyses to others to which we would not first submit ourselves (Turner 1991). Among the most artful and conscientious practitioners of reflexivity in science studies is Malcolm Ashmore (1989, in press), who takes social epistemology to task for first deconstructing the cognitive authority of science, and then apparently appropriating the very same authority for itself, namely, by suggesting various ways in which social epistemology can be "applied" to improve knowledge production in society. It would be fair to say that this has been the most frequent and substantial criticism made of social epistemology (e.g., Schmaus 1991; Dear in press). In Ashmore's hands, however, this criticism takes on a generality that enables us to see the philosophical sensibility that motivates the reflexive turn.

Although Ashmore (in press) officially lodges two objections to social epistemology, they are really the negative and positive sides of the same point. Both turn on the social constructivist cast of STS research, which reveals the inconclusive character of all claims to knowledge. Epistemic closure is always contingent and reversible—sometimes, so it seems, simply by providing an alternative account. If social epistemology draws on this research, then it cannot be for purposes of "application," because to apply something presupposes that one has something conclusive to apply, which is exactly what STS appears to deny. The positive side of this point is that STS actually sets out to do something quite antithetical to social epistemology, namely, to call into question the very idea of knowledge, understood as a privileged representation of some reality outside itself.

It is worth noting, at the outset, that the form of critique that Ashmore calls "reflexive" is not necessarily politically the most interesting or potent. (I will address that form at the end of this chapter.) Nevertheless, to his credit, Ashmore has presented an empirically informed, up-to-date version of the problem of philosophical skepticism, which does suggest complete generality. Ironically, however, his reflexive critique would seem to have bite only for someone who is *not already* a constructivist. After all, the classical skeptic is a disappointed realist, someone who seeks the truth but believes that, in the end, all appearances may be *false*. By contrast, Ashmore's constructivism is antirealist; it aims to show the indeterminacy of the true/false distinction, and hence, as noted above, the impossibility of social epistemology's "applying" anything from STS. Unfortunately, Ashmore's own reflexive strategy involves a second-order *application*

Opposing the Relativist

of a first-order concern. That is, if Ashmore wants to show that the stuff out of which something is constructed is itself constructed, then he must presume that the meaning of "constructed" remains univocal across the two contexts in which construction is said to occur. But this violates the constructivist point that there is no univocal meaning of "constructed," only the meanings constructed from context to context. One interpreter's sense of paradox may be another's idea of distinct contexts of utterance. The first would be a reflexive realist, like the classical skeptic; the other the unreflexive constructivist. Ashmore's "middle way," the reflexive constructivist, should be seen, not as standing between these two alternatives, but as its own orthogonal axis, a plea to change the genre in which knowledge is written about from fact to "fiction," the realm that philosophers typically identify as consisting mostly of sentences whose truth values are indeterminate. While such pleas may be eloquent or prudent, because they refuse to assert or deny, they cannot *refute* the possibility of making knowledge claims.

Another problem that arises from Ashmore's attempt to escape realism is that it threatens the empirical cast of social constructivism. In describing the "openness" of our epistemic situation, reflexive constructivists tend to run together such notions as "historically contingent," "essentially arbitrary," and "logically possible." It is easy to get the impression from reading this literature that all it would take to change one's view of the world would be to say something different (cf. Woolgar 1988b). I do not mean to deny that one's understanding can be changed substantially, but it cannot be changed into just anything else at any moment. Some possibilities for epistemic change are more immediate than others because of the changes that would need to be made in collateral practices. STS has provided much information about the nature of these possibilities, indeed, largely by revealing the networks of interdependence that are necessary to keep any social practice in place (e.g., Callon, Law, and Rip 1986; Latour 1987).

If we now stand back from Ashmore's particular critique, we see that reflexivity has been integral to most dynamic accounts of the history of science. Put most simply, a system is reflexive if it applies something it has learned about its environment to its own internal workings. STS differs from earlier accounts of science in terms of the general character of what it takes science to have learned. In many ways, these differences capture what is at stake in the *modernist-postmodernist dispute* that cuts across all the human sciences today. On the one hand, such nineteenth-century "modernist" theorists as Hegel and Comte took science to be discovering order in the world, an insight which can then be applied to regulate the development of science itself. On the other

Four: Some Worthy Opponents

hand, the "postmodernist" science practiced by STS has discovered disorder in the world, *especially* in the world where science is practiced (cf. Lyotard 1983). The reflexive histories told by Hegelians and positivists are of ever better methods that enable greater prediction and control of the environment. These are stories of increased *closure* and *inclusiveness*. By contrast, the reflexive histories told by STS are of ever greater discrepancies between universal principles and situated practices that are patched up in ever more opportunistic ways. These are stories of increased *openness* and *dispersion*. There are three dimensions along which the reflexive consequences of modernism and postmodernism may be compared, namely, in terms of the ways they *broaden*, *deepen*, and *limit* the scientific enterprise.

To *broaden* science reflexively is to apply to all of science what has been learned about one science. This is another way of stating the "unity of science" thesis associated with logical positivism. Crudely put, if a model or method enables order to be elicited in one domain of inquiry, then it should be extended to all domains; hence, the ubiquity of mechanistic models and experimental methods, once they were shown to succeed in physics. In this way, the positivists and other modernists thought that laws governing each domain would be forthcoming. The postmodernists also have their version of the unity of science thesis, which may be called *panconstructivism*. If the appropriateness of a given attribute to a given case must always be negotiated by social actors, then, in principle at least, there is nothing that prevents a nonhuman from being socially constructed as, say, a "scientist" or even a "person." Conversely, it is just as much a political act to withhold rights and responsibilities from a computer as it would be from a human. The only difference, according to the panconstructivist, is that whereas suppressed human voices often find someone to speak on their behalf, computers typically do not. However, a general strategy for granting nonhumans voices would be to treat the technological interfaces between ourselves and the nonhumans as media of communication instead of control. A cloud chamber would thus be a means for communicating with microphysical particles, and not simply for tracking their motions. One paradoxical consequence of this version of broad reflexivity is that postmodernists now face the prospect of investing everything with personhood–a veritable "sociology of things"–and, in the process, diminishing the value of being a person. Indeed, this consequence seems to be deliberate, insofar as postmodernists have often remarked on the discriminatory consequences that ontologically inflated criteria of personhood have had not only for nonhumans, but also for humans who failed to meet those criteria by not looking or acting in the right ways.

In short, devaluing personhood might be one of the best things to happen to people!

To *deepen* science reflexively is to divide a domain into parts that are then analyzed by the same principles that were originally used to study that domain. This process is familiar to modernists as the division of cognitive labor into special disciplines. This process typically involves adapting general principles, techniques, and instruments to ever more specific objects. For example, an experimental task that was originally used to test the problem-solving ability of humans in general can be refined in various ways to capture differences between, say, scientists and nonscientists, or men and women. The experimental method is not abandoned; rather, it is intensified. For their part, postmodernists become reflexively deep by intensifying the openness of their inquiry, typically by highlighting the discrepancies in perspective that are already latent in any situation that is defined by more than one person. This goes beyond, say, the ethnographer noting discrepancies in behavior. Indeed, it is here that the ethnographic method itself comes under attack for privileging the ethnographer's account of an episode at the expense of silencing the perspectives of those who participated in defining the episode. In that regard, the classical ethnographer is no less authoritarian than the experimentalist who dismisses her subjects' accounts of their behavior during an experiment. The reflexive remedy is to articulate the alternative perspectives without any attempt at resolving their differences. Ironically, this evenhandedness of representation often costs postmodernist research some credibility since it implies that the postmodernists are contributing to a body of factual knowledge, rather than to a "new literary form," as some of them (e.g., Mulkay 1985) describe their work.

Finally, from the standpoint of rhetoric, the most instructive feature of the reflexivity literature (Woolgar 1988a) is its attempt to *limit* the scientific enterprise by challenging the aspiring theorist to formulate her position in such a way that she is covered by it to the exact same extent as the people about whom she is theorizing. Thus, in its limiting mode, reflexivity demands that the theorist think more democratically than she would otherwise do. After all, theories are usually developed on the assumption that the theorist occupies a privileged vantage point, as indicated by the special language she introduces, a language that makes either too little or too much sense to the people under study. In the former case, theory fails to descend from the ivory tower (Fish 1989); in the latter, theory strips people of their illusions, often with little to replace them (e.g., when Marxism or Freudianism are taken to heart). By contrast, reflexively adequate theories should

Four: Some Worthy Opponents

be ones that both the theorist and the theorized can both live with and learn from.

But is social epistemology so irreducibly social, that it does not have anything valuable to say to individuals? Critical Legal Studies theorist Roberto Unger (1986, 1987; cf. Fuller 1988b) has fully grasped the deep issue at stake here, one that plagues any "scientifically" based social theory. If the theorist uses science to show that ordinary people radically misperceive the causes of their plight–causes which are, to a large extent, beyond their immediate control–then in what sense can such a revelation free people from their chains? More than any other social theory of the modern period, Marxism has been dogged by this problem of calibrating explanations of the past with guidance for the future. Deterministic materialism cannot plausibly yield utopian idealism, no matter how cunning the reason or how bloody the revolution (Aune 1990). Moreover, the problem is not merely at the level of ontology, but also at the level of epistemology. Given human cognitive limitations, mastering the intricacies of the Marxist account of capitalist oppression will likely force one out of the arena of everyday life in which revolutionary practice ultimately takes place. Once the account has been mastered, the theorist realizes that ordinary language is a grand mystification to be deconstructed or simply avoided, but certainly not to be used to persuade the masses. Indeed, such theory-induced political incapacity can result in contempt for the very classes whom theorists are supposed to understand and even identify with. This was certainly the fate of the Frankfurt School, the brand of Marxism that has managed to flourish most consistently in the academy (Fay 1987).

The reflexivist diagnosis that Unger, a native Brazilian, and other Third World theorists give of Marxism's failure as a political practice is that Marx formulated his theory by working in libraries or at home, by himself or in collaboration with Engels. In short, Marx merely theorized *about* people but did not theorize *with* or *for* them. If the latter two prepositions had come more into play, Marxism would have been a totally different theory, one well adapted to its intended audience, namely, workers lacking any formal training in either Hegelian dialecticsor Ricardian political economy. Sometimes theorists think that the reflexive adequacy associated with rhetorical effectiveness would amount to "dumbing down" a theory in order to "manipulate" the audience. When lodging such a complaint, theorists will often conflate signs of a theory's incomprehensibility with signs of its radicalness–so much so that a theorist may not deem her theory fit for ordinary ears ill equipped to hear the truth (Bloom 1987). Under those circumstances, reflexivity sounds like a call to either self-destruction or bland moderation in one's theoretical utterances. However, the truth of the matter is that reflexivity calls for overturning the politics of theorizing by making theoretical language less authoritarian and more negotiable—that is, for theorizing to be done from a third- to a second-person perspective (Fuller 1988a: chap. 11). Thus, social epistemology begins its descent from the Platonic heavens down to the agora where it belongs.

In sum, are there some concrete courses of action that the social epistemologist can take to reveal her sensitivity to the dimensions of reflexivity just outlined? The social epistemologist wants to remain committed to the modernist ideal of an empirically informed, theoretically progressive social science, but at the same time the possibility of an integrated science studies depends on her addressing the postmodernist challenge posed by STS's sense of reflexivity. Here are some brief suggestions, covering each of the three senses of reflexivity:

> *Broaden*: Instead of, say, conceptually prejudging the issue of whether computers can be scientists, decide the issue empirically by plopping, say, an expert system in a scientific setting and seeing how often, and under what circumstances, people who are recognized as scientists come to rely on the computer's judgment. As in the case of measuring the credibility of human scientists, the key indicator here is less a matter of whether the scientists consult the computer and more a matter of whether they actually follow its advice—especially in situations where there is competing advice from a recognized colleague.
>
> *Deepen*: Multiple perspectives undermine a piece of research only if they are allowed to diverge indefinitely, rather than forced to enter into dialogue with one another, specifically to encourage each perspective to articulate in its own terms the differences that it can detect in the others. The ultimate goal would be a more inclusive discourse that found a place for each distinctive position. Thus, the multiplicity of perspectives on a piece of research should be encouraged—but so too equally many attempts at their integration.
>
> *Limit*: If social epistemology is to have moral import, then its theories must empower the people who believe them. In the first instance, this means that the theories need to be comprehensible to their intended audiences. Failure to address this re-

Four: Some Worthy Opponents

flexive concern led the Frankfurt School to be cynical about Marxism's ultimate ability to liberate the masses. After all, realism has worked so effectively as a scientific ideology in large part because of its cognitive simplicity: it projects real world objects from theoretical terms, and it promises that, in the long term, all the theories will come together to explain everything by the fewest principles possible.

10 OPPOSING THE ANTITHEORIST

Ethnographers of the knowledge-producing tribes must find the recent humanist assertions that "theory has no consequences" a puzzling practice, much like an elaborate native ritual which seems to serve no purpose other than to inhibit the tribe's reproductive capabilities. Yet a cynic amongst the ethnographers may well suspect that by disavowing any claim to either knowledge or power, these antitheoretical humanists are merely dodging fights with real social scientists and professional politicians, against whom they would not even make a respectable show. I think that the cynic has something more than just her suspicions to go on. I will begin with the definitive statement of the "theory has no consequences" thesis: Stanley Fish's (1989) recent collection of essays *Doing What Comes Naturally*, and diagnose its impoverished view of theory, which is unfortunately shared by much of the STS community today. Informing my diagnosis is a positive, rhetorically oriented view of theorizing as the establishment of a normatively corrective presumption. The second half of this chapter is devoted to an exposition of my own viewpoint, including the challenges it poses to extant accounts in legal and scientific philosophy.

A Fish Story about Theory

Over the last quarter century, Fish's sphere of influence has expanded from Milton scholarship, where he established a beachhead for reader-response criticism, to literary criticism more generally, where his interpretive conventionalism marked him as an independently minded fellow traveler of the French deconstructionists, and most recently, to the law, where Fish is often taken to be the godfather of the antifoundationalist movement in American legal scholarship known as "Critical Legal Studies" (CLS). Fish's writing serves as an

Four: Some Worthy Opponents

ideal medium for disseminating antitheory throughout the humanities. Unlike his two main competitors, Richard Rorty and Richard Bernstein, Fish does not presume that philosophy is the lingua franca of the humanities—indeed, he presumes hardly any specialized learning at all, only an intuitive sense of the interpretive process. Thus, rather than polemicizing against The Tradition to an audience of interested spectators, Fish typically makes a direct appeal to common sense. Consequently, it might be said that Fish reads a little *too* easily. However, I hope to add some friction to Fish's smooth manner by being almost exclusively critical and diagnostic. Fish and his fellow travelers need to realize that at this stage of the dialectic they share enough of the deepest presuppositions of their "positivist" foes to lead the cynic to think that their antifoundationalist rage is tinged with a sense of betrayal. Moreover, the apparent finality of the "theory has no consequences" argument turns on the fact that the antitheorists have failed to assimilate some of positivism's own attempts at self-correction.

The most objectionable feature of Fish's voluminous writings is his presumption that practice (or *doing* X) is one thing and *theory* (or *talking about* X) is something else entirely. In fact, to hear Fish tell it, failure to respect this distinction is the source of most of the problems surrounding the cognitive status of the humanities today. Fish invokes the distinction to show that "critical self-consciousness," a capacity much vaunted by radical theorists, illuminates our prejudices but does little to eliminate them. In other words, comprehensive knowledge of X leaves the performance of X unchanged. A close-up look at one example, drawn from Fish's critique of Harvard Law professor and CLS guru Roberto Unger (1986, 1987; cf. Fuller 1988b), will highlight what lies in the balance of these dialectical maneuvers.

According to Fish, Unger's "transformative vision of politics" is ultimately of no political consequence. After granting Unger that overarching social structures arise and persist only in virtue of local political contingencies, Fish then cautions aspiring revolutionaries against taking comfort in Unger's premise, for "the political efforts still have to be made, and the assertion that they *can* be made is not one of [those efforts]" (1989: 431). One would expect such advice to issue from the pen of some stuffy ordinary language philosopher, keen on reinforcing the use-mention distinction—not from the pen of a dazzling postmodernist critic. Yet Fish's point is familiar enough: Being a theorist, Unger operates on a logical level that is once removed from the actual practice of politics. He is more commentator than participant, which would seem to imply that Unger's potential impact on politics, qua theorist, is decidedly limited. I say "would seem" because Fish's line of reasoning—its familiarity notwithstanding—does not really follow at all. In brief, Fish is bewitched by a metaphoric conflation of conceptual and causal "determination" that has captivated most philosophers since Hobbes. The metaphor is launched by regarding the move from X *to talk about* X—the ascent to the "meta" level—as a means for constructing an abstract representation of X. Like concrete representations such as painting a picture of X, abstract representations require that one stands at a certain distance away from the represented object, so as to gain full view of its form and its position in a field of other objects. But given the constraints of normal eyesight, this fuller view involves a loss of detail, not to mention a diminished ability to act on specific parts of the object. If this is how one thinks, it should come as no surprise that talk always seems to be *mere* talk, and theorizing seems to be a practice, like teaching, designed expressly for those who can't do.

To follow the metaphor further, an abstract representation "underdetermines" the objects it represents. The relation of underdetermination alludes to a tradeoff that is made between a representation and its objects. In order to stand for many things in many settings, a representation must be constructed at a sufficient distance from its objects so that most of their details drop out. Thus, when Unger invokes such abstractions as "destabilization rights," "institutional reconstruction," and "deviationist doctrine," he is representing a host of heterogeneous practices, the differences between which fade away against the commonalities that he wants to bring into focus. But do these abstractions actually enable Unger to make people see how they "deviate" in particular cases? Fish presumes that the answer is no, on the grounds that theory hovers too far above the world, and hence constrains the possibilities for practice too loosely. But Fish fails to take into account that theory and practice, representation and its objects, exist on the same plane, in the same world—however much theories and representations may advertise themselves otherwise. To paraphrase Ian Hacking (1983): No representation without intervention!

In sum, *pace* Fish, theory *does* have consequences for the simple reason that language is part of the causal order. Only something *in* the world can be *about* the world (cf. Fuller 1988a: chap. 2). Unger's abstractions, insofar as they are embodied as utterances and embedded in social contexts, have *many* consequences. This point should be taken as trivially true, not denied, as Fish curiously tends to do. The interesting question is whether Unger's abstractions have the consequences that he

Four: Some Worthy Opponents

either intends or would find desirable. That question is better treated *empirically* in a survey of what audiences do with Unger's utterances than *conceptually* in an analysis of the meaning of his bare words. To see the contrast here, consider the (entirely plausible) case of individuals who are inclined to take action only if they think that everyone else will as well. Lofty abstractions that speak to a common plight will probably do more to motivate such people than speech tailored to the particulars of each person's situation. From a rhetorical standpoint, such a unified message is doubly effective. It not only restrains people from abandoning the group effort as their particular needs are satisfied, but also encourages them to trust that the speaker aspires to do more than simply gain political advantage by mollifying special interest groups. Such dual rhetorical mastery lay behind the "philosophically inspired" French Revolution of 1789 and the Russian Revolution of 1917. Regarded as a set of abstract concepts, the slogan "Liberty, Equality, Fraternity" epitomizes the airy indeterminacy for which positivists from Auguste Comte to A. J. Ayer have derided metaphysicians. Yet, regarded as a piece of rhetoric operating in the world, the slogan could hardly have been more to the point, as it caused its disparate audiences to move in a focused and largely desired fashion.

If conceptually indeterminate speech can be–indeed, has been–an effective vehicle for bringing about change in the world, why has this point eluded Fish? I suspect that, in spite of his own deftness in deploying and detecting rhetoric, Fish is ultimately a *disenchanted logical positivist*. In the spirit of Ayer's verificationist principle, Fish seems to think that if the empirical consequences of a concept are not specified in the concept's definition, then the concept has no such consequences "by right or nature" (Fish 1989: 28). But whereas Ayer thought that scientific theories were unique in their verifiability, Fish believes that no theory fits the bill:

> Again, I am not denying that theory can have political consequences, merely insisting that those consequences do not belong by right or nature to theory, but are contingent upon the (rhetorical) role theory plays in the particular circumstances of a historical moment. (1989: 28)

This telling passage appears as Fish is debunking the idea that feminist theory has anything to do with the success of feminism as a political movement. But who else, other than a disenchanted positivist, would want to drive a wedge between theory and its consequences, only to show that the latter do not follow from the former? (Moreover, I wonder *where* exactly Fish would drive the wedge: Once the words leave the theorist's mouth? Once they enter the audience's ears?) Because the key concepts used by theorists of Marxism, feminism, and constitutional law are defined primarily in terms of other concepts in those theories, Fish concludes that the impact of each theory is limited to the licensing of moves–especially the shifting of the burden of proof–in the language games that center on the discussion of those theories. It follows that, for Fish, feminism, say, is little more than rules for conducting conversations in certain academic settings–conversations, to be sure, that Fish will go to great lengths to defend under the rubric of "professionalism," but which he will not allow to wander beyond their "conventional" boundaries. Again, Fish takes words too much at face value.

If Fish had wanted to find evidence for the efficacy of feminist theorizing, he should have looked at the subtle but substantial shaping of the academic mind that has been wrought by the enforcement of nonsexist language in the style manuals of most disciplines. For example, the detailed guidelines of the American Philosophical Association or the American Psychological Association constitute a corrective presumption to modes of thought that continue to pervade contemporary society. Legislating the substitution of "she" for "he" will not necessarily alter the predilections of male chauvinist philosophers, but it will place the burden on them to demonstrate why the feminine pronoun should *not* be used in particular contexts. Even in cases where the burden is successfully shouldered, as when referring to a group of people that did not include women, the chauvinist will still need to have given thought to the exclusion of women in that case and perhaps to why that was the case–and, perchance, to whether it ought to have been the case. Feminists are familiar with the cognitive detour that the chauvinist is forced to make in this situation as part of their "consciousness-raising" tactics, but it is no less appropriate to other forms of theorizing that aspire to greater influence than can be expected from the rounds of hermetic academic discourse (cf. Nelson 1986: especially Treichler 1986; Cooper [1989: 14-21] surveys and evaluates attempts from various times and places to shape social behavior through "language planning," including the enforcement of nonsexist writing). The last part of this chapter is thus devoted to a more general exploration of the kind of corrective presumption that nonsexist style manuals exemplify.

Four: Some Worthy Opponents

What Exactly Does "Theory Has No Consequences" Mean?

Having now been exposed to some of Fish's seductive arguments, we would do well to step back and consider in some detail the ambiguity of the claim "Theory has no consequences." At least three things can be meant by this claim, even if we take "theory" in its most ordinary sense (which, as a matter of fact, Fish does not). As it turns out, these three senses can also be found in recent disavowals of theory in STS, and so are especially worthy of our consideration:

> 1. *Theory cannot, by definition, have any consequences.* This is an extension of the logical thesis–normally associated with the later Wittgenstein–that the definition of a concept does not determine its range of application. At most, the definition supplies the concept's relation to other concepts in a common framework. Thus, formulating a theory (i.e., a system of concepts) and specifying the contexts where it may be properly used are logically distinct activities. Indeed, the latter activity crucially relies on situated judgment calls on "hard cases" that were not anticipated in the original formulation of the theory. Thus, armed with a complete mastery of *Structuralist Poetics* (Culler 1975) but no knowledge of the history of applied structuralism, I would be liable to the same misconceptions as befell the medieval physicists who tried to reconstruct the Greek experimental tradition with only the texts of Archimedes on hand. In STS, this capacity of theory appears in the guise of the Duhem-Quine argument against falsifiability and scientific realism, i.e., any theory can be logically saved in the face of any negative experimental outcome, but the theory cannot be credited in light of any positive outcome.

> 2. *Theory does not, in fact, have any consequences.* This is an empirical thesis that a social scientist might show about the lack of influence of theoretical pursuits on other social practices, including perhaps not only the practices that theories have been designed to influence but even the subsequent pursuit of theory itself. A vulgar Marxist materialist may find such a view attractive, as it would render intellectual discourse

Opposing the Antitheorist

entirely epiphenomenal. However, in these stark terms, the thesis flies in the face of our historical intuitions about the efficacy of certain theories, such as those offered by the figures of the French Enlightenment and, indeed, Marxists themselves. Still, the thesis may be stated more sophisticatedly to show that whenever a theory has seemed to have social impact, that impact has been due not to the theory per se but to something contingently associated with it, such as the status of the particular theorist. In STS, this point marks the turn to experiment as "the motor of scientific progress," in terms of which theory appears to be a post hoc rationalization of laboratory practice.

3. *Theory ought not, as a matter of principle, to have any consequences.* This is a normative thesis that might be formulated if theory were thought to have, or could have, the wrong sorts of social consequences. Such concerns were clearly voiced by Edmund Burke and other conservative opponents to the political impact of the French Enlightenment. Similar reservations have been recently expressed by Allan Bloom (1987), who argues that speculative theorizing can easily turn into dangerous ideologizing once it is unleashed from the cloistered colonies of cool-headed academics into the frenzy of the public sphere. Aside from preventing the gratuitous agitation of the masses, academics may want to contain the effects of theory out of a self-imposed intellectual modesty, which would both reward the efforts of data collectors, archivists, and the other "underlaborers" who supply whatever "real" content theories have, and make that content available for more direct critical scrutiny. In STS, a similar sentiment seems to inform Bruno Latour's (1988) critique of scientific theories as "acting at a distance" from the phenomena they purport to explain. By this he means the ability of a theoretical explanation to suppress important differences in the items it subsumes, all in the name of uniform standards of knowledge.

Needless to say, these three readings stand in a curious tension. For example, the prescription made in (3) seems to presuppose that there have been occasions in which (2) has been false, while the semantic character of (1) seems to render the claims made in either (2) or (3) beside the point. Nevertheless, all three are tightly woven into the fabric of Fish's argument, as in the following:

Four: Some Worthy Opponents

> Then this is why theory will never succeed: it cannot help but borrow its terms and its content from that which it claims to transcend: the mutable world of practice, belief, assumptions, point of view, and so forth. And, by definition, something that cannot succeed cannot have consequences, cannot achieve the goals it has set for itself by being or claiming to be theory, the goals of guiding and/or reforming practice. Theory cannot guide practice because its rules and procedures are no more than generalizations from practice's history (and from only a small piece of that history), and theory cannot reform practice because, rather than neutralizing interest, it begins and ends in interest and raises the imperatives of interest–of some local, particular, partisan project–to the status of universals. (Fish 1989: 321)

Another way of seeing that Fish mixes the three readings of his thesis to suit his immediate dialectical purpose is to follow what he has to say about his main example of theory, Chomskyan linguistics. At first, Fish pronounces on the conceptual impossibility of Chomsky's project, but then realizing that *something* that Chomsky has done has in fact become very influential, Fish turns to debunking theory's causal role. Finally, lest the reader think that, no matter how it happened, Chomsky's success was not such a bad thing after all, Fish warns against allowing theory to divert attention from genuine scholarly and critical practice. Thus, the moments of Fish's dialectic pass from (1) to (2) to (3), and in so doing, he implicitly invites, respectively, the philosopher, the social scientist, and the politician to assess the merits of his claim.

Fish's Positivistic Theory of "Theory"

In locating the conceptual space occupied by theory, the key contrast that Fish has in mind is one familiar to mathematicians and computer scientists, namely, between *algorithmic* and *heuristic* procedural rules. As Fish (1989: 317) sees it, the theorist aspires to the algorithmic in that she would like to discover rules that can function as a guide to a humanistic discipline's practice in all cases by being sufficiently explicit and neutral for any practitioner, regardless of her

Opposing the Antitheorist

particular interests, to follow the rules to the same result. A rule with heuristic status, by contrast, can guide practice only in certain cases which cannot be determined in advance of practice but only, in retrospect, once practice has in fact been successfully guided by the rule. In short, the desired distinction is between "the foolproof method" and "the rule of thumb."

In the course of *Doing What Comes Naturally*, Fish smuggles additional conceptual baggage into this distinction: the foolproof method turns out to be one that, at least in principle, can be derived a priori, which is to say, prior to all practice, and that is thus unaffected by the history of the practice it purports to govern; the rule of thumb turns out to be valid a posteriori in a rather particular way, namely, relative to the entire history of a practice and not simply to a practitioner's own experience in trying to apply the rule. These conceptual moves are properly characterized as "smuggling" because not only are they logically independent of Fish's opening moves (that is, a foolproof method need not be knowable a priori and a rule of thumb is not normally thought of as a kind of social convention), but they also ensure the success of his central argument; for Fish argues (a) that the pursuit of theory cannot succeed *because* all purported instances of theory are really generalizations from actual practice, and (b) that this failure of theory does not in the least impair the ability of practices to govern themselves locally through rules of thumb. The force of these two conclusions would not be so strong if their truth was not virtually deducible from Fish's idiosyncratic definition of the two kinds of rules. As a result, were the reader to wonder why foolproof methods could not be grounded empirically or why rules of thumb must have the force of social conventions, he would be at a loss for an answer in Fish.

Still, for all their question-begging character, Fish's tactics have precedent—indeed, in the very movement Fish claims to be opposing. For Fish's definition of "theory" as a "foolproof method" is nothing short of a positivist reduction. Before the rise of positivism in the nineteenth century, "theory" was generally used to describe privileged standpoints from which phenomena might be systematically inspected. A theory was thus typically "speculative" and "metaphysical," and the attitude of the theorist was one of contemplative detachment. It would be fair to say that this is still the ordinary sense of the word "theory" (cf. R. Williams 1976: 266-68). However, this was precisely the sense of theory attacked by the positivists, starting even with Comte, who argued that the only way in which one could tell whether one's theory was any good was by putting it to the experimental test: Does the

Four: Some Worthy Opponents

theory allow the inquirer to obtain what he wishes from the phenomena? A theory that could give a positive answer to this question on a regular basis was "theory enough" for the positivist, for it would then constitute a foolproof method for conducting one's inquiry. Like the metaphysical sense of theory, the positivist sense had to be articulated in a technical discourse. But instead of referring to subtle underlying entities which unified the array of phenomena under study in ways not always transparent to the casual observer, the positivist referred to operational procedures which any inquirer could implement and to which he could be held accountable by some larger community.

Indeed, in positivism's more virulent twentieth-century form (which followed a brush with pragmatism), a camp follower like A. J. Ayer (1959) would argue that what distinguishes "scientific theory" from other forms of theory is *not* its ability to permit us a deeper understanding of reality, but its ability to permit us a more substantial control over phenomena; hence, positivist philosophy of science is typically described as antirealist and instrumentalist." The implication of this view for theory in the social sciences, then, is clear: the better theory is the one better able to predict and control the behavior of people. Interestingly, ethical theories turn out to be crude theories of this kind, the only difference being that ethics clothes its interest in controlling behavior in the metaphysical language of "values." Fish's own remarks about "the consequences of theory" dovetail nicely with Ayer's here, since Fish believes that whatever impact theory has on humanist practice is due not to its truth directing the way to better interpretations, but to its force directing interpreters into certain desired forms of discourse. However, Fish ultimately fails to see his ties to positivism because by infusing his conception of "theory" with a priori status, he conflates the older metaphysical sense of the term, which he clearly rejects, with the newer positivistic sense, which he seems–at least in practice–to embrace.

Since Fish fails to see himself as an instance of the positivism he opposes, it is perhaps not surprising that he also misses the point of the positivist's longing for a "value-neutral" method, which Fish dismisses as patently impossible to achieve. "Value-neutrality," a term popularized by Max Weber after it had already provoked a generation of polemics among German economists at the end of the nineteenth century, was used hardly ever to characterize the activity of "constructing" a method (of, say, economic analysis) but more often to characterize the activity of testing or justifying it. Already, then, this fact about the term's usage concedes the point which Fish still thinks needs to be contested, namely, that all theory construction is laden with the values of the theorist, which are determined by the local nature of his own practices.

What is open to debate, however, is whether a theory can be tested or justified in a value-neutral manner. For example, philosophically inclined practitioners of the life sciences (e.g., John Eccles, Peter Medawar, Stephen Jay Gould, but especially David Faust 1985) have been persuaded by Karl Popper's view that the scientific community collectively achieves value-neutrality for a theory by having the theory's tester be someone other than the person who first proposed the theory—presumably someone who does not have a stake in the theory. Not only is this a nice myth, but it also seems to fit the sociological data, which suggest that the personalities of scientists polarize into two types, roughly the speculators and the experimentalists (cf. Mitroff 1974). As long as the scientific community has a healthy mixture of these personalities, all of whom see themselves as engaged in the same inquiry, then Popper's picture appears quite workable. In any case, Fish does not address the possibility that value-neutrality is simply the mutual cancellation of individual values on the collective level.

Now the issue of whether a theory can be justified in a value-neutral manner is somewhat trickier to resolve, largely for terminological reasons. The logical positivists frequently spoke of "theory-neutral observation," but in so doing they were using "theory" in the metaphysical sense mentioned earlier and "observation" to describe something already placed in a technical language. For purposes of criticizing Fish, then, it would be less misleading for the reader to substitute "value-neutral theory." The motivation behind the positivists' desire for such neutrality was that the two most recently heralded physical theories, relativity and quantum, were each supported by scientists of an idealist (Eddington for relativity and Bohr for quantum) and a realist (Einstein for both) bent. Yet regardless of where they stood on the idealist/realist debate, the scientists could account for the same facts from within their respective metaphysical positions. The positivists tried to generalize this insight to the idea that a theory is more justified as it can be deduced from higher-order theories, especially ones which would otherwise be mutually incompatible. Thus, the ultimately justified theory, the positivist's notorious "observation language," stands out by its ability to be deduced from all other theories and, therefore, remains justified regardless of which of those other theories end up getting rejected. And so, in practical terms, the possibility that Fish neglects here is that value-neutrality may simply be a theory's ability to be endorsed by people having otherwise conflicting values.

Four: Some Worthy Opponents

Toward a More Self-Critical Positivist Theory of "Theory"

At this point, it might prove instructive to increase the conceptual stakes in our critique of Fish. Instead of revealing Fish's errors by the positivist standards to which he secretly seems to aspire, we shall now turn to criticizing the standards themselves. This tactic is not as unfair as it may seem, since the positivists were among the first to realize the inadequacies of their own account of theory. Indeed, their reservations arose from further thought about the alleged value neutrality, or metaphysical indifference, of empirically testable theories.

Carnap (1967, pp. 332-39) originally set the stage when he appealed to relativistic and quantum mechanics as evidence for the "pseudoproblematic" status of philosophical (or, in Fish's sense, "theoretical") disputes, since such metaphysically divergent physicists as Bohr and Einstein could continue to add to the body of empirical knowledge while their philosophical differences remained unresolved. However, if true, the truth of Carnap's claim was very much an unintended consequence of the many famous exchanges held between the idealist and realist physicists, since Bohr, Einstein, et al. interpreted their own empirical inquiries as attempts to vindicate their respective metaphysics—an impossible task by positivist lights. Still, evidence could be found for a deliberate application of Carnap's thesis, and to great consequence—but in the less glamorous science of experimental psychology. Indeed, in his 1915 presidential address to the American Psychological Association, John B. Watson called for the abandonment of the entire introspectionist paradigm, on the grounds that its evidence was gathered simply by training subjects in the particular experimenter's response protocols, thereby explaining the hopelessly diverse array of data. Yet Watson quickly added that introspection's failure was behaviorism's gain, since if nothing else, the diversity of data proved that learning was a robust phenomenon worthy of study in its own right, independent of the content that the subject was taught (cf. Lyons 1986). Needless to say, behaviorism was destined to become the most important antitheoretic scientific research program of the century.

After the late 1930s, when behaviorism became the academically most powerful school of psychology in the United States and the logical positivists had themselves emigrated to this country, psychology was the science to which positivists most often referred for examples about the eliminability of theory. Not surprisingly, they quickly found themselves pondering Fish-like thoughts, which culminated in what Carl Hempel dubbed the *Theoretician's Dilemma*:

> If the terms and principles of a theory serve their purpose they are unnecessary [since they merely summarize the known data]; and if they do not serve their purpose they are surely unnecessary. But given any theory, its terms and principles either serve their purpose or they do not. Hence, the terms and principles of any theory are unnecessary. (Hempel 1965: 186)

To his credit, Hempel solved the dilemma by abandoning a crucial positivist assumption. It is the assumption, rather prominent in Fish, that algorithms and heuristics mark a distinction *in kind*. That is, positivists generally suppose that some rules work all the time and can therefore function as proof procedures, whereas other rules are more open-ended and work only occasionally. Against this, Hempel argued that heuristics are merely *imperfectly known algorithms*. In other words, if a rule works only occasionally, it cannot be immediately inferred that the phenomena to which the rule applies are indeterminate or, in some fundamental way, escape rule governance. Rather, it simply means that the right rule has yet to be found. Another way of making Hempel's point is that positivists act as if a theory becomes somehow less theoretical, and somehow less worthy of scientific attention, if it turns out to be *false*. This view is clearly mistaken, since a major role for theory is not merely to save but to *extend* the range of phenomena, which necessarily involves an element of risk, given that we never know in advance whether our extensions will be correct. What, then, informs these extensions of theory? According to Hempel, as well as Popper and Quine, none other than the sorts of "local" considerations that Fish seems to think vitiate the epistemic status of theory. For all its intriguing character, *there is nothing in the least self-contradictory about the possibility that certain universal truths may be discoverable only under quite particular historical circumstances*. This is a clear oversight in Fish's argument that did not equally elude the positivists–or, as we saw in the last chapter, Karl Mannheim, whose sociology of knowledge took this as the guiding methodological insight (cf. Meja and Stehr 1990).

Four: Some Worthy Opponents

The Universality, Abstractness, and Foolproofness of Theory

A related confusion into which Fish seems to have fallen as a result of making aprioricity essential to theory concerns the relation of *universality* to *fallibility* and *corrigibility*. Contrary to the spirit of Fish's definition, universality does not necessarily imply infallibility and incorrigibility. Admittedly, in the history of the Western tradition, several philosophers have followed Plato in believing that whatever is universal cannot be false and thus need not be changed. However, those philosophers (and it is now controversial whether even Descartes should be included among them) have also tended to think that universals are apprehended by a special mental faculty, which Plato called *nous*, whose workings are infallible in virtue of bypassing the potentially deceptive route of sensory experience. It is curious that Fish seems to think that this is still taken to be a live option among contemporary theorists in the humanities who are no doubt familiar with the conclusions of *The Critique of Pure Reason*. Moreover, both Charles Sanders Peirce and Popper are noted for having argued that as long as a universal principle is treated as a hypothesis under test, there is no contradiction in saying that it might be false and hence revisable. Isaac Levi has since gone further to separate issues of corrigibility from those of fallibility: a discipline may legitimately decide to revise a universal principle, even if it has not been shown false, simply because its interests have changed (cf. Levi 1985).

When Fish seems to think that certain theorists believe they are possessed with "pure reason" or some other form of "intellectual intuition," he may be expressing latent skepticism about the sort of knowledge that can be gained from abstraction. Certainly, this would make sense of his dismissive remarks about Chomsky's project of universal linguistics. Again to cite precedent, Aristotle talked about the abstraction of a universal in two sorts of ways: on the one hand, he described it as the extraction of what is common to a set of particulars (*korismos*); on the other hand, he described it as what is left after the particularizing features of a particular are removed (*aphairesis*). Generally speaking, if a philosopher has talked about abstraction in the former way, she has tended to be sanguine about its efficacy, while if she has talked about it in the latter way, she has tended to be skeptical. Joining Fish in the list of skeptics are William of Ockham, Bishop

Berkeley, and F. H. Bradley. All happen to believe that a conceptual distinction is legitimate only if it can be cashed out as an empirically real difference. (Even Bradley the Absolute Idealist buys this line, insofar as he believes that since there are no empirically real differences, it follows that there are no legitimate conceptual distinctions.) With this history in mind (cf. Weinberg 1968), let us now return to Chomsky (cf. Fish 1989: 314-18).

If, as Chomsky argues, our linguistic performance is a degenerate expression of our linguistic competence, then an anti-abstractionist like Fish will demand that there be some way of empirically eliminating the degenerate elements of *our* linguistic performance–not the performance of machine simulations–so as to reveal this underlying competence. Fish claims that experiments of this kind are bound to fail because language works *only* because sentences are always situated in a context of utterance, which consists of those very elements Chomsky calls "degenerate." In his aversion to abstraction Fish may, thus, be likened to the physiologist who thinks that examining dead bodies defeats the whole point of studying the human organism, which is, after all, to discover how life works. (Cf. Bernard 1964, for the absurdity of this argument in physiology.) And while both sides of this analogy have some prima facie plausibility, can either side, especially Fish's, hold up under stiff scrutiny?

From within Chomsky's camp, there is an instructive way of diagnosing the source of Fish's anti-abstractionism which should make us think twice about what exactly is being criticized here. Jerry Fodor (1981: 100-126) distinguishes two reasons why a scientist–let us say a cognitive scientist–is interested in abstraction or "idealization." First, she might want to model the optimally rational thinker, in which case his object of study is indeed something closer to a computer simulation than a real human being. However, she might instead want to model real suboptimally rational thinkers, in which case foolproof computational methods will have to be supplemented by other rules that don't work nearly so well and, as a result, account for the numerous errors that real thinkers make. Both Chomsky and Fish confuse these two interests in abstraction: Chomsky claims to be interested in modeling real speakers but his techniques suggest that he is really modeling ideal speakers; Fish catches on to this fact but mistakenly concludes that Chomsky's is the way of all abstractive projects and that, therefore, all should be rejected. Again the error here has probably been inherited from Plato, if we understand the Platonic concept of "Form" to imply that any particular is a degenerate version of just *one* universal, or, in more Aristotelian terms, that each particular consists of one essence and

Four: Some Worthy Opponents

many nonessential features. After all, Chomsky does not merely bracket considerations of context from his search for linguistic universals, but actually does not believe that context has much to contribute to a general understanding of language. And certainly, Fish is reacting at least as much to this devaluation of context as *nonessential* as to the fact that Chomsky restricts his interests in the essential features of language to whatever can be captured in a competence grammar. Both therefore err in thinking that if there are universals to be found in a given set of particulars, then there are *at most* one, when in fact a more realistic representation of the particular may be gotten by supposing that they are governed by *several* universals–in the case of language, principles of *pragmatics* as well as ones of syntax and semantics.

A final set of confusions into which both theorists and antitheorists are prone to fall concerns the sense in which a method can be "foolproof." In many ways these are the subtlest confusions, and, in any case, they are the ones which most naturally lead to a discussion of the recent antitheorizing in legal studies. To get at them, we shall introduce a distinction in types of rules first raised by John Rawls (1955) in an attempt to revamp Kantian normative theory, namely, between *regulative* and *constitutive rules*. A regulative rule is one drafted by a legislature and which appears as a statutory law: for example, "All wrongdoers must be punished." Notice that this rule is stated as a universal principle but does not mention which cases count as instances of the principle. The latter problem is the business of adjudication, which works by applying constitutive rules. These rules determine how and which particular cases should be constructed under the principle–say, that "Jane Doe is a wrongdoer"–usually on the basis of tacit criteria for which a judicial opinion provides ex post facto justification.

Among the many advantages of Rawls's distinction is that it allows us to make sense of the idea that a rule can be universally applicable without itself specifying the universe of cases to which it may be legitimately applied. To put it more succinctly, the distinction shows us that *a theory does not entail its practical applications*. This conclusion often turns out to be the point of many of the later Wittgenstein's examples of mathematical practice: namely, that my knowledge of, say, the Peano axioms and all the theorems of arithmetic–the sorts of universal principles that mathematicians study and formalize–is never sufficient to determine which arithmetic principle I am applying when I am trying to complete a particular number series. Thus, when the mathematical realist claims that there is a fact of the matter as to how the series goes, she offers small comfort to the person counting, who is interested in finding out exactly *which* fact it is.

Opposing the Antitheorist

Likewise, Fish is probably guilty of confusing Rawls's two types of rules when he claims that "foolproof methods" and "rules of thumb" are incompatible pursuits for the humanist. For even if there were a computer algorithm specifying the steps by which one correctly interprets a poem, one would still need some other rule–perhaps a rule of thumb, perhaps another algorithm–for identifying cases in which it would be relevant to apply the algorithm. In other words, knowledge of how to interpret poems still does not tell us how to recognize poems in the first place. Moreover, again contra Fish, Rawls's two types of rules are sufficiently independent of one another that it would be coherent for the practitioners of a humanistic discipline to agree on procedures for identifying poems without agreeing on procedures for interpreting them, or even vice versa. In the latter case, where rules exist for interpretation but *not* identification, one can imagine the discipline agreeing to "If x is a poem, then x is read in this manner . . ." and still disagreeing over whether a given x is in fact a poem. The example that seems to arise most often in the antitheory literature is E. D. Hirsch's (1967) strategy of "general hermeneutics," against which Fish inveighs.

Convention, Autonomy, and Fish's "Paper Radicalism"

Before ending our catalogue of the conceptual problems facing the leading force of antitheory in the humanities, a word should be said about a term to which Fish attaches great importance in connection with how the practices of a discipline are authorized: *convention*. In political and linguistic philosophy, conventions are contrasted with *contracts* and *grammars* in that a convention is a practice that emerged largely without design yet continues to be maintained in virtue of the beneficial consequences accrued by the individuals adhering to it. While Fish usually understands "convention" in this way, he also sometimes means it in the sense of "conventionalism," which is a doctrine about what confers validity or legitimacy on a theoretical statement, namely, that it follows from some explicit earlier agreement about definitions and assumptions. This second sense of convention arises especially when Fish wants to devalue the kind of legitimacy supposedly claimed for a theory by its proponents and, therefore, stresses the similarity between theories and games. However, as

Four: Some Worthy Opponents

Hilary Putnam (1975: 153-92) first observed, conventionalism's metaphysical implications are really much stronger than someone like Fish thinks, because an implicature of claiming "p follows by convention" is the claim "p follows *by virtue of nothing else*," which seems to commit the conventionalist to what Putnam calls "negative essentialism." Thus, in terms of the sort of knowledge to which Fish must have access in order to legitimately argue that disciplinary procedures are *nothing but* game rules, he is being just as "metaphysical" as his opponent, the theorist, who believes that such procedures really represent a part of how things are.

Fish's frequent appeals to the "conventional" character of interpretation play a somewhat unexpected role in the delimitation of theory's powers. Ordinarily, claiming that a practice is conventional amounts to denying its naturalness and, hence, to suggesting that the practice may be changed. But this is not what Fish is really doing. On the contrary, he invokes the conventional in order to signal that practices are explicitly bounded, self-regulating fields of action that do not become less authoritative simply because their origin and maintenance have been subject to a variety of local contingencies. By retooling the rhetoric of the conventional in this way, Fish manages to cater to two opposing constituencies at once. On the one hand, he appeals to the anti-intellectualist streak in American thought, which suspects that the professoriate poses a threat to our folk mores once academic discourse is permitted to stray beyond its natural habitat in the ivory tower (cf. M. White 1957; Hofstadter 1974). On the other hand, Fish is also able to champion esoteric humanists within the academy, who, armed with the insight that all forms of knowledge are conventional, can now honestly claim that their activities are no less legitimate than those of grant-guzzling natural scientists.

The political bottom line for Fish may be summed up as *I'm OK, you're OK—as long as we each know our place*. Separate and *therefore* equal. The original sin is overextension, and theory is the devil's artifice. This particularly aggressive brand of relativism reveals Fish's true sophistic colors. Specifically, Fish's profound ambivalence toward the Left in *Doing What Comes Naturally* marks him as the ultimate *foul weather* friend. To stay true to the sophistic ideal of making the lesser argument appear the greater, he can support the Left only when they are on the defensive, but never when they are on the offensive. A good way of positioning myself vis-à-vis Fish is in terms of contrasting anthropological strategies of "cultural diversity." One is to preserve established differences, and the other is to promote endless hybridiza

Opposing the Antitheorist

tion. Fish opts for the former, I for the latter. The concomitant notions of autonomy are likewise opposed. I associate autonomy with combinability, while Fish links it with *purity*–indeed, a "retreat to purity," to harken back to a humanist tendency first identified in *Social Epistemology* (Fuller 1988a: chap. 8).

The appeal to the purity, or autonomy, of research is a topos common to scholars across the arts and sciences. Often this appeal is expressed in the phrase "pursuing knowledge for its own sake." The expression conjures up two images of how knowledge might be pursued. On the one hand, the forthright inquirer might follow the truth wherever it leads, regardless of the amount of resources consumed in the process. In the extreme case of certain high-energy physics projects today, the cost of pursuing knowledge as an end in itself entails that virtually everyone and everything be incorporated as means, usually labor and capital, toward realizing that end. On the other hand, the pure pursuit of knowledge might signal a call to modesty, as the inquirer restricts her efforts to what knowledge can reasonably be expected to control, namely, the production of more knowledge. This is the more humanistic route, which traditionally has been informed by two quite opposing considerations. Some, like Allan Bloom and other latter-day Platonists, are concerned about the effects on people (mostly students) whose minds are unprepared to receive knowledge in an unadulterated form, mainly because they have not been themselves directly involved in the knowledge production process. These humanists profess the cultivation of "sensibility" in their thoughts and "decorum" in their actions. The most interesting modern expression of this sentiment is the institution of "academic freedom" (German *Lehrfreiheit*) as a self-regulated guild right: Speak as you will, but only in your field. In contrast to all this, other humanists, like Fish, aim to deflate just these pretensions, that somehow they need to watch what they say and do because of their potential impact on society; hence, the denial of theory's consequentiality. Yet this camp, too, is committed to the modest sense of purity. It believes that the moral imperative of tolerance always outweighs any claim to epistemic privilege: The pure inquirer restrains her own totalizing impulses, in order to enable the flourishing of others who have just as much a claim on the truth as she does.

The irony behind such magnanimity is that humanists started singing the praises of relativism and pluralism only after successive historical failures at gaining control over politicians, artists, publics, and each other. The increasing independence of general hermeneutics

Four: Some Worthy Opponents

from the specifics of biblical and legal interpretation is just part of this process of turning adversity into virtue (cf. Gadamer 1975, Grafton 1990). Put most cynically, humanistic knowledge must be pursued as an end in itself because it certainly hasn't been a reliable means to any other end! My point here is not simply to suggest that a certain measure of self-deception may be involved in attempts, like Fish's, to defend scholarly pluralism. After all, clever sophist that he is, Fish could well concede the diagnosis yet deny the cure. He might counter: Even if the politics of academic tolerance is a defensive reaction against an embarrassing historical track record, that still does not speak against the "consequences" of being tolerant. But as I have tried to show here, treating academic disciplines as well-bounded language games blinds one to the consequences that disciplinary discourses have outside their intended fields of application. Relativism makes sense as an epistemological doctrine only if a community can be identified relative to which knowledge claims can be held accountable. The relativist presumes that the consequences of acting on a knowledge claim can be contained to just the members of that community. We know, however, that knowledge claims are continually imported and exported across disciplinary boundaries, which themselves shift over time, changing their relation not only to other fields but to society at large. Given such complex circulation patterns, it would come as no surprise to learn that some knowledge claims end up having their most significant impact on people outside their discipline of origin, in ways neither intended nor anticipated by their originators. The evaluation and mediation of these effects is the normative challenge that awaits the humanistic inquirer who is willing to treat knowledge production as more consequential than Fish would permit us to do.

I do not want to exaggerate the powers of theory to change the world, but simply to urge that these powers, such as they are, be regarded as a matter for empirical inquiry and practical control. No one denies that, in the twentieth century, the pretensions of theory–especially of Marxist origin–have been deflated in a variety of ways (cf. Crook 1991). But is this something that should have been anticipated, just given what Fish would call the "nature" of theory? Or would it not be less cynical, as well as rhetorically sounder, to suppose that a good theory requires not only that its speaker be persuaded to keep speaking but also that its audience be moved to act appropriately? In other words, the failure of theories may be due more to the failure of theorists as persuaders than to the failure of theory as such. What, then, is the appropriate rhetorical habitat for a theory? The answer lies in the realm of *presumption*, an account of which I will now provide.

Consequential Theory: An Account of Presumption

Presumptions are normative instruments for injecting some make-believe into an all too real world, with the long-term hope that reality may become more like make-believe. A scientific community may never know whether a given theory is really true, but by granting the theory paradigmatic status, members of the community are forced to act as if it were so, which causes them to frame their positions in terms of that presumption. Since a presumption is established on explicitly normative grounds, it is often difficult to claim that, just because a community presumes certain things, it follows that most of the community's members actually believe those things to be true. In fact, if the presumption is doing any real normative work, and hence correcting people's prior beliefs, then individual members of a community should suspect the truth or appropriateness of the presumption in particular cases, but nevertheless believe that the presumption should be upheld so as to force the relevant countervailing arguments to be mustered. The presumption of innocence in Anglo-Saxon law seems to work this way, but it also captures the attitude that disciplinary practitioners have toward a "widely held" theory in their field.

One way to understand this attitude is in the Durkheimian sense of reinforcing a collective identity, for the presumptively true theory in a field exemplifies what it is to excel at the methodological standards that confer epistemic legitimacy on the field as a whole. This point stands, even if the theory is eventually superseded and currently suspected by a large portion of field. Moreover, it may be argued (pro Popper and contra Polanyi) that what distinguishes science from a community of faith is precisely that leading theories are presumed rather than believed. As a result, scientists are professionally mandated to treat presumptions not as positive accomplishments in their own right but as way stations to be superseded on the road to inquiry.

A critic typically takes aim at a presumption, whether it be of a defendant's innocence (in the case of a trial prosecutor) or a belief's orthodoxy (in the case of an innovative scientist). The critic functions as an agent of *rationality* insofar as she clearly distinguishes the presumption's ability to be defended (its real probative strength) from its ability to prevent attack (its mere conventionality). The critic succeeds

Four: Some Worthy Opponents

in overturning the presumption (and thus in fully "bearing" the burden of proof) if she can show that the presumption's unassailability has served only to mask its indefensibility. A situation of this sort arises if the presumption can be defended on no grounds other than the fact that it has traditionally been presumed. Among the more obvious candidates for overturned presumptions would, therefore, be various "folk" beliefs that may have been warranted when first introduced, but now have only habit on their side against our current background knowledge.

A presumption overturned in one case is not overturned once and for all. In other words, the effects of criticism appear to be purely local, confined to the single challenged case. (This is why one false prediction does not refute a theory–a point that Lakatos realized, but Popper did not.) For example, if a particular defendant is proven guilty, the general presumption of innocence remains unaffected. Indeed, the presumption would not be overturned even if defendants were always proven guilty, since the reason why innocence is presumed in trials is conceptually unrelated to the police's success rate at apprehending guilty parties. Likewise, if we extend this legalistic model to epistemic matters, to prove in one case that folk psychology does not offer the best explanation for someone's behavior does not diminish the presumption in favor of folk psychological explanations. However, it might seem that in epistemic matters the frequency with which the presumption is overturned *should* play a role in determining the presumption's fate–that the law should resemble more closely how science is intuitively thought to operate. Thus, Nicholas Rescher (1977) has developed an account of presumption closely tied to a claim's probability, with each defeat of the claim increasing its burden of proof.

I, too, believe that the difference between legal and epistemic presumptions has been exaggerated. In my view, both are *normative correctives* to widespread beliefs, which may, in the long term, cause those beliefs to change, but which do not depend on that prospect for their validity (cf. Fuller 1988a: chap. 4). Not surprisingly, then, the critic has her work cut out for her! Let us consider how this applies in both the legal and the scientific cases (cf. Fuller 1985).

PRESUMPTION IN LEGAL MATTERS

Richard Whately (1963: pt I, chap. 3, sec. 2), the nineteenth-century Anglican bishop and rhetorician, clearly modeled the modern theory of presumption on what he took to be the conservative grounds for presuming innocence in Anglo-Saxon legal procedure. It is common for rhetoricians nowadays to interpret this conservatism as a strategy for risk-

Opposing the Antitheorist

averse institutional action (cf. Goodnight 1980: 304-32). Thus, the judge would presume the defendant innocent in order to minimize the worst possible trial outcome–a safeguard against a hasty judicial decision that would needlessly ruin the defendant's life, were it subsequently shown to be based on a faulty understanding of the facts. But even given this interpretation of presumption's "conservative" function, it does not follow that presuming *innocence* is the most conservative course of action, especially if we are judging that action according to its consequences rather than its intentions, and moreover, if we expand the scope of the parties potentially under risk to include not only the individual brought to trial but the society at large. Both sorts of judgment are empirical in character, not merely the products of conceptual analysis performed upon "innocence" and "conservatism." It may be, for example, that the presumption of innocence has the effect of encouraging individuals to be more reckless in their actions–to do the sorts of things that superficially resemble crimes–because they know that even if they are brought to trial the burden will be with the plaintiff to show that the superficial resemblance is anything more than that. Even more likely is the increase in risk that the rest of the society will have to absorb as a result of the fact that the presumption of innocence will permit unconvicted felons to roam free. Thus, the social function served by a presumption of innocence probably has little to do with whatever risk-averse impulses may have prompted the good bishop (cf. Goodnight 1980: 322).

However, the presumption of innocence may discourage illegal activities in a somewhat different way. For example, in American civil procedure, there are two senses in which the burden of proof must be borne in a case. Whereas the plaintiff must always bear the burden of *persuasion* in demonstrating the defendant guilty of the alleged wrongdoing, the defendant may have the *burden of producing evidence* that shows that the plaintiff has not interpreted the defendant's actions in the most natural manner. The idea behind the defendant's bearing the burden of production is that if the defendant is indeed innocent, then he or she will likely have access to some fact that recontextualizes the case sufficiently to defeat the plaintiff's charge (Conrad et al. 1980: 840-43).

In fact, the problem so far with our analysis of the presumption of innocence is that the considerations that we just raised to show its possible risk-enhancing consequences are the very ones invoked by French jurisprudents in justifying the presumption of *guilt* as the appropriate stance for the judge to take toward the defendant (cf. Abraham 1968: 98-103, for a comparison of the role that presumption plays in Anglo-

Four: Some Worthy Opponents

American accusatorial and European inquisitorial legal systems). In other words, *both* presumptions—of innocence and of guilt—have been legitimated on the same conservative basis, even though there are probably no empirical grounds for believing that either presumption especially contributes to a well-ordered society. In that case, why should the legal system presume anything at all about the defendant?

As it turns out, considerations of risk do not play nearly as big a role in Whately's thinking about presumption as the need to ground the persistence of a form of social life in a principle of sufficient reason: that is, there is a prima facie reason for believing that anything that has been the case ought to continue being the case. It might not be clear exactly what social good is served by the persistent social form, but simply the fact that the form has persisted is evidence for its serving some such good. Thus, the defendant is presumed innocent of *this* wrongdoing because he was innocent of other wrongdoings prior to his appearance in court. Consequently, given the good inductive evidence against the defendant's being guilty, the Anglo-Saxon judge is instructed to proceed cautiously in his inquiries so as to ensure against an unnatural understanding of what took place. The work of Alfred Sidgwick (1884: brother of the utilitarian Henry Sidgwick) marks the transition from Whately's "sufficient reason" analysis of presumption to the more modern "risk" analysis. Sidgwick argues that an inductive regularity probably points to some well-founded phenomenon that it would be risky to act against. This line of argument is generally based on John Stuart Mill's work on the relation between induction and utility. Still, the principle of sufficient reason can cut either for or against the defendant, depending on the form of social life the persistence of which the presumption is supposed to underwrite. It need not be the image of a particular defendant as a law-abiding citizen; instead, it could be, as in French juristic reasoning, the image of law enforcement agencies as consistently doing a job that needs to be done, which would then justify a presumption of guilt.

Whately's intentions notwithstanding, we should hesitate before embracing the sufficient reason interpretation of presumption. After all, the traditional appeal of sufficient reason approaches has been their psychologically compelling character. "Things just don't happen for no good reason," we are prone to say—but about *what* exactly, in the average legal proceeding? Are we not more likely to think that the defendant, and not the law enforcement agencies, has done something socially deviant (criminal or otherwise) to bring about the need for a trial in the first place? Taking psychology as our guide, then, sufficient reason would seem to weigh on the side of a presumption of guilt rather

Opposing the Antitheorist

than a presumption of innocence. That is, if we assume that presumption must operate to *conserve* our intuitions about our fellow persons rather than to *correct* them. If we go the latter route, as I now suggest, then we must turn our gaze from the *short-term* role that the presumption of innocence plays in *impeding* the judge's *actions* against *particular* defendants to its *long-term* role in *revising* the judge's (and society's) *attitudes* toward defendants in *general*.

If nothing else, the presumption of innocence implies that the law enforcement agencies which bring an individual to trial are likely to be in error, and that the defendant, unless proven otherwise, has acted within the confines of the law. The level of fallibility attributed to the legal system on this view not only runs against our ordinary intuitions but is also quite foreign to the considerations which lead, once again, the French to presume guilt of the defendant. Yet it would be equally misleading to say that the French presumption merely reinforces the intuitions that the Anglo-Saxon presumption seeks to correct. For in the French system, the presumption of guilt licenses the judge to suppose that, regardless of whether the defendant is indeed in the wrong, something strange has been afoot worthy of further examination. What follows, then, is an exhaustive inquiry into the facts of the case, which continues until the judge feels that he has achieved an accurate understanding of what took place and can therefore subsume the case under the appropriate law. Indeed, the investigative powers of the judge are so extensive that he may freely suspend the rights of citizens (e.g., by wiretapping or opening their mail) in pursuit of crucial bits of evidence. By so conferring a greater value on the thoroughness than on the swiftness of the legal proceedings, French law supports an attitude of objectivity, impartiality, and ultimately the sort of certainty classically associated with an unhampered search for the truth. In the long term, then, the presumption of guilt manages to tinge the workings of legal conventions with the aura of scientificity, which has the effect of commanding greater respect of the citizenry than such conventions might otherwise do.

As I have earlier suggested, the presumption of innocence in Anglo-Saxon law serves a strikingly different purpose, namely, as a constant reminder of the *mere* conventionality and hence likely fallibility of law enforcement agencies. This acts against the natural psychological tendency, on the part of both the judge and the onlooking citizenry, to make too easy an identification between being a defendant and being a wrongdoer, which has the long-term effect of increasing the judge's capacity for fairness. The capacity for fairness, however, is not tied in any empirically clear way to increased social stability or even more

Four: Some Worthy Opponents

correct verdicts. In that case, the presumption of innocence is perhaps best interpreted as simply a mental discipline undertaken by the judge and onlookers to correct what is taken to be an inherently bad psychological tendency that can, in various indirect ways, inhibit the administration of justice and, more generally, the wholesomeness of the defendant's subsequent interaction with his or her fellows.

The upshot of this part of the discussion, then, is that instead of seeing presumption as a conservative force in legal reasoning and hence an *object of criticism*, our reinterpretation of the Anglo-Saxon presumption of innocence illustrates a sense in which presumption may function as a *tool for criticizing* and revising beliefs that are widespread among judges and other legal functionaries, and yet not conducive to promoting the goals of the legal system. At this point, I would like to extend this new view of presumption to epistemic matters, where it helps to spell out much of what is involved in radical conceptual change in science.

PRESUMPTION IN EPISTEMIC MATTERS

In keeping with the new view of presumption, radical conceptual change directly brings about a change in the orthodox beliefs of the scientific community, and only indirectly a change in the actual beliefs of individual members of that community. The difference suggested here between *orthodox beliefs* and *individual beliefs* is reflected in the different answers that would be given to the following two questions:

> 1. What should the members of a community take to be the dominant beliefs of their community?
>
> 2. What should each member of a community believe for herself?

Indeed, I maintain that radical conceptual change is possible because the answer to question (2) places no necessary constraints on the answer to question (1), while the answer to question (1) can be deployed as part of a strategy for altering the answer to question (2).

Question (1) considers how members of the community think that the burden of proof should be distributed among their beliefs. For even if the vast majority of members of the community happen to hold a certain belief, that is no indication of whether they would allow it to pass in open forum without strenuous argument. As in the case of legal functionaries wanting to inhibit their natural tendency toward believing that all defendants are probably lawbreakers, there are *methodological* and *ideological* reasons why the members of a scientific community might have an interest in keeping certain of their widely held beliefs from achieving the status of an orthodoxy by holding those beliefs accountable to standards of proof that they clearly cannot meet (cf. Harman 1986: 50-52, for the only recent attempt in analytic philosophy to deal with the difference between beliefs "held for oneself" and those "held for others").

Among the methodological reasons may be that the belief, though widely held, is held only on the basis of indirect evidence or on the purely pragmatic grounds that such a belief is implicit in the assumption that the standing beliefs of the community form a maximally coherent set. Elevating a belief of this sort to the status of an orthodoxy, tempting as it may be, would inhibit any further testing and thereby prevent the scientific community from discovering whatever element of falsehood it may contain. Indeed, Popperians would be especially suspicious of such a belief, since its official acceptance promises to close critical inquiry on the set of beliefs it renders maximally coherent. Another methodological reason for restricting the set of orthodox beliefs is to keep domains of inquiry separate. For example, it would not be surprising to learn that most natural scientists believe not only in the existence of God but also in the occurrence of supernatural causation at some point in the history (most likely at the origin) of the universe. Yet the extent of the burden of proof that such a belief must bear (especially as measured by the number of alternative orthodox explanations that must be first ruled out) ensures that it will never again become part of the scientific orthodoxy. Indeed, Jeffrey Stout (1984) has epitomized the Scientific Revolution in the seventeenth century as marking, not the beginning of a decline in the belief in God, but a decline in the social recognition of such a belief as rational, thus shifting the burden of proof from nonbelievers to believers. Less exotic examples of the same phenomenon include the presumption against folk psychological explanations on the part of social scientists as a means of keeping their disciplines separate from common sense, and the presumption against explanatory appeals to the unconscious and class interests on the part of classical humanists as a means of keeping their disciplines separate from the social sciences.

As for the ideological reasons, a commonly held belief may remain unorthodox because it instills a "bad" attitude toward scientific inquiry. For example, few scientists would probably deny that sociologists have accurately captured the extent to which scientific research agendas are opportunistic, indeed to the point that scientists openly

Four: Some Worthy Opponents

admit to manipulating ex post facto the significance of research findings to fit currently popular theoretical debates. Yet, these scientists would also probably resist the suggestion that they henceforth justify knowledge claims in terms of their opportunistic agendas, even if the sociologists turn out to succeed in showing that "the logic of opportunism" better explains (predicts) their behavior than appeals to the allegedly univocal relation in which evidence stands to theory (cf. Knorr-Cetina 1981). Indeed, more than just setting high probative standards, the scientific community has erected many purely *conceptual* barriers that serve to make the sociologist's stance difficult even to articulate. Many of these barriers appear as distinctions that have been canonized by positivist philosophy of science, especially reasons versus causes and theoretical versus practical. While these distinctions are drawn precisely enough so that each pair of terms is jointly exhaustive, their range of applicability is sufficiently malleable so that anything "inherently" a feature of scientific reasoning can always be made to appear on the "reasons" or "theoretical" sides of the distinction. In this way, scientists are able to project to themselves and to the nonscientific community an image, and ultimately an attitude, of detachment from thoughts of career advancement and other forms of self-interest.

So far we have looked at the use of presumption (and, correlatively, burden of proof) in scientific reasoning as a means of *arresting* change that the canon of orthodox beliefs would naturally undergo if all commonly held beliefs were granted orthodoxy. However, perhaps the more interesting use of presumption is in *facilitating* a change in the canon, granting orthodoxy to beliefs that have yet to be widely held by members of the scientific community. For if there is an analogue to the presumption of innocence in Anglo-Saxon law, it is in the attempt to facilitate such a change in the scientific community. However, my model for the structure of this change will be drawn from a determinedly nonscientific source: Pascal's Wager on the existence of God. In a crucial respect, Pascal's problem is similar to that faced by a scientific revolutionary such as Galileo: both want to believe something that, for the moment at any rate, is unwarranted. It would seem, therefore, that they must scotch either their criteria of rationality (which says to have only warranted beliefs) or their desired belief (as rationality would demand). But there is a third way out: *they can cause themselves to arrive at a situation in which their desired belief is warranted.*

Now, as Bernard Williams (1973: chap. 9) has noted, this third situation can arise either because the belief is indeed true (the change in situation would thus have resulted from some improvement in our cognitive powers) or because appearances have been manipulated so as to make the belief seem true (the change in situation would thus have resulted from some form of deception). Clearly, the former is preferable to the latter, but if Feyerabend's (1975) account of Galileo is to be believed, the latter will occasionally do as well. Williams goes on to make the interesting point that a simple sentence-uttering mechanism may have knowledge without having beliefs, in that whereas what one knows can be read off what one says, what one believes cannot be similarly read off, since it is up to the individual to decide what he will say given what he believes. This allows a possibility for *intentional* falsehood that is lacking in the machine. If we regard the change in presumption that occurs during a scientific revolution as a decision to say what one does not necessarily believe, at least in the short term, then Williams provides us with a way of identifying a "collective will" of the scientific community which adopts a presumption in favor of certain knowledge claims that its members do not yet individually believe.

In any case, Pascal or Galileo will have to reorganize their environments in such a way that the sort of reasons that would be needed to warrant his desired belief could become available. Again, this may simply involve fabricating some evidence that is tailor-made to the belief, or it may require an extended crucial experiment that is especially designed so that if the evidence warranting the belief does not arise under those circumstances, then the belief is probably false. Whereas the first situation is set up to be foolproof, the second clearly is not, since it is dependent on the state of the world regardless of one's own beliefs. And whatever Galileo may have had in mind, Pascal thought of his wager, with its attendant requirement that he conduct a thoroughgoing Christian life, as a crucial experiment in the above sense. Thus, Pascal feels the risk of the wager, as the strength of God's signs to him vary on a day-to-day basis. Interestingly, this second strategy of presumption formation also approximates Charles Sanders Peirce's use of the term "presumption" (otherwise called "abduction" or "hypothesis"), which, as Ann Holmquest (1986) has observed, operates for Peirce as the motor of scientific progress.

Of course, the distinction between merely fabricating evidence and positioning oneself to acquire genuine evidence can be easily erased from the prospective believer's mind, if she is also able to cause herself to forget her interest in wanting to hold the belief, as Jon Elster (1979: chap. 2) suggests in his account of presumption as "precommitment." Annette Baier (1985: chap. 4) offers an interesting slant on this topic, in

Four: Some Worthy Opponents

that she accepts that changing one's mind consists of rethinking the evidential relations of what one already knows, and is thus not tied to a specific piece of evidence or argument, as ordinary belief revision is. However, Baier argues that this rethinking occurs by remembering what was earlier forgotten–à la Plato's *Meno*–instead of forgetting what was recalled. Needless to say, either proposal–but especially Elster's–is a tall order, since forgetting is something done not deliberately but only as a by-product of some other activity. A good candidate for an activity of this sort is the restructuring of discourse that is necessary for articulating any presumptively new relation between language and the world: the introduction of new terms, new meanings for old terms, and new inferential moves within the language in general. As one plays this new language game, it becomes so natural that the player becomes convinced that she has been implicitly playing all her life. This describes not only Pascal's adopted Christian lifestyle (in, e.g., Pensées 252), but also the Whiggish characterization in terms of which revolutionary scientists tend to see their predecessors once they have presumed a new paradigm.

Moreover, there is good social psychological evidence that the mere articulation of the new language game, on a regular and elaborate enough basis by enough people, will have the long-term effect of changing the beliefs of individual scientists. The evidence, drawn admittedly from studies of how political factions tend to gain dominance, suggests that at first any splinter group (say, an inchoate paradigm) is presumed by the public (say, the scientific community at large) to be a minority voice that would not have needed to speak up had its views been adequately represented by the dominant party. As Whately himself put it, under those circumstances, only he who asserts must defend. However, as time goes on, the presumption starts to shift away from the dominant party if it refuses to answer the claims made by the splinter group. For in that case, the public tends to interpret the silence as tacit acceptance of the claims and hence ideological capitulation to the splinter party, which, in turn, often ends up leading the majority of voters to move to its side as well, creating a bandwagon effect." The spiral of silence" (Noelle-Neumann 1982) is the expression that public opinion researchers use to characterize this frequent phenomenon, but readers of Kuhn may recognize it as the "Planck Principle" which baldly claims that a new paradigm triumphs once voices of opposition from the old are silenced. If this link is apt, then we see the beginnings of a theory of long-term rational conceptual revision that incorporates much of what is distinctive in Kuhn's work into a general account of presumption. Fuller (1985) made some opening moves in that direction.

POSTSCRIPT

The World of Tomorrow, as Opposed to the World of Today

In the world of tomorrow, breakthroughs in the natural sciences are regarded as triumphs of applied sociology and political economy, rather than of, say, theoretical physics, chemistry, or biology. It is presumed that a distinctive knowledge product reflects an innovative form of social interaction among knowledge producers and their publics. As for the languages of the special sciences–physics, chemistry, biology, and so on–they are taken to cut the world up spatially rather than conceptually. However, the relevant sense of space is more that filled by a transnational corporation, whose parts are distributed throughout the globe, than that filled by a nation-state, which occupies one well-bounded place. In other words, the metaphor of "disciplinary boundaries" suggests a misleading sense of space in the world of tomorrow. For, these are not absolute spaces, preexistent realms of being waiting to be discovered by science; rather, they are relative spaces constituted by structured social interaction. Whereas, in today's world a public television program recounts a scientific breakthrough by focusing on the construction of an experiment or technique that enabled some discovery to take place, in tomorrow's world the focus is instead on the arguments, both in person and in print, that were used to convince various constituencies that the artifact constructed in the lab warranted specific responses that turned out to empower certain people at the expense of others. In tomorrow's world, this differential empowerment, this redistribution of resources, is what, in today's world, would be called the "empirical content" of science. It is the social analogue of the positivist-pragmatist definition of knowledge as the difference that accepting or rejecting a given claim would make to the world.

In today's world, language is a "thin" phenomenon, generally confined to the well-ordered noises that come from people's mouths, pens,

Postscript

or keyboards. This is counterposed to external reality, which has a mind of its own that often resists the attempts of language to represent it. In tomorrow's world, however, language is a "thick" phenomenon that exists only in and through social interaction, of which formal syntax and phonology are simply abstractions–convenient for some purposes but misleading for most. Here language does not merely "represent" the structure of reality, but is already embedded *as* the structure of reality. In other words, "structure" is less an imaginative projection and more a mnemonic recovery. Those aspects of tomorrow's world that will be called "external" refer to cognitive liabilities, namely, whatever we cannot predict, recall, or otherwise structurally incorporate without great effort. The remedy is to restructure our environment so as to enable the perception of new things and the ignorance of others.

In the world of today, self-styled "holists" in the philosophy and sociology of science say that certain theories are preferred to others because they demand a change in fewer of the beliefs that we already hold. The effort implicitly conserved by this preference is that of conception or imagination. However, in the world of tomorrow, the relevant quantities conserved are labor and capital, as theorists become more prone to argue about the social and material dislocation that would attend the implementation of alternative theories: Who would be dis/enabled to speak authoritatively? For what? When and where? And to what effect, for whose good? To put it most crassly, any theory can be *made* true, if we are willing to pay the price.

The love affair that Western thought has had with the idea of truth as something that is "discovered" or "revealed" finally comes to an end in the world of tomorrow. Today's talk of knowledge as gradually emerging through a process of "decontextualization" sounds odd to tomorrow's ears, just as eighteenth-century talk of the chemical process of "dephlogistication" sounds strange to today's. The oddity in both cases lies in the image of something becoming more substantial as it suffers a loss. In the one case, the loss of phlogiston supposedly added to the weight of a burnt piece of metal; in the other, loss of context-specificity supposedly adds to the validity of a knowledge claim. But in tomorrow's world, gains of both sorts will be seen, quite reasonably, as due to gains: just as dephlogistication turned into oxidation, decontextualization will turn into standardization. Whereas, in today's world the scales fall from one's eyes when one, faces the truth, in tomorrow's world one must learn to see the world aright. Thus, emancipation comes to be known as a subtle form of imposition, thereby enabling the transaction costs of knowledge production to come into view, which is to say, epistemologists become used to asking who had to pay how much for knowledge that is nevertheless advertised as being the property of all.

Today it is common for civic-minded people to worry about the impact of science on social policy. In particular, they fear "necessities" and "essences"–that science will arrive at some ultimate facts about ourselves and the world that will trump democratic values of liberty, equality, and progress: Are intellectual differences attributable to racial ones? Is personality determined in infancy? Questions like these encourage otherwise enlightened liberals to argue that some things are better left unknown, or that science needs to be held tightly in check by our common humanity. The world of tomorrow does not deny the need for a humane science, but it will be more receptive to a scientifically changed conception of humanity. Scientophobia is a thing of the past, as people come to realize that the determinateness of reality–that there are facts of the matter–does not imply a cosmic sense of determinism. In fact, it implies *the very opposite*–especially once we take seriously the idea that claims to truth or falsehood are impossible without linguistic and other technologies capable of enforcing the true/false distinction.

The possibilities for human action are expanded enormously once the bounds of the "human" are taken to exceed the capacities of the unadorned body and, more specifically, once the innate is no longer seen as unchangeable. This point is especially instrumental in overcoming the threat to our political sensibilities currently posed by the specter of, say, a genetic basis of intelligence. The model for handling such possibilities tomorrow is today's attitudes toward myopia. That there is a wide, and probably innate, variability in people's visual abilities has led neither to the devaluation of the social contributions made by the nearsighted, nor to remedial courses for improving myopic vision; rather, eyeglasses have been made generally available at a nominal cost. In short, by technologically extending the body, today's brute biology is converted to tomorrow's consumer economics. This does not make matters any less controversial: After all, who pays for producing and distributing the prosthetic devices that directly benefit only a portion of the population? But economization makes the questions more tractable by opening them to negotiation.

Nowadays, it is common to see alternative research programs as "competing" to explain, or otherwise "save," roughly the same range of phenomena. This makes the history of science seem like a series of winner-takes-all jousting matches. As for the losers, they either scramble for cover in the enemy camp or disappear altogether. No such zero-sum games are to be found in the histories of science written in tomorrow's

Postscript

world. Rather, the cost of conducting the joust is itself taken more seriously, as each rival research program is portrayed as trying to outdo the other in its ability to incorporate its rival's interests without losing its own original focus. In tomorrow's world, the model for epistemic change is no longer war–"scientific revolutions" will lose their *Sturm und Drang* quality–but democratic party politics, in which we all win or lose together. Grantsmanship becomes the art of coalition formation. As a result, potentially affected third parties, who in today's world would be overlooked by the grant proposers, may tomorrow become decisive in swinging grant money from one team to another. These third parties may be openly courted, and, indeed, research funding may start to be seen as a form of "campaigning" that envelopes the entire populace in discussions over the consequences of pursuing competing lines of research. Science policymakers take such maneuvers into account as they periodically rearrange the scientists' incentive structure, so that they are motivated to team up with members of different disciplines or research traditions. Much of the concern and interest that is currently generated by how the individual scientist conducts her research is shifted, in tomorrow's world, to a focus on the patterns by which the products of such research are combined and distributed. In this respect, the local autonomy that scientists and their well-wishers jealously guard today is gladly granted in tomorrow's world, because it is finally appreciated that what really matters is what happens once science ventures forth from the laboratory.

One factor that facilitates tomorrow's image of epistemic pursuit is a closer link between the material scarcity that gives rise to budgets (i.e., we cannot afford to fund every project) and the cognitive dissonance that gives rise to theory choices (i.e., not every theory of a given domain can be true). Left to their own devices, with limitless time, money, and energy, scientists can entertain a variety of incompatible theories indefinitely. Hard decisions–the stuff of which paradigm shifts are made–do not naturally arise in the pursuit of pure inquiry, but must be occasioned by the intrusion of a world of action upon pure thought. For example, there are two ways of looking at the recent superabundance of funds for science. Today's way is to say that more funds leads to better science, largely because of the waste that is needed in order to foster the serendipitous character of good research. By contrast, tomorrow's way looks askance at superabundance, as promoting inefficiency by offering no incentive for scientists to prioritize their research or to consider how they might work with others to mutual advantage. From today's standpoint, tomorrow's way looks utopian insofar as tight budgets alone will not solve the problem of conceptualizing knowledge production in cost-benefit terms. There would seem to remain the problem of identifying the relevant epistemic outputs and assigning values to them. Yet, from tomorrow's standpoint, today's worry appears beside the point, as fuzzy outputs are retrospectively seen as indicative of an inquiry without clear decision points. That biology and physics, say, evince incommensurable values that make them impossible to be compared is simply an artifact of their drawing from separate pools of funds, which prevents them from ever coming into direct competition. Specify the parameters of the decision that needs to be made–by whom, for whom, between what, for how long–and the relevant sorts of outputs and value dimensions will come into focus. In any case, the outputs in tomorrow's world are probably not quite the ones that policymakers currently fall back on, such as an author's citation counts, since these reflect scientists' attempts at maximizing their opportunities in the existing disciplinary structure–hardly the forum for registering their views on the worth of that structure, which is what policymakers need to know in tomorrow's world.

In today's world, the idea that science is a public trust has merely presumptive status: that is, people presume without proof that they are somehow served by science. This has delivered unto science a passive consumer culture that has no formal mechanism to account for what scientists do. Not so in tomorrow's world, where knowledge production absorbs a still greater share of human and material resources than it does now. Under these new circumstances, science is more integrated into the political structure, so that the lay public is routinely called (in the manner of jury duty today) to participate in research projects, during which the public has a say in at least the interpretation, and quite possibly the conduct, of the research. Scientific expertise in tomorrow's world is not treated with the uncritical respect that it is often accorded today. Rather, appeals to expertise are seen primarily as means to end debate and, as such, to be tolerated as expedient in the short term but to be suspected in the long. Much in the original spirit of positivism and pragmatism, "method" in tomorrow's world is regarded as something *opposed to*, not in league with, expertise–insofar as method implies a publicly accessible procedure for evaluating testimony, whereas expertise suggests the elite authorization of testimony. Because this point is realized, the social scientific understanding of knowledge production is no longer feared as supplanting natural science expertise with a yet more pernicious form of scientism. The success of a "science of science" depends on whether the inner workings of knowledge production are fathomable by people who are clearly nonexpert, and hence in a position to empower society at large with knowledge of these workings. As

Postscript

might be expected, one consequence of this penetration of science's internal mechanism in tomorrow's world is a shift in what counts as a "hands-on" understanding of science. Nowadays, such an understanding is conveyed by expert scientists in their disciplinary jargons, whereas, in the world of tomorrow, jargon is regarded as an abstraction–an abstraction that hovers above the micromechanics of text production and social interaction that captures science "as it actually happens."

APPENDIX

Course Outlines for Science and Technology Studies in a Rhetorical Key

TECHNICAL COMMUNICATION FOR SCIENTISTS

STS research is "constructivist" in the sense that the objects of scientific inquiry–and, indeed, the separateness of science from other social practices–are constructed by introducing and enforcing certain ways of writing, or "conventions." But even when these conventions seem to have universal status, their application to a specific case needs to be negotiated with the audience you are addressing. For example, scientists know that they must operationalize their concepts in order for colleagues to test their hypotheses. It remains an open question, however, how you manage to convey this in the paper you are about to write. The results determine the group of people who are empowered to hold what you say accountable. If you write in the standard technical prose of your discipline, then only other people trained in that field will be able to evaluate what you say. The repeated occurrence of this phenomenon gives the science-society boundary the sharpness that it has today. However, as with other forms of discrimination, it is not clear that this one is warranted.

As a matter of fact, more people than just you and your colleagues have a stake in the conduct of your research, although the way in which scientists typically write obscures that fact from both yourself and those people. Jargon is not the stuff of which the public interest is made, yet the activities hiding behind that jargon are maintained largely through taxes and corporate sponsorship. You may look at this challenge to write accessibly (or "accountably") either as an *obligation* or as an *opportunity*. In one important sense, the two are linked. For, given the increasing percentage of public and private funds devoted to research, and the increasingly public character of the consequences of such research, it is only a matter of time (perhaps the occurrence of one

Appendix

high-tech "accident" too many) before greater accountability will be *demanded*. Rather than appearing to be forced to do something that you would otherwise not do, why not now take the opportunity to develop a style that enables the interested lay reader to ask critical questions of what you are up to?

In this course, we will stress the opportunities that accountable writing can open up for you. No doubt you come to this course thinking that this will involve diluting the scientific content of your writing and further corrupting the dilution with the extrascientific concerns that are needed to attract the lay reader's attention. Perhaps the chief goal of this course is to disabuse you of these preconceptions by persuading you of the constructivist premise of STS research. In other words, whenever you say that writing for a larger public forces you to "dilute the content" of your prose, you are simply expressing resistance to the idea of having certain groups in society—on whose goodwill you *already* depend—ask you critical questions. And, what you may initially see as the public's "extrascientific" concerns are in fact attempts to draw the science-society boundary somewhat differently from the way it was drawn in the past.

One thing you need to realize at the outset is that you and the public have more in common than you realize. In particular, neither of you spends much time thinking about science in society, or the political dimension of knowledge more generally—that is, until it affects you personally. As a result, you share some rather naive views. Since it is generally easier to spot one's own faults in what others do, we will start by critically examining published works that foster the naive view, even in their attempt to make science more publicly accessible. These will serve as "bridging texts" that will increase your awareness of the interdependency of science and society. We will try to simulate a technique developed by the psychologist Jean Piaget to get children to see contradictions in their thought in such a way that they are then able to develop a more comprehensive framework within which those contradictions can be resolved. Whereas Piaget had his subjects perform experiments designed to make the contradictions vivid for them, you will be asked to locate paradoxical turns in a text's argument that arise as a result of its promoting conflicting images of science. Typically, an unwittingly placed word or expression will reveal the paradox. For example, most of the talk about the "self-governing," or "autonomous," character of scientific research has been generated during the period when it has been increasingly subject to federal and corporate sponsorship. You will then be asked to rewrite the text so as to bring out the conflicting images of science that this suggests, the point being to make the conflict visible for public discussion.

Appendix

Projected Textbook: James Collier and David Toomey, two English composition instructors who have been converted to STS, are preparing the text that would realize the goals of this course (Collier and Toomey 1994). Below is their current working table of contents. Each chapter is devoted to an aspect of the technical writing process–communicator roles, communication strategies, and tools. To each aspect correspond readings in the relevant STS issues raised (indicated here in parentheses):

Scientific and Technical Communication for the Twenty-First Century

Introduction: Understanding the Relation of Science, Technology and Society

Part I: The Roles of Technical Communicators

1. Learner (New Images of Science and Technology)
2. Reader (Reading Science)
3. Writer (Technical Communication: Beyond Form and Content)
4. Participant (Democracy, Science and Information)

Part II: The Strategies of Technical Communicators

5. Considering the Audience (Gender, Science and Technology)
6. Conducting research (Scientific Journalism)
7. Arguing and Persuading (The Ethics of Scientific Practice)
8. Formatting, Designing and Using Graphics (Intellectual Property)

Part III: The Tools of Technical Communicators

9. Letters, Memos and Electronic Mail (Technology in an Age of Limits)
10. Job-finding Materials (The Workplace in the 21st century)
11. Functional, physical, and process descriptions (Science Writing for the Layperson)
12. Instructions (Scientific and Technological Controversies)
13. Proposals (Corporate and Private Funding of Science and Technology)
14. Technical and analytical reports (Governmental Science)
15. Scientific Articles and Abstracts (Scientific Publication and Scholarship)
16. Oral Presentations (Communication Within the Laboratory)

Appendix

PHILOSOPHY OF SCIENCE

Philosophy used to be the discipline that tried to explain everything within one system. In those days—which continued well into this century—philosophy was principally identified with metaphysics, which was, in turn, distinguished from the more limited missions of the special sciences. The turn from "philosophy" to "philosophy of science" began when the positivists projected this traditional systematic function of philosophy onto the idea of "unified science," in which the search for underlying principles was replaced by a scheme for translating and reducing the phenomena of all the special sciences into one master science. In this scheme, philosophers would not themselves explain anything, but would instead articulate standards of explanation and pass judgment on the adequacy of particular explanations. In effect, the positivists put the philosopher in the role of referee of the knowledge process, monitoring the flow of information between the disciplines, e.g., to ensure that one discipline was not relying on ideas or data that some other discipline had rendered problematic, obsolete, or in some other way unwarranted.

Nowadays, in our "postmodern" world, it is fashionable to think that there is nothing valuable about either metaphysical explanation or its positivist successor, reductionism. Consequently, most contemporary accounts of explanation deny that there is any overarching need for explanation—even in science—aside from particular requests that people have for knowing why certain things happen (cf. van Fraassen 1980: chap. 5). The point of this course is to counteract this "settle for less" mentality that has beset recent discussions of explanation—and much else in the philosophy of science.

STS practitioners ought to take an interest in this topic because the most distinctive conceptual moves made by STS researchers (e.g., the Edinburgh school, critical Marxism, constructivism, actor-network theory) have involved showing that seemingly disparate phenomena are in fact instances of the same deep and general principles: e.g., that there really is no sharp difference between science and the rest of society, initial appearances to the contrary. One does not need to be Bruno Latour (1987) to believe that explanations necessarily have both a political and an epistemic character, and that the two are not easily separated. For example, in defending a contemporary version of the metaphysical search for deep explanations, Robert Nozick (1982) considers why so much intellectual and sometimes even political power is gained by being able to redescribe disparate features of reality in terms

Appendix

of One Big Picture. You will be asked to be on the alert for this duality in the weeks that follow. My own slant is that explanations are claims to intellectual property, so that the successful claimant is socially acknowledged as having authority over the "disposition" of the thing explained (or *explanandum*, in positivist lingo). Thus, if someone else wishes to make use of the *explanandum*, she implicitly holds herself accountable to the person(s) socially acknowledged as having provided an explanation for it.

For the term paper, you should take something that is normally explained by one discipline and explain it in terms of the theories and data of another discipline. It does not matter whether it is, say, sociology explaining physics or vice versa. In any case, you would need to redescribe the phenomenon explained so that it can be discussed more fluently in the second discipline. What you come up with might look quite strange to a practitioner of the first discipline. In fact, you should document the resistance you meet in trying to make the translation between the two disciplines, both in terms of finding the relevant words and principles and in terms of persuading practitioners of the first discipline that your proposed explanation actually illuminates something that interests them.

HISTORY OF SCIENCE

Take an explanation of some historical episode that you find reasonably convincing, or at least that you are willing to defend for purposes of this paper. After describing this account in detail, and especially why you find it stronger than competing accounts, then go on to explore how you would persuade the people accounted for by this explanation that it does, indeed, make the most sense of what they were up to. Imagine that this act of persuasion requires that you go back to the original moment in time, so that you can work only with information that the people had back then. Clearly, then, you won't be able to simply give them the reasons why *you* bought the explanation, since your reasons were informed by later research, to which they do not yet have access. In that case, what conceptual/empirical obstacles would you have to overcome in order to show that your account of them makes sense as a "natural extension" of what they already believe? Would you have to change some of their fundamental beliefs, and could you do that in a relatively nonobtrusive manner by working with other things they know?

Suppose you were interested in persuading Newton of Frank Manuel's (1969) psychoanalytic explanation of his work. While you

Appendix

wouldn't be able to appeal to the interplay of such Freudian mechanisms as ego, superego, and id, you could nevertheless appeal to such seventeenth century analogues as the intellect, the will, and the passions. However, translating the etiological side of the Freudian explanation would require some ingenuity, as early childhood encounters with parents had yet (in the seventeenth-century) to acquire the significance for adult behavior that Freud bestowed on them. Yet, even here, it shouldn't be too hard to find a seventeenth-century belief that could serve as a touchstone from which to start to convince Newton that, say, unresolved tensions about his mother decisively influenced his scientific work. Perhaps your best bet would be to take advantage of Newton's intimate familiarity with biblical doctrines of sin.

This exercise will force you to integrate the following concerns: What is a good historical explanation? Can the historian "dialogue with the past" in some interesting sense, or is the expression just idle humanist rhetoric? How do you determine what people at a given time knew? Can the difference between your own "third-person" and your interlocutor's "first-person" accounts of the event be reconciled by some "second-person" acts of persuasion, or must you as historian choose between the two perspectives?

SCIENCE POLICY

Since STS will survive as a field only by constant outreach to other disciplines and the general public (e.g., in teaching or policy jobs), you must learn to argue your case in a clear and incisive manner. Toward this end, the course will depart somewhat from the usual heavy emphasis on writing. Instead, the class will prepare debates on the merits of some controversial issue in STS, for which the materials covered in this course will provide a general framework, but little specific guidance. Here is a list of possible "resolutions" from which to choose:

> –Since "objectivity" is illusory, STS should be explicitly oriented toward a political agenda.
> –Feminism has the theoretical resources to radically revise our understanding of science.
> –Almost everything interesting about modern science can be explained in terms of the larger cultural issues dominating our time.
> –What passes for "science" these days is so big that it is better seen as a kind of transnational corporation than a knowledge producing enterprise.

Appendix

–STS gains strength from not having a clear disciplinary identity.
–STS research can refute certain claims that philosophers have put forth about science.
–STS researchers should study the natural and social sciences in the same way.
–Knowledge is powerful only because a few people have it.
–STS researchers generally understand the nature of science better than practicing scientists do.
–It takes a special psychological makeup to become a successful scientist.

In typical academic debate, resolutions are presented in fairly vague terms, and the affirmative's opening move is to give the resolution a more precise interpretation. In that case, the negative side must address that interpretation of the resolution. In practice, this means that you will find a partner and select a resolution, one person arguing the affirmative case, and the other the negative case. Although you will be arguing for opposing viewpoints, you should do your research collaboratively, so that you can make each other's arguments more effective. In fact, both members of the same team will receive *the same grade* for their debate. Think of these debates as staged events. The affirmative side will have 20 minutes, the negative 10 minutes for rebuttal, then 20 minutes to provide her own position, and then the affirmative gets the final 10 minutes to rebut the negative's position, with another 15 minutes devoted to unrehearsed questions from the audience (which may include invited members of the faculty). The professor will remain quiet during the event, debriefing each team afterward on the strengths and the weaknesses of the arguments presented. This exercise is designed to simulate three features of real world encounters that STS researchers increasingly face:

> 1. an audience potentially receptive to what you have to say, but initially uninformed about the issues involved (a point will be made of *not* assigning the class any specific readings in advance of a given debate)

> 2. a time constraint that will force you to say less than half as much as you would, were you preparing an adequate term paper on the topic of the debate

> 3. the need to take a clear stand, instead of vacillating between

Appendix

>positions in a more or less thoughtful manner, as "proper" academic writing all too often encourages (moreover, by taking a clear stand that might be demonstrably wrong, your audience will be encouraged to engage you).

In addition to these virtues, one general sensibility that this course aims to instill is what the Greek rhetoricians called *kairos*, or "timeliness," an art that seems to have been lost as soon as rhetoric moved out of the forum and into the classroom (cf. Kinneavy 1986).

Because academic writing tends to be increasingly treated (by both its authors and its readers) as purely archival, university life provides few incentives for communicators to urge the timeliness of their arguments. Unfortunately, this lost art is crucial to persuading people in policy settings, where you need to show not only that your case has merits but, more important, that a certain course of action ought to be taken–and soon–in light of those merits. For an academically trained person just entering the policy arena, the challenge is to insinuate one's abstract concerns (for empowering disadvantaged groups, for instilling global consciousness, etc.) in concrete issues that are *already* on the minds of policymakers. Lobbyists do this all the time by convincing legislators to set up a free-standing agency to deal with problems of the sort that the lobbyist has successfully highlighted in a well-publicized case (Ornstein and Elder 1978). The lobbyist succeeds when she shows that the case at hand exemplifies certain general concerns that deserve systematic treatment. Students in this course will be required to engage in the "casuistic" thinking and research that lobbying requires (Jonsen and Toulmin 1988 is the best philosophical history of this topic). In practice, this means learning to integrate academic and journalistic sources, as well as scientific and political agendas, in forging a persuasive argument.

REFERENCES

INDEX

REFERENCES

Abbott, Andrew. 1988. *The System of Professions.* University of Chicago: Chicago.
Abraham, Henry. 1968. *The Judicial Process.* Oxford University Press: Oxford.
Ackerman, Robert. 1985. *Data, Instrument, Theory.* Princeton University Press: Princeton.
Adorno, Theodor, ed. 1976. *The Positivist Dispute in German Sociology.* Heinemann: London.
Agassi, Joseph. 1985. *Technology.* Kluwer: Dordrecht.
Agger, Ben. 1989. *Fast Capitalism.* University of Illinois Press: Urbana.
Aitchison, Jean. 1981. *Language Change: Progress or Decay?* Universe Books: New York.
Albrow, Martin, and King, Elizabeth, eds. 1990. *Globalization, Knowledge and Society.* Sage: London.
Almond, Gabriel. 1989. *A Discipline Divided: Schools and Sects in Political Science.* Sage: London.
Almond, Gabriel, and Verba, Sidney. 1963. *The Civic Culture.* Princeton University Press: Princeton.
Althusser, Louis. 1989. *Philosophy and the Spontaneous Philosophy of the Scientist Scientists.* Verso: London.
Amabile, Teresa. 1983. *The Social Psychology of Creativity.* Springer Verlag: New York.
Amundson, Ron. 1982. "Science, Ethnoscience, and Ethnocentrism." *Philosophy of Science* 49: 236-50.
Anderson, John. 1986. *The Architecture of Cognition.* Cambridge University Press: Cambridge.
Argote, L., and Epple, D. 1990. "Learning Curves in Manufacturing." *Science* 247: 920 24.
Arkes, Hal, and Hammond, Kenneth, eds. 1986. *Judgment and Decision Making.* Cambridge University Press: Cambridge.
Arnauld, Antoine. 1964. *The Art of Thinking.* Bobbs-Merrill: Indianapolis.
Aronowitz, Stanley. 1988. *Science as Power.* University of Minnesota Press: Minneapolis.
Ash, Mitchell. 1980. "Academic Politics in the History of Science: Experimental Psychology in Germany: 1879-1941." *Central European History* 13: 255-86.
Ashmore, Malcolm. 1989. *The Reflexive Thesis.* University of Chicago Press: Chicago.
Ashmore, Malcolm. "Social Epistemology and Reflexivity." Forthcoming in *Argumentation*.
Aune, James. 1990. "Cultures of Discourse: Marxism and Rhetorical Theory." In Williams and Hazen 1990.

References

Averch, Harvey. 1985. *A Strategic Analysis of Science and Technology Policy.* Johns Hopkins University Press: Baltimore.
Averch, Harvey. 1989. "New Foundations for Science and Technology Policy." Paper delivered at the conference on "The mutual relevance of science studies and science policy," Virginia Tech.
Ayer, A. J. 1936. *Language, Truth and Logic.* Gollancz: London.
Ayer, A. J., ed. 1959. *Logical Positivism.* Free Press: New York.
Baars, Bernard. 1986. *The Cognitive Revolution in Psychology.* Guilford Press: New York.
Bachrach, Peter, and Baratz, Morton. 1962. "The Two Faces of Power." *American Political Science Review* 56: 947-52.
Baier, Annette. 1985. *Postures of the Mind.* University of Minnesota Press: Minneapolis.
Barnes, Barry. 1974. *Scientific Knowledge and Sociological Theory.* Routledge: London.
Barnes, Barry. 1982. *T. S. Kuhn and Social Science.* Columbia University Press: New York.
Barnes, Barry, and Bloor, David. 1982. "Relativism, Rationalism, and the Sociology of Knowledge." In Martin Hollis and Steven Lukes, eds., *Rationality and Relativism.* MIT Press: Cambridge.
Barnes, Barry, and Edge, David, eds. 1982. *Science in Context.* Open University Press: Milton Keynes UK.
Bartley, W. W. 1984. *The Retreat to Commitment.* 2d ed. Open Court Press: La Salle IL.
Barwise, John, and Perry, Jon. 1983. *Situations and Attitudes.* MIT Press: Cambridge.
Basalla, George. 1988. *The Evolution of Technology.* Cambridge University Press: Cambridge.
Bazerman, Charles. 1988. *Shaping Written Knowledge.* University of Wisconsin Press: Madison.
Bazerman, Charles. 1989a. *The Informed Reader.* Houghton Mifflin: Boston.
Bazerman, Charles. 1989b. *The Informed Writer.* 3d ed. Houghton Mifflin: Boston.
Bechtel, William, ed. 1986. *Integrating Scientific Disciplines.* Martinus Nijhoff: Dordrecht.
Bechtel, William. 1988. *Philosophy of Science: An Overview for Cognitive Science.* Lawrence Erlbaum Associates: Hillsdale NJ.
Beiser, Frederick. 1987. *The Fate of Reason: German Philosophy from Kant to Fichte.* Harvard University Press: Cambridge.
Bell, Daniel. 1960. *The End of Ideology.* Free Press: New York.
Bell, Daniel. 1973. *The Coming of Post-Industrial Society: A Venture in Social Forecasting.* Basic Books: New York.
Ben-David, Joseph. 1984. *The Scientist's Role in Society.* 2d ed. University of Chicago Press: Chicago.
Berkowitz, Leonard, and Donnerstein, Edward. 1982. "Why External Validity Is More Than Skin Deep." *American Psychologist* 37: 245-57.
Bernard, Claude. 1964. *The Study of Experimental Medicine.* Dover: New York.
Bernstein, Richard. 1983. *Beyond Objectivism and Relativism.* University of Pennsylvania Press: Philadelphia.
Bhaskar, Roy. 1979. *The Possibility of Naturalism.* Harvester: Brighton UK.

References

Bijker, Wiebe; Hughes, Thomas; and Pinch, Trevor, eds. 1987. *The Social Construction of Technological Systems.* MIT Press: Cambridge.
Billig, Michael. 1987. *Arguing and Thinking.* Cambridge University Press: Cambridge.
Bitzer, Lloyd. 1968. "The Rhetorical Situation." *Philosophy and Rhetoric* 1: 1-14.
Blaug, Mark. 1978. *Economic Theory in Retrospect.* 3d ed. Cambridge University Press: Cambridge.
Block, Fred. 1990. *Post-Industrial Possibilities.* University of California Press: Berkeley.
Bloom, Allan. 1987. *The Closing of the American Mind.* Simon and Schuster: New York.
Bloomfield, Brian, ed. 1987. *The Question of Artificial Intelligence.* Croom Helm: London.
Bloor, David. 1976. *Knowledge and Social Imagery.* Routledge: London.
Bloor, David. 1979. "Polyhedra and the Abominations of Leviticus." *British Journal of the History of Science* 13: 254-72.
Bloor, David. 1982. "Durkheim and Mauss Revisited: Classification and the Sociology of Knowledge." *Studies in History and Philosophy of Science* 13: 267-98.
Bloor, David. 1983. *Wittgenstein: A Social Theory of Knowledge.* Blackwell: Oxford.
Bocock, Robert. 1986. *Hegemony.* Tavistock: London.
Boehme, Gernot. 1977. "Cognitive Norms, Knowledge Interests, and the Constitution of the Scientific Object." In E. Mendelsohn et al., eds., *The Social Production of Scientific Knowledge.* Reidel: Dordrecht.
Boehme, Gernot, and Stehr, Nico, eds. 1986. *The Knowledge Society.* Reidel: Dordrecht.
Bok, Derek. 1982. *Beyond the Ivory Tower.* Harvard University Press: Cambridge.
Booth, Wayne. 1979. *Critical Understanding.* University of Chicago Press: Chicago.
Boring, Edwin. 1957. *A History of Experimental Psychology.* 2d ed. Appleton Century Crofts: New York.
Botha, Rudolf. 1989. *Challenging Chomsky.* Blackwell: Oxford. Bottomore, Tom, and Nisbet, Robert, eds. 1977. *A History of Sociological Analysis.* Basic Books: New York.
Boulding, Kenneth. 1968. *Beyond Economics.* University of Michigan Press: Ann Arbor.
Bourdieu, Pierre. 1986. *Distinction.* Harvard University Press: Cambridge.
Brannigan, Augustine. 1981. *The Social Basis of Scientific Discoveries.* Cambridge University Press: Cambridge.
Brannigan, Augustine. 1989. "Artificial Intelligence and the Attributional Model of Scientific Discovery." *Social Studies of Science* 19: 601-12.
Brannigan, Augustine, and Wanner, Richard. 1983. "Historical Distributions of Multiple Discoveries and Theories of Scientific Change." *Social Studies of Science* 13: 417-35.
Brenner, Reuven. 1987. *Rivalry.* Cambridge University Press: Cambridge.
Brown, Harold. 1989. "Towards a Cognitive Psychology of What?" *Social Epistemology* 3: 129-38.
Brown, James Robert, ed. 1984. *The Rationality Debates: The Sociological Turn.* Reidel: Dordrecht.

References

Brunswik, Egon. 1956. *Perception and the Representative Design of Psychological Experiments*. University of California Press: Berkeley.
Burke, Kenneth. 1969. *The Grammar of Motives*. University of California Press: Berkeley.
Burnham, John. 1988. *How Superstition Won and Science Lost*. Rutgers University Press: New Brunswick NJ.
Button, Graham, ed. 1991. *Ethnomethodology and the Human Sciences*. Cambridge University Press: Cambridge.
Buxton, William, and Turner, Stephen. 1992. "Edification and Expertise." In Terence Halliday and Morris Janowitz, eds., *Sociology and Its Publics*. University of Chicago Press: Chicago.
Byrne, Richard, and Whiten, Andrew, eds. 1987. *Machiavellian Intelligence*. Oxford University Press: Oxford.
Callon, Michel, and Latour, Bruno. 1981. "Unscrewing the Big Leviathan." In Knorr-Cetina and Cicourel 1981.
Callon, Michel; Law, John; and Rip, Arie. 1986. *Mapping the Dynamics of Science and Technology*. Macmillan: London.
Campbell, Donald. 1969. "Ethnocentrism of Disciplines and the Fishscale Model of Omniscience." In Muzarif Sherif, ed., *Interdisciplinary Relationships in the Social Sciences*. Aldine Press: Chicago.
Campbell, Donald. 1974. "Evolutionary Epistemology." In P. Schilpp, ed. *The Philosophy of Karl Popper*. Open Court Press: La Salle IL.
Campbell, Donald. 1988. *Methodology and Epistemology for Social Science*. University of Chicago Press: Chicago.
Campbell, Donald. 1989. "Fragments of the Fragile History of Psychological Epistemology and Theory of Science." In Gholson et. al. 1989.
Campbell, Donald, and Stanley, Julian. 1963. *Experimental and Quasi-Experimental Designs for Research*. Rand McNally: Chicago.
Carnap, Rudolf. 1967. *The Logical Structure of the World*. University of California Press: Berkeley.
Cartwright, Nancy. 1983. *How the Laws of Physics Lie*. Oxford University Press: Oxford.
Cassirer, Ernst. 1953. *Substance and Function*. Dover: New York.
Catania, Charles, and Harnad, Stevan, eds. 1988. *The Selection of Behavior*. Cambridge University Press: Cambridge.
Cherniak, Christopher. 1986. *Minimal Rationality*. MIT Press: Cambridge.
Cherwitz, Richard, ed. 1990. *Rhetoric and Philosophy*. Lawrence Erlbaum Associates: Hillsdale NJ.
Chisholm, Roderick. 1977. *Theory of Knowledge*. Prentice-Hall: Englewood Cliffs NJ.
Chomsky, Noam. 1959. Review of *Verbal Behavior* by B. F. Skinner. *Language* 35: 26-58.
Chomsky, Noam. 1972. *Language and Mind*. 2d ed. Harcourt Brace Jovanovich: New York
Chomsky, Noam. 1980. *Rules and Representations*. Columbia University Press: New York.
Chubin, Daryl, project director. 1991. *Federally Funded Research: Decisions for a Decade*. Office of Technology Assessment: Washington DC.
Chubin, Daryl, and Chu, Ellen, eds. 1989. *Science Off the Pedestal*. Wadsworth: Belmont CA.

References

Chubin, Daryl, and Hackett, Edward. 1990. *Peerless Science*. SUNY Press: Albany.
Churchland, Paul. 1979. *Scientific Realism and the Plasticity of Mind*. Cambridge University Press: Cambridge.
Churchland, Paul. 1984. *Matter and Consciousness*. MIT Press: Cambridge.
Churchland, Paul, and Hooker, Clifford, eds. 1985. *Images of Science*. University of Chicago Press: Chicago.
Clark, Noel, and Stephenson, Geoffrey. 1989. "Group Remembering." In Paulus 1989.
Clifford, James, and Marcus, George, eds. 1986. *Writing Cultures*. University of California Press: Berkeley.
Cohen, I. Bernard. 1985. *Revolutions in Science*. Harvard University Press: Cambridge.
Cohen, L. Jonathan. 1986. *The Dialogue of Reason: A Defense of Analytic Philosophy*. Oxford University Press: Oxford.
Coleman, James. 1961. *The Adolescent Society*. Free Press: New York.
Coleman, James. 1990. *The Foundations of Social Theory*. Harvard University Press: Cambridge.
Collier, James, and Toomey, David. 1994. *Scientific and Technical Communication for the 21st Century*. HarperCollins: New York.
Collini, Stefan; Winch, D.; and Burrow, J. 1983. *That Noble Science of Politics*. Cambridge University Press: Cambridge.
Collins, Harry. 1981. "What Is TRASP?" *Philosophy of the Social Sciences* 11: 215-24.
Collins, Harry. 1985. *Changing Order*. Sage: London.
Collins, Harry. 1989. "Computers and the Sociology of Scientific Knowledge." *Social Studies of Science* 19: 613-24.
Collins, Harry. 1990. *Artificial Experts*. MIT Press: Cambridge.
Collins, Randall. 1979. *The Credential Society*. Academic Press: New York.
Collins, Randall. 1988. *Theoretical Sociology*. Harcourt Brace and Jovanovich: New York.
Collins, Randall, and Ben-David, Joseph. 1966. "Social Factors in the Origins of a New Science: The Case of Psychology." *American Sociological Review* 34: 451-65.
Conrad, J. J.; Friedenthal, J. H.; and Miller, A. R. 1980. *Civil Procedure: Cases and Materials*. 3d ed. West: St. Paul MN.
Cooper, Robert. 1989. *Language Planning and Social Change*. Cambridge University Press: Cambridge.
Corlett, J. Angelo. 1991. "Some Connections between Epistemology and Cognitive Psychology." *New Ideas in Psychology* 9: 285-306.
Cozzens, Susan, and Gieryn, Thomas, eds. 1990. *Theories of Science in Society*. Indiana University Press: Bloomington.
Crease, Robert, and Samios, Nicholas. 1991. "Managing the Unmanageable." *Atlantic* (January) 263: 80-87.
Crook, Stephen. 1991. *Modernist Radicalism and Its Aftermath*. Routledge: London.
Culler, Jonathan. 1975. *Structuralist Poetics*. Cornell University Press: Ithaca.
Culler, Jonathan. 1982. *On Deconstruction*. Cornell University Press: Ithaca.
Cutcliffe, Stephen. 1989. "The Emergence of STS as an Academic Field." In P. Durbin, ed., *Research in Philosophy and Technology*. JAI Press: Greenwich CT.

References

Czubaroff, Jeanine. 1989. "The Deliberative Character of Strategic Scientific Debates." In H. Simons, ed., *Rhetoric in the Human Sciences.* Sage: London.
D'Amico, Robert. 1989. *Historicism and Knowledge.* Routledge and Kegan Paul: London.
Danziger, Kurt. 1990. *Constructing the Subject.* Cambridge University Press: Cambridge.
Davidson, Donald. 1984. *Inquiries into Truth and Interpretation.* Oxford University Press: Oxford.
Deane, Phyllis. 1989. *The State and the Economic System.* Oxford University Press: Oxford.
Dear, Peter. 1987. *Mersenne and the Learning of the Schools.* Cornell University Press: Ithaca.
Dear, Peter. "We May Give Advice, But We Do Not Inspire Conduct." Forthcoming in *Annals of Scholarship.*
Debray, Regis. 1981. *Teachers, Writers, Celebrities.* Verso: London.
De Mey, Marc. 1982. *The Cognitive Paradigm.* Reidel: Dordrecht.
Dennett, Daniel. 1984. *Elbow Room: The Varieties of Free Will Worth Wanting.* MIT Press: Cambridge.
Dennett, Daniel. 1987. *The Intentional Stance.* MIT Press: Cambridge. Denzin, Norman. 1970. *The Research Act.* Aldine: Chicago.
Derrida, Jacques. 1976. *Of Grammatology.* Johns Hopkins University Press. Baltimore.
Dewey, John. 1922. *Human Nature and Conduct.* Random House: New York.
Dewey, John. 1946. *The Public and Its Problems.* Gateway: Chicago.
Dewey, John. 1958. *Experience and Nature.* Dover: New York.
Dewey, John. 1960. *The Quest for Certainty.* G. P. Putnam and Sons: New York.
Dibble, Vernon. 1964. "Four Types of Inference from Documents to Events." *History and Theory* 4: 203-21.
Dickson, David. 1984. *The New Politics of Science.* Pantheon: New York.
Doise, William. 1986. *Levels of Explanation in Social Psychology.* Cambridge University Press: Cambridge.
Dolby, R. G. A., and Cherry, Christopher. 1989. "Symposium on the Possibility of Computers Becoming Persons." *Social Epistemology* 3: 321-48.
Douglas, Mary. 1986. *How Institutions Think.* Syracuse University Press: Syracuse.
Douglas, Mary, and Wildavsky, Aaron. 1982. *Risk and Culture.* University of California Press: Berkeley.
Downes, Stephen. 1990. "The Prospects for a Cognitive Science of Science." Ph.D. dissertation, Virginia Tech.
Dreyfus, Hubert, and Dreyfus, Stuart. 1986. *Mind over Machine.* Free Press: New York.
Ehrlich, Paul. 1978. *The Population Bomb.* Rev. ed. Ballantine Books: New York.
Elgin, Catherine. 1988. "The Epistemic Efficacy of Stupidity." *Synthese* 74: 297-311.
Elster, Jon. 1979. *Ulysses and the Sirens.* Cambridge University Press: Cambridge
Elster, Jon. 1980. *Logic and Society.* John Wiley and Sons: Chichester UK.
Elster, Jon. 1983. *Sour Grapes.* Cambridge University Press: Cambridge.
Elster, Jon. 1984. *Explaining Technical Change.* Cambridge University Press: Cambridge.
Elster, Jon. 1985. *Making Sense of Marx.* Cambridge University Press: Cambridge.

References

Elster, Jon. 1989. *Solomonic Judgments*. Cambridge University Press: Cambridge.
Engels, Friedrich. 1934. *The Dialectics of Nature*. Progress Publishers: New York.
Engestrom, Y.; Brown, K.; Engestrom, R.; and Koistinen, K. 1990. "Organizational Forgetting." In D. Middleton and D. Edwards, eds., *Collective Remembering*. Sage: London.
Ericsson, K. Anders, and Simon, Herbert. 1984. *Protocol Analysis: Verbal Reports as Data*. MIT Press: Cambridge.
Etzkowitz, Henry. 1989. "Entrepreneurial Science in the Academy." *Social Problems* 36: 14-29.
Ezrahi, Yaron. 1990. *The Descent of Icarus*. Harvard University Press: Cambridge.
Faust, David. 1985. *The Limits of Scientific Reasoning*. University of Minnesota Press: Minneapolis.
Fay, Brian. 1987. *Critical Social Science*. Cornell University Press: Ithaca.
Febvre, Lucien. 1982. *The Problem of Unbelief in the Sixteenth Century*. Harvard University Press: Cambridge.
Feyerabend, Paul. 1975. *Against Method*. New Left Books: London.
Feyerabend, Paul. 1979. *Science in a Free Society*. New Left Books: London.
Feyerabend, Paul. 1981. "Two Models of Epistemic Change: Mill and Hegel." In Feyerabend, ed., *Problems of Empiricism*. Cambridge University Press, Cambridge.
Fields, Christopher. 1987. "The Computer as Tool." *Social Epistemology* 1: 5-25.
Fine, Arthur. 1984. "The Natural Ontological Attitude." In Leplin 1984.
Fine, Arthur. 1986. "Unnatural Attitudes: Realist and Instrumentalist Attachments to Science." *Mind* 95: 149-79.
Fisch, Menachem, and Schaffer, Simon, eds. 1991. *William Whewell: A Composite Portrait*. Cambridge University Press: Cambridge.
Fish, Stanley. 1989. *Doing What Comes Naturally*. Duke University Press: Durham.
Fisher, David. 1990. "Boundary Work and Science." In Cozzens and Gieryn 1990.
Fleck, Ludwig. 1980. *The Genesis and Development of a Scientific Fact*. University of Chicago Press: Chicago.
Fodor, Jerry. 1975. *The Language of Thought*. Crowell: New York.
Fodor, Jerry. 1981. *Representations*. MIT Press: Cambridge.
Fodor, Jerry. 1987. *Psychosemantics*. MIT Press: Cambridge.
Forman, Paul. 1971. "Weimar Culture, Causality, and Quantum Theory, 1918-1927." *Historical Studies in the Physical Sciences* 3: 1-115.
Forrester, John, ed. 1985. *Critical Theory and Public Life*. MIT Press: Cambridge.
Foucault, Michel. 1970. *The Order of Things*. Random House: New York.
Foucault, Michel. 1975. *Archaeology of Knowledge*. Harper and Row: New York.
Freidson, Eliot. 1986. *Professional Powers*. University of Chicago Press: Chicago.
Fuller, Steve. 1983. "A French Science with English Subtitles." *Philosophy and Literature* 7: 3-14.
Fuller, Steve. 1984. "The Cognitive Turn in Sociology." *Erkenntnis* 74: 439-50.
Fuller, Steve. 1985. "Bounded Rationality in Law and Science." Ph.D. dissertation, University of Pittsburgh.
Fuller, Steve. 1988a. *Social Epistemology*. Indiana University Press: Bloomington.
Fuller, Steve. 1988b. "Playing without a Full Deck: Scientific Realism and the Cognitive Limits of Legal Theory." *Yale Law Journal* 97: 549-80.
Fuller, Steve. 1989a. *Philosophy of Science and Its Discontents*. Westview: Boulder. Rev. 2d ed., published by Guilford Press, New York, in 1992.

References

Fuller, Steve. 1989b. "Philosophy of Science since Kuhn: Readings on the Revolution That Has Yet to Come." *Choice* 26: 595-603.

Fuller, Steve. 1990. "Does It Pay to Go Postmodern If Your Neighbors Do Not?" In G. Shapiro, ed., *After the Future: Postmodern Times and Places*. SUNY Press: Albany.

Fuller, Steve. 1991. "The Proprietary Grounds of Knowledge." *Journal of Social Behavior and Personality* 6: 105-28.

Fuller, Steve. 1992a. "Social Epistemology and the Research Agenda of Science Studies." In Pickering 1992.

Fuller, Steve. 1992b. "Knowledge as Product and Property." In N. Stehr and R. Ericson, eds., *The Cultures of Knowledge and Power*. Routledge: London.

Fuller, Steve. 1992c. "Epistemology Radically Naturalized: Recovering the Normative, the Experimental, and the Social." In Giere 1992.

Furner, Mary. 1975. *Advocacy and Objectivity*. University of Kentucky Press: Lexington.

Gadamer, Hans. 1975. *Truth and Method*. Seabury Press: New York.

Galbraith, James. 1988. "The Grammar of Political Economy." In Klamer et al. 1988.

Galbraith, John Kenneth. 1974. *The New Industrial State*. Harmondsworth UK: Penguin.

Galison, Peter. 1987. *How Experiments End*. University of Chicago Press: Chicago.

Galison, Peter. 1992. "Image and Logic." Manuscript.

Galison, Peter, and Hevly, Bruce, eds. 1992. *Big Science*. Stanford University Press: Palo Alto.

Gardner, Howard. 1973. *The Quest for Mind*. Random House: New York.

Gardner, Howard. 1987. *The Mind's New Science*. 2d ed. Basic Books: New York.

Geertz, Clifford. 1973. *Interpreting Cultures*. Harper and Row: New York.

Geertz, Clifford. 1980. "Blurred Genres." *American Scholar* 49: 165-79.

Geertz, Clifford. 1983. *Local Knowledge*. Basic Books: New York.

Gellner, Ernest. 1979. *Spectacles and Predicaments*. Cambridge University Press: Cambridge.

Gellner, Ernest. 1989. *Plough, Sword, and Book*. University of Chicago Press: Chicago.

Georgescu-Roegen, Nicholas. 1970. *The Entropy Law and the Economic Process*. Harvard University Press: Cambridge.

Gettier, Edmund. 1963. "Is Justified True Belief Knowledge?" *Analysis* 23: 121-23.

Gholson, Barry; Houts, Arthur; Shadish, William; and Neimeyer, Robert, eds. 1989. *Psychology of Science: Contributions to Metascience*. Cambridge University Press: Cambridge.

Gibson, J. J. 1979. *The Ecological Approach to Visual Perception*. Houghton Mifflin: Boston.

Giddens, Anthony. 1984. *The Constitution of Society*. University of California Press: Berkeley.

Giddens, Anthony. 1989. *The Consequences of Modernity*. Stanford University Press: Palo Alto.

Giddens, Anthony, and Turner, Jonathan, eds. 1987. *Social Theory Today*. Stanford University Press: Palo Alto.

Giere, Ronald. 1988. *Explaining Science*. University of Chicago Press: Chicago.

Giere, Ronald. 1989a. "Scientific Rationality as Instrumental Rationality." *Studies in History and Philosophy of Science* 20: 377-84.

References

Giere, Ronald. 1989b. "Computer Discovery and Scientific Interests." *Social Studies of Science* 19: 638-43.
Giere, Ronald, ed. 1992. *Cognitive Models of Science*. Minnesota Studies in the Philosophy of Science, vol. 15. University of Minnesota Press: Minneapolis.
Gieryn, Thomas. 1983. "Boundary Work and the Demarcation of Science from Nonscience." *American Sociological Review* 48: 781-95.
Gilbert, Nigel, and Mulkay, Michael. 1984. *Opening Pandora's Box*. Cambridge University Press: Cambridge.
Glymour, Clark. 1987. "Android Epistemology and the Frame Problem." In Pylyshyn 1987.
Goldenberg, Sheldon. 1989. "What Scientists Think of Science." *Social Science Information* 28: 467-81.
Golding, Martin. 1974. *Philosophy of Law*. Prentice-Hall: Englewood Cliffs NJ.
Goldman, Alvin. 1985. "Epistemics: The Regulative Theory of Cognition." In Kornblith 1985.
Goldman, Alvin. 1986. *Epistemology and Cognition*. Harvard University Press: Cambridge.
Goldman, Alvin. 1989. "Strong and Weak Justification." In J. Tomberlin, ed., *Philosophical Perspectives*, vol. 2. Ridgeview Publishing: Atascadero CA.
Goldman, Alvin. 1992. *Liaisons: Philosophy Meets the Cognitive and Social Sciences*. MIT Press: Cambridge.
Goodin, Robert. 1980. *Manipulatory Politics*. University of Chicago Press: Chicago.
Goodin, Robert. 1990. "Liberalism and the Best Judge Principle." *Political Studies* 32: 181-95.
Gooding, David; Pinch, Trevor; and Schaffer, Simon, eds. 1989. *The Uses of Experiment*. Cambridge University Press: Cambridge
Goodnight, G. Thomas. 1980. "The Liberal and Conservative Presumptions: On Political Philosophy and the Foundations of Public Argument." *Proceedings of the Summer Conference on Argumentation*, ed. J. Rhodes and S. Newell. SCA: Falls Church VA.
Goodwin, Craufurd. 1988. "The Heterogeneity of Economists' Discourse." In Klamer et al. 1988: 207-20.
Gorman, Michael. 1989. "Beyond Strong Programmes: How Cognitive Approaches Can Complement SSK." *Social Studies of Science* 19: 643-52.
Gorman, Michael. 1991. "Simulating Social Epistemology in the Psychology Lab." In Giere 1992.
Gorman, Michael and Carlson, Bernard. 1989. "Can Experiments Be Used to Study Science?" *Social Epistemology* 3: 89-106.
Gorman, Michael; Gorman, Margaret; and Latta, R. 1984. "How Disconfirmatory, Confirmatory, and Combined Strategies Affect Group Problem Solving." *British Journal of Psychology* 75: 65-79.
Gould, Stephen Jay. 1981. *The Mismeasure of Man*. Norton: New York.
Gouldner, Alvin. 1957. "Cosmopolitans and Locals." *Administrative Science Quarterly* 2: 281-306, 444-80.
Grafton, Anthony. 1990. *Forgers and Critics*. Princeton: Princeton University Press.
Grafton, Anthony, and Jardine, Lisa. 1987. *From Humanism to the Humanities*. Duckworth: London.

References

Graham, Loren; Lepenies, Wolf; and Weingart, Peter, eds. 1983. *Functions and Uses of Disciplinary Histories*. D. Reidel: Dordrecht.
Greenburg, Daniel. 1967. *The Politics of Pure Science*. New American Library: New York.
Greenwood, John. 1989. *Explanation and Experiment in Social Psychological Science*. Spring-Verlag: New York.
Gross, Alan. 1990. *The Rhetoric of Science*. Harvard University Press: Cambridge.
Gross, David. 1989. "Critical Synthesis on Urban Knowledge: Remembering and Forgetting in the Modern City." *Social Epistemology* 3: 3-22.
Gruber, Howard. 1981. *Darwin on Man*. University of Chicago Press: Chicago.
Gunnell, John. 1986. *Between Philosophy and Politics*. University of Massachusetts Press: Amherst.
Guthrie, W. K. C. 1969. *A History of Greek Philosophy: The Fifth-Century Enlightenment*. Cambridge University Press: Cambridge.
Habermas, Jürgen. 1985. *The Theory of Communicative Action*, vol. 1. Beacon Press: Boston.
Habermas, Jürgen. 1987. *The Philosophical Discourse of Modernity*. MIT Press: Cambridge.
Hacking, Ian. 1975. *The Emergence of Probability*. Cambridge University Press: Cambridge.
Hacking, Ian. 1983. *Representing and Intervening*. Cambridge University Press: Cambridge.
Hacking, Ian. 1984. "Five Parables." In R. Rorty, J. Schneewind, and Q. Skinner, eds., *Philosophy in History*. Cambridge University Press: Cambridge.
Hacking, Ian. 1990. *The Taming of Chance*. Cambridge University Press: Cambridge.
Hall, A. Rupert. 1963. *From Galileo to Newton: 1630-1720*. Collins: London.
Hamowy, Ronald. 1987. *The Scottish Enlightenment and the Theory of Spontaneous Order*. Southern Illinois University Press: Carbondale.
Hanson, Russell. 1958. *Patterns of Discovery*. Cambridge University Press: Cambridge.
Haraway, Donna. 1989. *Simians, Cyborgs, and Women*. Routledge: London.
Hardin, Garrett. 1959. *Nature and Man's Fate*. New York: New American Library.
Harding, Sandra. 1986. *The Science Question in Feminism*. Cornell University Press: Ithaca.
Harding, Sandra. 1991. *Whose Science? Whose Knowledge?* Cornell University Press: Ithaca.
Harman, Gilbert. 1983. "Rational Action and the Extent of Intentions." *Social Theory and Practice* 9: 123-41.
Harman, Gilbert. 1986. *Change In View*. MIT Press: Cambridge.
Harré, Rom. 1979. *Personal Being*. Blackwell: Oxford.
Harré, Rom. 1981. "Philosophical Aspects of the Micro-Macro Problem." In Knorr-Cetina and Cicourel 1981.
Harré, Rom. 1989. "Metaphysics and Methodology: Some Prescriptions for Social Psychological Research." *European Journal of Social Psychology* 19: 437-47.
Harré, Rom, and Secord, Paul. 1979. *The Explanation of Social Behavior*. 2d ed. Blackwell: Oxford.
Harris, Benjamin. 1988. "A History of Debriefing in Social Psychology." In Morawski 1988.

References

Harris, Marvin. 1968. *The Rise of Anthropological Theory*. Thomas Crowell: New York.
Hartman, Joan, and Messer-Davidow, Ellen, eds. 1991. *(En)gendering Knowledge*. University of Tennessee Press: Knoxville.
Harvey, David. 1986. *The Condition of Postmodernity*. Blackwell: Oxford.
Haskell, Thomas, ed. 1984. *The Authority of Experts*. Indiana University Press: Bloomington.
Hayek, Friedrich. 1973. *Law, Legislation, and Liberty*. University of Chicago Press: Chicago.
Hebb, Donald. 1949. *The Organization of Behavior*. Wiley: New York.
Hedges, Larry. 1987. "How Hard Is Hard Science, How Soft Is Soft Science?" *American Psychologist* 42: 443-55.
Heelan, Patrick. 1983. *Space-Perception and the Philosophy of Science*. University of California Press: Berkeley.
Hegel, Georg Wilhelm Friedrich. 1892. *Lectures on the History of Philosophy*. Routledge and Kegan Paul: London.
Held, David. 1987. *Models of Democracy*. Stanford University Press: Palo Alto.
Hempel, Carl. 1965. *Aspects of Scientific Explanation*. Free Press: New York.
Hewstone, Miles. 1989. *Causal Attribution*. Blackwell: Oxford.
Heyes, Cecelia. 1989. "Uneasy Chapters in the Relationship between Psychology and Epistemology." In Gholson et al. 1989.
Hirsch, E. D. 1967. *Validity in Interpretation*. Yale University Press: New Haven.
Hirschman, Albert. 1977. *The Passions and the Interests*. Princeton University Press: Princeton.
Hirschman, Albert. 1982. *Shifting Involvements*. Princeton University Press: Princeton.
Hirschman, Albert. 1989. "Reactionary Rhetoric." *Harper's Magazine* (May) 263: 63-70.
Hirschman, Albert. 1991. *The Rhetoric of Reaction*. Harvard University Press: Cambridge.
Hofstadter, Richard. 1974. *Anti-Intellectualism in American Life*. Random House: New York.
Holland, John; Holyoak, Keith; Nisbett, Richard; and Thagard, Paul. 1986. *Induction*. MIT Press: Cambridge.
Hollinger, David. 1990. "Free Enterprise and Free Inquiry." *New Literary History* 21: 897-919.
Holmquest, Anne. 1986. "Rhetoric, Signs, and the Theory of Discovering Knowledge: The Relevance of Charles Peirce's Theory of Presumption to Charles Darwin's Method of Discovery." Ph.D. dissertation, University of Iowa.
Holton, Gerald. 1978. *The Scientific Imagination*. Cambridge University Press: Cambridge.
Horowitz, Irving Louis. 1986. *The Communication of Ideas*. Oxford University Press: Oxford.
Hoselitz, Bert, ed. 1970. *A Readers Guide to the Social Sciences*. 2d ed. Free Press: New York.
Hoskin, Keith, and Macve, Richard. 1986. "Accounting and the Examination: A Genealogy of Disciplinary Power." *Accounting, Organizations, and Society*. 11: 105-36.

References

Houts, Arthur, and Gholson, Barry. 1989. "Brownian Notions." *Social Epistemology* 3: 139-46.
Hovland, Carl; Janis, Irving; and Kelley, Harold. 1965. *Communication and Persuasion*. 2d ed. Yale University Press: New Haven.
Howe, Henry, and Lyne, John. 1992. "Gene Talk." *Social Epistemology* 6: 109-63.
Huber, Peter. 1988. *Liability: The Legal Revolution and Its Consequences*. Basic Books: New York.
Huber, Peter. 1990. "Pathological Science in Court." *Daedalus* 119. 4: 97-117.
Hull, David. 1988. *Science as a Process*. University of Chicago Press: Chicago.
Irvine, John, and Martin, Ben. 1984. *Foresight in Science*. Francis Pinter: London.
Jansen, Sue Curry. 1988. *Censorship*. Oxford University Press: Oxford.
Johnson Laird, Philip. 1988. *The Computer and the Mind*. Harvard University Press: Cambridge.
Jonsen, Albert, and Toulmin, Stephen. 1988. *The Abuse of Casuistry*. University of California Press: Berkeley.
Kahneman, Daniel. 1973. *Attention and Effort*. Prentice Hall: Englewood-Cliffs NJ.
Kahneman, Daniel; Slovic, Paul; and Tversky, Amos, eds. 1982. *Judgments under Uncertainty: Heuristics and Biases*. Cambridge University Press: Cambridge.
Karp, Walter. 1988. "In Defense of Politics." *Harper's Magazine* (May) 276: 41-49.
Keith, William. 1986. "Believing and Acting: Toward a Rational Theory of Persuasion." Ph.D. dissertation, University of Texas.
Keith, William. 1990. "Cognitive Science on a Wing and a Prayer." *Social Epistemology* 4: 343-56.
Keith, William. "Argument Practices." Forthcoming in *Argumentation*.
Keller, Evelyn Fox. 1985. *Reflections on Gender and Science*. Yale University Press: New Haven.
Keohane, Robert, ed. 1986. *Neorealism and Its Critics*. Columbia University Press: New York.
Keohane, Robert. 1988. "The Rhetoric of Economics as Viewed by a Student of Politics." In Klamer et al. 1988.
Keynes, John Maynard. 1936. *The General Theory of Employment, Interest, and Money*. Harcourt Brace: New York.
Killingsworth, Jimmie, and Palmer, Jacqueline. 1991. *Ecospeak*. Southern Illinois University Press: Carbondale.
Kinneavy, James. 1986. "Kairos: A Neglected Concept in Classical Rhetoric." In J. Moss, ed., *Rhetoric and Praxis*. Catholic University Press: Washington DC.
Kitchener, Richard. 1986. *Piaget's Theory of Knowledge*. Yale University Press: New Haven.
Klamer, Arjo; McCloskey, Donald; and Solow, Robert, eds. 1988. *The Consequences of Economic Rhetoric*. Cambridge University Press: Cambridge.
Klapp, Orin. 1991. *The Inflation of Symbols*. Transaction Books: New Brunswick NJ.
Klein, Julie. 1990. *Interdisciplinarity*. Wayne State University Press: Detroit.
Knorr-Cetina, Karin. 1981. *The Manufacture of Knowledge*. Pergamon Press: Oxford.
Knorr-Cetina, Karin, and Cicourel, Aaron, eds. 1981. *Advances in Sociological Theory*. Routledge and Kegan Paul: London.
Knorr-Cetina, Karin, and Mulkay, Michael, eds. 1983. *Science Observed*. Sage: London.
Kornblith, Hilary, ed. 1985. *Naturalizing Epistemology*. MIT Press: Cambridge.

References

Kruglanski, Arie. 1991. "Social Science Based Understandings of Science." *Philosophy of the Social Sciences* 21: 223-31.
Kuhn, Thomas. 1970 (1962). *The Structure of Scientific Revolutions*. 2d ed. University of Chicago Press: Chicago.
La Follette, Marcel, ed. 1983. *Creationism, Science, and the Law*. MIT Press: Cambridge.
Lakatos, Imre. 1978. *Proofs and Refutations*. Cambridge University Press: Cambridge.
Lakatos, Imre. 1979. *Methodology of Scientific Research Programmes*. Cambridge University Press: Cambridge.
Lakatos, Imre, and Musgrave, Alan, eds. 1970. *Criticism and the Growth of Knowledge*. Cambridge University Press: Cambridge.
Lakoff, George. 1987. *Women, Fire, and Dangerous Things*. University of Chicago Press: Chicago.
Lane, Robert. 1990. *The Market Experience*. Cambridge University Press: Cambridge.
Langley, Pat; Simon, Herbert; Bradshaw, Gary; and Zytkow, Jan. 1987. *Scientific Discovery*. MIT Press: Cambridge.
Lasswell, Harold. 1948. *Power and Personality*. Norton: New York.
Latour, Bruno. 1981. "Insiders and Outsiders in the Sociology of Science; or, How Can We Foster Agnosticism?" *Knowledge and Society* 3: 199-216.
Latour, Bruno. 1987. *Science in Action*. Harvard University Press: Cambridge.
Latour, Bruno. 1988. "The Politics of Explanation." In Woolgar 1988a.
Latour, Bruno. 1989. *The Pasteurization of France*. Harvard University Press: Cambridge.
Latour, Bruno, and Woolgar, Steve. 1986 (1979). *Laboratory Life: The Social Construction of Scientific Facts*. 2d ed. Princeton University Press: Princeton.
Laudan, Larry. 1977. *Progress and Its Problems*. University of California Press: Berkeley.
Laudan, Larry. 1981. *Science and Hypothesis*. Reidel: Dordrecht.
Laudan, Larry. 1983. "The Demise of the Demarcation Problem." In R. Laudan, ed., *Working Papers on the Demarcation of Science and Pseudoscience*. Virginia Tech: Blacksburg.
Laudan, Larry. 1984. *Science and Values*. University of California Press: Berkeley.
Laudan, Larry. 1987. "Progress or Rationality: The Prospects for a Normative Naturalism." *American Philosophical Quarterly* 24: 19-31.
Laudan, Larry. 1989. "The Rational Weight of the Past." In M. Ruse, ed., *What Philosophy of Biology Is: Essays in Honor of David Hull*. Reidel: Dordrecht.
Laudan, Larry. 1990a. *Science and Relativism*. University of Chicago Press: Chicago.
Laudan, Larry. 1990b. "Aim-less Epistemology?" *Studies in History and Philosophy of Science* 21: 315-22.
Laudan, Larry; Donovan, Arthur; Laudan, Rachel; Barker, Peter; Brown, Harold; Leplin, Jarrett; Thagard, Paul; and Wykstra, Stephen. 1986. "Testing Theories of Scientific Change." *Synthese* 69: 141-223.
Laudan, Rachel; Laudan, Larry; and Donovan, Arthur. 1988. "Testing Theories of Scientific Change." In A. Donovan et al., eds., *Scrutinizing Science*. Kluwer: Dordrecht.

References

Law, John, ed. 1986. *Power, Action, and Belief*. Routledge and Kegan Paul: London.

Laymon, Ronald. 1991. "Idealizations, Externalities, and the Economic Analysis of Law." In J. Pitt and E. Lugo, eds., *The Technology of Discovery and the Discovery of Technology*. Society for Philosophy and Technology: Blacksburg VA.

Layton, Edward. 1977. "Conditions of Technological Development." In I. Spiegel-Roesing and D. Price, eds., *Science, Technology, and Society*. Sage: London.

Lemon, Lee, and Reis, Marion, eds. 1965. *Russian Formalist Criticism*. University of Nebraska Press: Lincoln.

Lepage, Henri. 1978. *Tomorrow, Capitalism*. Open Court: La Salle IL.

Leplin, Jarrett, ed. 1984. *Scientific Realism*. University of California Press: Berkeley.

Levi, Isaac. 1985. *Decisions and Revisions*. Cambridge University Press: Cambridge.

Levine, John. 1989. "Reaction to Opinion Deviance in Small Groups." In Paulus 1989.

Lévi-Strauss, Claude. 1964. *Structural Anthropology*. Harper and Row: New York.

Lichtenberg, Judith, ed. 1990. *Democracy and the Mass Media*. Cambridge University Press: Cambridge.

Lippman, Thomas. 1992. "Lab for Controversial Supercollider Project Sprouts in Texas Soil." *Washington Post*, March 21 A3.

Lowe, Adolph. 1965. *On Economic Knowledge*. Harper and Row: New York.

Lowenthal, David. 1987. *The Past Is a Foreign Country*. Cambridge University Press: Cambridge.

Luhmann, Niklas. 1979. *The Differentiation of Society*. Columbia University Press: New York.

Lynch, William. 1989. "Arguments for a Non-Whiggish Hindsight: Counterfactuals and the Sociology of Knowledge." *Social Epistemology* 3: 361-65.

Lyons, William. 1986. *The Disappearance of Introspection*. MIT Press: Cambridge.

Lyotard, Jean-François. 1983. *The Postmodern Condition*. University of Minnesota Press: Minneapolis.

McCloskey, Donald. 1985. *The Rhetoric of Economics*. University of Wisconsin Press: Madison.

McCloskey, Donald. 1987. *Econometric History*. Collier Macmillan: London.

McCloskey, Donald. 1991. *If You're So Smart...* University of Chicago Press: Chicago.

MacCorquodale, Kenneth. 1970. "On Chomsky's Review of Skinner's *Verbal Behavior*." *Journal of the Experimental Analysis of Behavior* 13: 83-99.

McDowell, John. 1982. "The Obsolescence of Knowledge and Career Publication Profiles." *American Economic Review* 72: 752-68.

McGee, Michael Calvin. 1980. "The 'Ideograph': A Link between Rhetoric and Ideology." *Quarterly Journal of Speech* 66: 1-16.

McGee, Michael Calvin, and Lyne, John. 1987. "What Are Nice Folks Like You Doing in a Place Like This?" In Nelson et al. 1987.

MacIntyre, Alasdair. 1984. *After Virtue*. Notre Dame Press: South Bend.

MacKenzie, Richard, and Tullock, Gordon. 1981. *The New World of Economics*. 3d ed. Richard Irwin: Homewood IL.

References

McLuhan, Marshall. 1962. *The Gutenberg Galaxy*. University of Toronto Press: Toronto.
McLuhan, Marshall. 1964. *Understanding Media: The Extensions of Man*. McGraw Hill: New York.
Maffie, James. 1990. "Recent Work in Naturalized Epistemology." *American Philosophical Quarterly* 27: 46-60.
Mahoney, Michael. 1989. "Participatory Epistemology and the Psychology of Science." In Gholson et al. 1989.
Maier, Robert, ed. 1989. *Norms in Argumentation*. Foris: Dordrecht. Malefijt, Anne. 1974. *Images of Man*. Alfred Knopf: New York.
Manicas, Peter. 1986. *A History and Philosophy of the Social Sciences*. Blackwell: Oxford.
Mannheim, Karl. 1936. *Ideology and Utopia*. Routledge and Kegan Paul: London.
Mannheim, Karl. 1940. *Man and Society in an Age of Reconstruction*. Routledge and Kegan Paul: London.
Manuel, Frank. 1969. *A Portrait of Sir Isaac Newton*. Harvard University Press: Cambridge.
Marcus, George, and Fischer, Michael. 1986. *Anthropology as Cultural Critique*. University of Chicago Press: Chicago.
Margolis, Howard. 1987. *Patterns, Thinking, and Cognition*. University of Chicago Press: Chicago.
Marshall, Alfred. 1920. *Principles of Economics*. 8th ed. Macmillan: London.
Martindale, Don. 1960. *The Nature and Types of Sociological Theory*. Houghton Mifflin: Boston.
Mayr, Otto. 1986. *Authority, Liberty, and Automatic Machinery in Early Modern Europe*. Johns Hopkins University Press: Baltimore.
Meja, Volker, and Stehr, Nico, eds. 1990. *Knowledge and Politics*. Routledge: London.
Merleau-Ponty, Maurice. 1962. *The Phenomenology of Perception*. Routledge and Kegan Paul: London.
Merleau-Ponty, Maurice. 1963. *The Structure of Behavior*. Beacon Press: Boston.
Merquior, J. G. 1986. *From Prague to Paris*. Verso: London.
Merton, Robert. 1973. *The Sociology of Science*. University of Chicago Press: Chicago.
Meyerson, Emile. 1930. *Identity and Reality*. Allen and Unwin: London. Miller, Arthur. 1986. *Imagery in Scientific Thought*. MIT Press: Cambridge.
Miller, Arthur, and Davis, Michael. 1983. *Intellectual Property*. West: St. Paul MN.
Miller, Carolyn. 1992. "*Kairos* in the Rhetoric of Science." In S. Witte, R. Cherry, and N. Nakadate, eds., *A Rhetoric of Doing*. Southern Illinois University Press: Carbondale.
Miller, Carolyn. "Progress, Development, and Change: *Kairos* in the History of Technology." Forthcoming in *Argumentation*.
Minsky, Marvin. 1986. *The Society of Mind*. Simon and Schuster: New York.
Mirowski, Philip, ed. 1986. *The Reconstruction of Economic Theory*. Kluwer: Boston.
Mirowski, Philip. 1989. *More Heat than Light*. Cambridge University Press: Cambridge.
Mirowski, Philip. 1991. "Postmodernism and the Social Theory of Value." *Journal of Post-Keynesian Economics* 13: 565-582.

References

Mitroff, Ian. 1974. *The Subjective Side of Science*. Elsevier: Amsterdam.
Moore, Ronald. 1978. *Legal Norms and Legal Science*. University of Hawaii Press: Honolulu.
Morawski, Jan, ed. 1988. *The Rise of Experimentation in American Psychology*. Yale University Press: New Haven.
Mukerji, Chandra. 1990. *A Fragile Power*. Princeton University Press: Princeton.
Mulkay, Michael. 1979. "Knowledge and Utility: Implications for the Sociology of Knowledge." *Social Studies of Science* 9: 69-74.
Mulkay, Michael. 1985. *The Word and the World*. George Allen and Unwin: London.
Nagel, Thomas. 1987. *The View from Nowhere*. Oxford University Press: Oxford.
Nelkin, Dorothy. 1987. *Selling Science*. W. H. Freeman: New York.
Nelson, Cary, ed. 1986. *Theory in the Classroom*. University of Illinois Press: Urbana.
Nelson, John. ed. 1986. *Tradition, Interpretation, and Science*. SUNY Press: Albany.
Nelson, John. 1987. "Stories of Science and Politics." In Nelson et al. 1987.
Nelson, John; Megill, Allan; and McCloskey, Donald, eds. 1987. *The Rhetoric of the Human Sciences*. University of Wisconsin Press: Madison.
Nersessian, Nancy. 1984. *Faraday to Einstein: Constructing Meaning in Scientific Theories*. Martinus Nijhoff: Dordrecht.
Newell, Alan, and Simon, Herbert. 1972. *Human Problem Solving*. Prentice Hall: Englewood Cliffs NJ.
Newmeyer, Frederick. 1980. *Linguistic Theory in America*. Academic Press: New York.
Nickles, Thomas. 1986. "Remarks on the Use of History as Evidence." *Synthese* 69: 253-66.
Nickles, Thomas. 1989. "Heuristic Appraisal in Science." *Social Epistemology* 3: 300-20.
Nielsen, Joyce. 1990. *Feminist Research Methods*. Westview: Boulder.
Noelle-Neumann, Elisabeth. 1982. *The Spiral of Silence*. University of Chicago Press: Chicago.
Nozick, Robert. 1982. *Philosophical Explanations*. Harvard University Press: Cambridge.
Ohmann, Richard. 1987. *The Politics of Letters*. Wesleyan University Press: Middletown.
O'Neill, Onora. 1990. "Practices of Toleration." In Lichtenberg 1990.
Ong, Walter. 1958. *Ramus, Method, and the Decay of Dialogue*. Harvard University Press: Cambridge.
Ong, Walter. 1986. *Orality and Literacy*. Methuen: London.
Ophir, Adi, and Shapin, Steven. 1991. "The Place of Knowledge." *Science in Context* 4: 3-21.
Ornstein, Norman, and Elder, Shirley. 1978. *Interest Groups, Lobbying, and Policymaking*. Congressional Quarterly Press: Washington DC.
Orr, C. Jack. 1990. "Critical Rationalism: Rhetoric and the Voice of Reason." In Cherwitz 1990.
Pappas, George, and Swain, Marshall, eds. 1978. *Justification and Knowledge*. Cornell University Press: Ithaca.
Pareto, Vilfredo. 1935. *Mind and Society*. Harcourt, Brace, and Jovanovich: New York.
Parfit, Derek. 1984. *Reasons and Persons*. Oxford University Press: Oxford.

References

Parsons, Talcott. 1937. *The Structure of Social Action*. Free Press: New York.
Parsons, Talcott. 1951. *The Social System*. Free Press: New York.
Pascal, Blaise. *Pensées*. Harper and Row: New York.
Paulus, Paul, ed. 1989. *Psychology of Group Influence*. 2d ed. Lawrence Erlbaum Associates: Hillsdale NJ.
Peirce, Charles Sanders. 1955. *Philosophical Writings of Peirce*. Ed. Justus Buchler. Dover: New York.
Petersen, Arne. 1984. "The Role of Problems and Problem Solving in Popper's Early Work on Psychology." *Philosophy of the Social Sciences* 24: 239-50.
Piaget, Jean. 1971. *Psychology and Epistemology*. Penguin: Harmondsworth UK.
Pickering, Andrew. 1984. *Constructing Quarks*. University of Chicago Press: Chicago.
Pickering, Andrew, ed. 1992. *Science as Practice and Culture*. University of Chicago Press: Chicago.
Polanyi, Michael. 1957. *Personal Knowledge*. University of Chicago Press: Chicago.
Polanyi, Michael. 1969. *Knowing and Being*. University of Chicago Press: Chicago.
Popper, Karl. 1950. *The Open Society and Its Enemies*. Princeton University Press: Princeton.
Popper, Karl. 1957. *The Poverty of Historicism*. Harper and Row: New York.
Popper, Karl. 1959. *The Logic of Scientific Discovery*. Harper and Row: New York.
Popper, Karl. 1966. *The Open Society and Its Enemies*. Oxford University Press: Oxford.
Popper, Karl. 1970. "Normal Science and Its Dangers." In Lakatos and Musgrave 1970.
Popper, Karl. 1972. *Objective Knowledge*. Oxford University Press: Oxford.
Porter, Theodore. 1986. *The Rise of Statistical Thinking: 1820-1900*. Princeton University Press: Princeton.
Porter, Theodore. "Standardization, Objectivity, and the Social Authority of Calculation." Forthcoming in *Annals of Scholarship*.
Prelli, Lawrence. 1989. *A Rhetoric of Science*. University of South Carolina Press: Columbia.
Price, Derek de Solla. 1986. *Little Science, Big Science, and Beyond*. Columbia University Press: New York.
Price, Don K. 1965. *The Scientific Estate*. Harvard University Press: Cambridge.
Proctor, Robert. 1991. *Value-Free Science?* Harvard University Press: Cambridge.
Putnam, Hilary. 1975. *Mind, Language, and Reality*. Cambridge University Press: Cambridge.
Pylyshyn, Zenon. 1979. "Imprecision and Metaphor." In A. Ortony, ed. *Metaphor and Thought*. Cambridge University Press: Cambridge.
Pylyshyn, Zenon. 1984. *Computation and Cognition*. MIT Press: Cambridge.
Pylyshyn, Zenon, ed. 1987. *The Robot's Dilemma*. Ablex: Norwood NJ.
Quine, W. V. O. 1953. "Two Dogmas of Empiricism." In W.V.O. Quine, ed., *From a Logical Point of View*. Harper and Row: New York.
Quine, W. V. O. 1960. *Word and Object*. MIT Press: Cambridge.
Quine, W. V. O. 1985. "Epistemology Naturalized." In Kornblith 1985.
Rachlin, Howard. 1989. *Judgment, Decision, and Choice*. Freeman: New York.
Rawls, John. 1955. "Two Concepts of Rules." *Philosophical Review* 64: 3-32.
Rawls, John. 1972. *A Theory of Justice*. Harvard University Press: Cambridge.

References

Reber, Arthur. 1987. "The Rise (and Surprisingly Rapid Fall) of Psycholinguistics." *Synthese* 72: 325-39.
Reichenbach, Hans. 1938. *Experience and Prediction*. University of Chicago Press: Chicago.
Rescher, Nicholas. 1977. *Dialectics*. SUNY Press: Albany.
Rescher, Nicholas. 1984. *The Limits of Science*. University of California Press: Berkeley.
Restivo, Sal. 1988. "Modern Science as a Social Problem." *Social Problems* 35: 206-25.
Ricci, David. 1984. *The Tragedy of Political Science*. Yale University Press: New Haven.
Robbins, Lionel. 1937. *An Essay on the Nature and Significance of Economic Science*. Macmillan: London.
Rogers, Everett. 1962. *The Diffusion of Innovations*. Free Press: New York.
Rooney, Ellen. 1991. *Seductive Reasoning*. Cornell University Press: Ithaca.
Root-Bernstein, Scott. 1989. *Discovering*. Harvard University Press: Cambridge.
Rorty, Richard. 1979. *Philosophy and the Mirror of Nature*. Princeton University Press: Princeton.
Rorty, Richard. 1982. *Consequences of Pragmatism*. University of Minnesota Press: Minneapolis.
Rorty, Richard. 1988. "Is Natural Science a Natural Kind?" In E. McMullin, ed., *Construction and Constraint*. Notre Dame Press: South Bend.
Rorty, Richard. 1989. *Contingency, Irony, and Solidarity*. Cambridge University Press: Cambridge.
Rosenau, Pauline. 1992. *Postmodernism and the Social Sciences*. Princeton University Press: Princeton.
Rosenberg, Shawn. 1988. *Reason, Ideology and Politics*. Princeton University Press: Princeton.
Roth, Paul. 1987. *Meaning and Method in the Social Sciences*. Cornell University Press: Ithaca.
Roth, Paul. 1991. "The Bureaucratic Turn: Weber contra Hempel in Fuller's *Social Epistemology*." *Inquiry* 34: 365-76.
Rouse, Joseph. 1987. *Knowledge and Power: Toward a Political Philosophy of Science*. Cornell University Press: Ithaca.
Rumelhart, Donald, and McClelland, James, eds. 1986. *Parallel Distributed Processing*. 2 vols. MIT Press: Cambridge.
Russell, David. 1991. *Writing in the Academic Disciplines: 1870-1990*. Southern Illinois University Press: Carbondale.
Salmon, Wesley. 1967. *The Foundations of Scientific Inference*. University of Pittsburgh Press: Pittsburgh.
Sampson, Geoffrey. 1980. *Making Sense*. Oxford University Press: Oxford.
Sartre, Jean-Paul. 1976. *Critique of Dialectical Reason*. New Left Books: London.
Sassower, Raphael; Bender, Frederic; and Levine, David. 1990. "Symposium on the Role of Scarcity in Economic Thought." *Social Epistemology* 4: 75-120.
Schaefer, Wolf, ed. 1984. *Finalization in Science*. Kluwer: Dordrecht.
Scharff, Robert. 1989. "Positivism, Philosophy of Science, and Self-Understanding in Comte and Mill." *American Philosophical Quarterly* 26: 253-68.
Schmaus, Warren. 1985. "Hypotheses and Analyses in Durkheim's Sociological Methodology." *Studies in the History and Philosophy of Science* 16: 1-30.

References

Schmaus, Warren. 1991. Review of Social Epistemology. *Philosophy of the Social Sciences* 21: 121-25.

Schultz, Duane. 1981. *A History of Modern Psychology*. 3d ed. Academic Press: New York.

Schumpeter, Joseph. 1942. *Capitalism, Socialism and Democracy*. Harper and Row: New York.

Science in Context. 1987- . Semiannual journal. Cambridge University Press: Cambridge.

Science, Technology, and Human Values. 1976- . Quarterly journal. Sage: Newbury Park CA.

Scott, Joan. 1987. "Women's History and the Rewriting of History." In C. Farnham, ed., *The Impact of Feminist Research in the Academy*. Indiana University Press: Bloomington.

Segall, Marshall; Campbell, Donald; and Herskovitz, Melville. 1966. *The Influence of Culture on Visual Perception*. Bobbs-Merrill: Indianapolis.

Segerstrale, Ullica. 1992. "Bringing the Scientists Back In: A Critique of Contemporary Sociology of Scientific Knowledge." In T. Brante, S. Fuller, and W. Lynch, eds., *Controversial Science*. SUNY Press: Albany NY.

Serres, Michel. 1982. *Parasite*. Johns Hopkins University Press: Baltimore.

Shannon, Thomas. 1990. *An Introduction to the World-Systems Perspective*. Westview Press: Boulder.

Shapere. Dudley. 1984. *Reason and the Search for Knowledge*. D. Reidel: Dordrecht.

Shapere, Dudley. 1987. "Method in the Philosophy of Science and Epistemology." In Nancy Nersessian, ed., *Science as Process*. Martinus Nijhoff: Dordrecht.

Shapin, Steven. 1991. " The Mind in Its Own Place." *Science in Context* 4: 191-218.

Shapin, Steven, and Schaffer, Simon. 1985. *Leviathan and the Air-Pump*. Princeton University Press: Princeton.

Shils, Edward, ed. 1968. *Criteria for Scientific Development*. MIT Press: Cambridge.

Shope, Robert. 1983. *The Analysis of Knowing*. Princeton University Press: Princeton.

Shotter, John. 1984. *Social Accountability and Selfhood*. Blackwell: Oxford.

Shrager, Jeff, and Langley, Pat, eds. 1990. *Computational Models of Scientific Discovery and Theory Formation*. Morgan Kaufman: San Mateo CA.

Shrum, Wesley, and Morris, Joan. 1990. "Organizational Constructs for the Assembly of Technological Knowledge." In Gieryn and Cozzens 1990.

Sidgwick, Alfred. 1984. *Fallacies, a View of Logic from the Practical Side*. Appleton: New York.

Siegel, Harvey. 1989. "Philosophy of Science Naturalized?" *Studies in History and Philosophy of Science* 20: 365-75.

Siegel, Harvey. 1990. "Laudan's Normative Naturalism." *Studies in History and Philosophy of Science* 21: 295-313.

Simmel, Georg. 1964. *The Sociology of Georg Simmel*. Ed. Kurt Wolff. Free Press: New York.

Simon, Herbert. 1957. *Models of Man*. John Wiley: New York.

Simon, Herbert. 1976. *Administrative Behavior*. Free Press: New York.

Simon, Herbert. 1981. *The Sciences of the Artificial*. 2d ed. MIT Press: Cambridge.

Simon, Herbert. 1991a. "Comments on the Symposium on Computer Discovery and the Sociology of Knowledge." *Social Studies of Science* 21: 143-48.

Simon, Herbert. 1991b. *Models of My Life*. Basic Books: New York.

References

Simon, Julian. 1990. *Population Matters*. Transaction Books: New Brunswick NJ.
Skinner, B. F. 1953. *Science and Human Behavior*. Free Press: New York.
Skinner, B. F. 1957. *Verbal Behavior*. Appleton-Century Crofts: New York.
Skinner, Quentin, ed. 1987. *The Return of Grand Theory in the Human Sciences*. Cambridge University Press: Cambridge.
Slezak, Peter. 1989a. "Scientific Discovery by Computer as Empirical Refutation of the Strong Programme." *Social Studies of Science* 19: 563-600.
Slezak, Peter. 1989b. "Computers, Contents, and Causes: Replies to My Respondents." *Social Studies of Science* 19: 671-95.
Slezak, Peter. 1990. "Man Not a Subject for Science?" *Social Epistemology* 4: 327-42.
Snow, C. P. 1964. *The Two Cultures and a Second Look*. Cambridge University Press: Cambridge.
Social Epistemology. 1987-. Quarterly journal. Taylor and Francis: London.
Social Studies of Science. 1971- . Quarterly journal. Sage: London.
Soja, Ed. 1988. *Postmodern Geographies*. Verso: London.
Sorell, Tom. 1991. *Scientism*. Routledge: London.
Sowell, Thomas. 1987. *A Conflict of Visions*. William Morrow: New York.
Star, Leigh, and Griesemer, James. 1989. "Institutional Ecology, Translations, Boundary Objects." *Social Studies of Science* 19: 387- 420.
Stehr, Nico, and Ericson, Richard, eds. 1992. *The Culture and Power of Knowledge*. Walter de Gruyter: Berlin.
Stehr, Nico, and Meja, Volker, eds. 1984. *Society and Knowledge*. Transaction Books: New Brunswick NJ.
Stephens, Mitchell. 1988. *A History of the News*. Viking: New York.
Stern, Fritz, ed. 1956. *The Varieties of History*. Cleveland: Meridian Books.
Stich, Stephen. 1983. *From Folk Psychology to Cognitive Science*. MIT Press: Cambridge.
Stich, Stephen. 1985. "Could Man Be an Irrational Animal?" In Kornblith 1985.
Stich, Stephen. 1990. *The Fragmentation of Reason*. MIT Press: Cambridge.
Stich, Stephen, and Nisbett, Richard. 1984. "Expertise, Judgment, and the Psychology of Inductive Inference." In Haskell 1984.
Stinchcombe, Arthur. 1990. *Information and Organizations*. University of California Press: Berkeley.
Stocking, George. 1968. *Race, Culture, and Evolution*. University of Chicago Press: Chicago.
Stout, Jeffrey. 1984. *The Flight from Authority*. Notre Dame Press: South Bend.
Swedberg, Richard. 1989. *Economics and Sociology*. Princeton University Press: Princeton.
Symposium on (Goldman's) Epistemology and Cognition. 1989. *Philosophia* 19: 2-3.
Sztompka, Piotr. 1990. "Conceptual Frameworks in Comparative Inquiry." In Albrow and King 1990.
Terveen, Loren. 1992. "In the Footprints of the Masters: Embedding Knowledge Acquisition in Organizational Activity." In B. Gaines, ed., *Working Notes of the AAAI Spring Symposium*. (Copies available from the Knowledge Science Institute, University of Calgary.)
Thagard, Paul. 1988. *A Computational Philosophy of Science*. MIT Press: Cambridge.
Thomas, W. I., and Thomas, D. S. 1928. *The Child in America*. Alfred Knopf: New York.

References

Thompson, Michael; Ellis, Richard; and Wildavsky, Aaron. 1990. *Culture Theory.* Westview Press: Boulder.
Tilly, Charles. 1991. "How (and What) Are Historians Doing?" In D. Easton and C. Schelling, eds., *Divided Knowledge.* Sage: Newbury Park, 1991.
Tong, Rosemarie. 1989. *Feminist Thought.* Westview: Boulder.
Toulmin, Stephen. 1958. *The Uses of Argument.* Cambridge University Press: Cambridge.
Toulmin, Stephen. 1972. *Human Understanding.* Princeton University Press: Princeton.
Toulmin, Stephen. 1990. *Cosmopolis: The Hidden Agenda of Modernity.* Free Press: New York.
Traweek, Sharon. 1988. *Beamtimes and Lifetimes.* Harvard University Press: Cambridge.
Traweek, Sharon. 1992. "Border Crossings." In Pickering 1992.
Treichler, Paula. 1986. "Teaching Feminist Theory." In C. Nelson 1986.
Turner, Ralph, ed. 1975. *Ethnomethodology.* Penguin: Harmondsworth UK.
Turner, Stephen. 1991. "Social Constructionism and Social Theory." *Sociological Theory* 9: 22-33.
Tversky, Amos, and Kahneman, Daniel. 1974. "Judgment under Uncertainty: Heuristics and Biases." *Science* 185: 1124-31.
Tweney, Ryan. 1989. "A Framework for the Cognitive Psychology of Science." In Gholson et al. 1989.
Tweney, Ryan. 1991. "On Bureaucracy and Science." *Philosophy of the Social Sciences* 21: 203-13.
Tweney, Ryan; Mynatt, Clifford; and Doherty, Michael, eds. 1981. *On Scientific Thinking.* Columbia University Press: New York.
Unger, Roberto. 1986. *The Critical Legal Studies Movement.* Harvard University Press: Cambridge.
Unger, Roberto. 1987. *Social Theory.* Cambridge University Press: Cambridge.
Van Fraassen, Bas. 1980. *The Scientific Image.* Oxford University Press: Oxford.
Verdon, Michael. 1982. "Durkheim and Aristotle: Some Incongruous Congruences." *Studies in the History and Philosophy of Science* 13: 333-52.
Vickers, Brian. 1987. *In Defense of Rhetoric.* Oxford University Press: Oxford.
Waddell, Craig. 1990. "The Role of Pathos in the Decision-making Process." *Quarterly Journal of Speech* 76: 381-400.
Waddell, Craig. "Public Indifference to the Population Explosion." Forthcoming in *Social Epistemology.*
Wallas, Graham. 1910. *Human Nature in Politics.* 2d ed. Constable: London.
Wallerstein, Immanuel. 1990. "Societal Development or Development of the World-System?" In Albrow and King 1990.
Wallerstein, Immanuel. 1991. *Unthinking Social Science.* Blackwell: Oxford.
Wasby, S. 1970. *Political Science—The Discipline and Its Dimensions.* Scribners: New York.
Webb, E. J.; Campbell, D. T.; Schwartz, R. D.; Sechrest, L. B.; and Grove, J. B. 1981. *Non-reactive Measures in the Social Sciences.* Houghton Mifflin: Boston.
Weber, Max. 1964. *Methodology of the Social Sciences.* Free Press: New York.
Weinberg, Julius. 1968. "Abstraction in the Formation of Concepts." In *Dictionary of the History of Ideas.* Charles Scribner: New York. 1: 1-9.
Weiss, Carol, ed. 1977. *Using Social Research in Public Policy Making.* Lexington Books: Lexington MA.

References

Weizenbaum, Joseph. 1976. *Computer Power and Human Reason*. Freeman: San Francisco.
Wenzel, Joseph. 1989. "Relevance–and Other Norms of Argument: A Rhetorical Exploration." In Maier 1989.
Westermarck, Edward. 1912. *Ethical Relativity*. Routledge and Kegan Paul: London.
Whately, Richard. 1963 (1828). *Elements of Rhetoric*. Southern Illinois University Press: Carbondale.
White, Harrison. 1981. "Where Do Markets Come From?" *American Journal of Sociology* 87: 517-47.
White, Hayden. 1972. *Metahistory*. Baltimore: Johns Hopkins University Press.
White, Morton. 1957. *Social Thought in America: The Revolt against Formalism*. Beacon Press: Boston.
Whitley, Richard. 1985. *The Social and Intellectual Organization of the Sciences*. Oxford University Press: Oxford.
Wicklund, Robert. 1989. "The Appropriation of Ideas." In Paulus 1989.
Willard, Charles. 1983. *Argumentation and the Social Grounds of Knowledge*. University of Alabama Press: Tuscaloosa.
Willard, Charles. 1990. "The Problem of the Public Sphere: Three Diagnoses." In Williams and Hazen 1990.
Willard, Charles. 1991. "Authority." *Informal Logic* 12: 11-22.
Willard, Charles. 1992. "Liberalism and the Problem of Competence." Manuscript.
Williams, Bernard. 1973. *Problems of the Self*. Cambridge University Press: Cambridge.
Williams, David, and Hazen, Michael, eds. 1990. *Argumentation Theory and the Rhetoric of Assent*. University of Alabama Press: Tuscaloosa.
Williams, Raymond. 1976. *Keywords*. Oxford University Press: Oxford.
Winant, Terry, and Ross, Stephen David. 1991. "Symposium on the Philosophical Politics of Ostracism and the Inarticulable." *Social Epistemology* 5: 317-34.
Wolff, Robert Paul; Moore, Barrington; and Marcuse, Herbert. 1969. *A Critique of Pure Tolerance*. Beacon Press: Boston.
Woolgar, Steve. 1985. "Why Not a Sociology of Machines?" *Sociology* 19: 557-72.
Woolgar, Steve, ed. 1988a. *Knowledge and Reflexivity*. Sage: London.
Woolgar, Steve. 1988b. *Science: The Very Idea*. Tavistock: London.
Wrong, Dennis. 1961. "The Oversocialized Conception of Man." *American Sociological Review* 26: 184-93.
Wuthnow, Robert. 1989. *Communities of Discourse*. Harvard University Press: Cambridge MA
Xenos, Nicholas. 1989. *Scarcity and Modernity*. Routledge and Kegan Paul: London.
Zagacki, Kenneth, and Keith, William. 1992. "Rhetoric, Topoi, and Scientific Revolutions." *Philosophy and Rhetoric* 25: 59-78.
Zilsel, Edward. 1942. "The Genesis of the Concept of Physical Law." *Philosophical Review* 51: 245-65.
Zolo, Daniel. 1989. *Reflexive Epistemology: The Philosophical Legacy of Otto Neurath*. Kluwer: Dordrecht.

INDEX

Adaptive preference formation, 110
Anthropologists, 80, 120-21, 190, 320, 326, 328
Anthropology, 97, 116-17, 119-21, 198, 258, 316, 320, 323, 326, 331; of science, 133
Antitheorizing, 362
Antitheory, 348, 363
Arationality assumption, 149, 193, 321
Aristotelian, 27, 109, 133, 153, 230, 289, 306, 361
Aristotle, 18, 21, 133, 153, 191, 200, 209, 211, 231, 360
Ashmore, Malcolm, 24, 96, 326, 340-41
Ayer, A. J., 188, 249, 350, 356

Barnes, Barry, 9, 10, 322-24
Behaviorism, 44, 119, 135, 139, 152, 161-62, 165, 170, 358
Behaviorists, 83, 151-52, 155-56, 162, 164, 167, 257
Big Science, xxi, 6, 46, 239-40, 245, 249, 283-84, 296, 334, 386
Bloor, David, 9, 145-51, 160, 162, 168-69, 193, 202-5, 220-21, 223, 258-60, 320, 322-26
Burke, Kenneth, 187, 189

Campbell, Donald, 49, 72, 78, 113, 135-36, 220, 243
Capitalism, 51, 127, 133, 172, 243, 297-98; entrepreneurial, 238; venture, 34
Chomsky, Noam, 79, 139, 151-68, 170, 354, 360-62; Chomskyan linguistics, 354
Churchland, Paul, 72, 186, 331, 335
Cognitive: democracy, 28; historiography of science, 172; history of science, 171; psychology, 21, 63, 72, 135, 159, 173, 195, 257, 340; revolution, 139, 151-56, 170, 177; revolutionaries, 155-56; science, xxi, 149-52, 169, 302
Cognitive Science Unit, 144
Cognitivism, 44, 155, 172
Cognitivist(s), 154-56, 161, 170-78
Collins, Harry, 9, 11, 110, 115, 121, 145-46, 181-82, 258-59, 266-67, 320-26
Collins, Randall, 72, 298-99
Computer, 63; agency, 184; algorithm, 363; as "black box," 176; as "boundary object," 63; cognitive capacities of, 62; digital, 71, 233; and human thought processes, representation of, 183; models, 146, 157; models of scientific discovery, 139, 159; programs, 268; revolution, 23, 146; scientists, 149, 354; simulation(s), 149, 152, 160, 180, 266, 297, 361; simulations of scientific discovery, 148; software engineering, 334
Computerization, 223
Computers, 139, 142-43, 149; as actants, 160; as members of our epistemic communities, 65; personal, 147; as scientists, 345; to model thought, 104, 125
Consequential Theory, 367
Constructivism, 180, 189, 302, 326, 340, 386
Convention(s), 184, 310, 325, 363-64, 383; authorizing, 280; classroom, 303; disciplinary, 280; literary, 206; narrative, 242; pursuit of, 307; reading and writing, xv; semantic, 17; short-term, 156; social, 262, 355
Conventional, 146, 166; academic disciplines, 23; character of differences separating academic disciplines, xx; character of disciplinary boundaries, 72, 85; character of interpretation, 364; scientific community as, 300; wisdom, 159
Conventionalism, 363; interpretive, 347; metaphysical implications, 364
Conventionalist, 364
Conventionality, 38
Conventionality Presumption, 25, 48, 313
Counterfactual, 243; conditions, 212; historical, 233; historiography, 248; history, 211; question(s), 218, 243
Counterfactually realizable situations, 78
Counterrelativism, 334

415

Index

Criticism, 14, 39, 56, 58, 368, 372; rational, 283; reader-response 162, 347; science 310; self-, 89

Democracy, xxi; contemporary democracies, 291-93, 294, 300; free inquiry in, 231; representative, xviii; and science, xxi, 280-82; scientific imagery in a, 237
Democratic, xviii, decision-making processes, 3; government, 31; liberalism, 290; rhetoric, 235, 306; science portrayed as, 307; theory, xviii; values, 237, 379
Democratic Presumption, 26
Democratization, xii; of education, 261
Democratizing, 183; intellect, 242; knowledge, xx, 249; mission of STS, 96; science, 266
Dennett, Daniel, 91, 151, 160, 163, 170, 176
Dewey, John, 5, 72, 89, 93-94, 220, 249, 262, 264, 266, 292, 302; naturalism, 99
Dialectic(s), 15, 61, 282, 326, 348; constraints, 86; Hegelian, 344; laws of, 34
Dialectical, 38; advantage, 165; ambivalence, 321; approach, 136; constraints, 86; disadvantage, 84, 149; error, 149; impasse, 86; and interpenetrability, 36; resources, 86; overcoming of existing disciplinary differences, 269; as responsive environment, 213; as synthesis, 269
Dialectical Presumption, 25
Dialecticians, 60
Discipline(s), xi, xii, xx, 4, 8, 25, 28, 33, 34, 36-44, 47-49, 51-52, 56, 60-62 64, 72-73, 79, 82, 91, 103-22, 125, 130-31, 133, 135, 157, 159, 162, 165, 186, 195, 202-3, 220-21, 233, 242, 268, 274, 285, 289, 296, 311, 313, 316, 326, 351, 354, 360, 363, 366, 373, 380, 383, 386-88; discourse, 104; history, 192; integration of, 313; special, 343; -specific problems, 298
Douglas, Mary, 168, 258, 330

Ecological, 280; -materialist, 120; validity, 161, 218;
Ecologically valid, 82
Ecologists, 249, 305
Ecology, xvii; as epistemology, 280; systems 120
Economic: 207-8; agent, 247; analysis, 356; consequences of embodied forms of knowledge, 51; historians, 199; modelling, 98; rationality, 124; reasoning, 130; and social history, 207; sociology, xxi, 297; value of research, 334; viability, 4
Economics, 10, 29, 44-45, 62, 97-98, 103, 105, 116, 119, 123, 130, 180; authority of, 130; consumer, 379; historical method in, 122; neoclassical, 159; of science, 133; techno-, 233; welfare, 91
Economist(s), xiii, 27, 47, 62, 93, 98, 124, 126, 130-33, 195; classical political, 47; German, 356; neoclassical, 20; sense of rationality, 124
Education, xvi, 31, 108, 114, 195, 221, 240, 260, 282, 288, 303, 385; ends of, 30; funding, 260; graduate schools of, 5; higher, 231; policy, 43, 260-61; science, 31, 215
Elster, Jon, 55, 110, 211, 248, 279, 291, 328, 339, 375
End(s): epistemically appropriate, 323; of inquiry, 140, 284; natural, 230; of knowledge, xi, 92-93, 309; of science, 87, 93, 223-24; societal, xv; ultimate, 16; variety of, 30; various, 328
Enlightenment, xv, 281, 311; as empowerment, 291, French, 353; liberals, 209; politics, 311; Post-Enlightenment age, 107; scientific, 283; Scottish, 265
Epistemology, xvi, xviii, 21-22, 26, 69, 72, 74-75, 88-89, 101, 106, 135, 144, 180, 249, 279, 285, 309, 324, 327; analytic, 6, 72, 75; android, 183; classical, 58, 64, 76, 83, 90, 92, 187, 213; evolutionary, 90, 243; function for rhetoric, 33; genetic, 172; lack of interpenetration, 74; naturalized, 70, 84, 87; realist, 330; scientific, 118
Ethics, 21, 26, 249, 279, 325, 356; emotive theory of, 188
Ethnographers, 95, 347
Ethnographic accounts of laboratory life, 95
Ethnography, 80, 82, 95; inquiry, 80; method, 81
Ethnomethodologists, 113-14, 311
Ethnomethodology, 80
Ethos, xv, xix, 253, 283
Experiment(s): 80-81, 109, 113, 115, 138, 213, 353, 377; crucial, 235, 312, 375; historians of, 12; natural science, 247; replicating, 257; in trust, 80
Experimental: cultures, 110; design, 218-19; method(s), 38, 80-81, 97, 136, 190, 204, 212, 219, 342-43; intervention, 48, 80; knowledge, 136; physics, 73, 117-18; psychologists, 149, 166, 267; psychology, 79, 83, 155, 161, 358; replication, 267;

Index

Experimental (*continued*)
   social psychology, 64
Experimenter and subject, 38, 125

Faust, David, 23, 76, 202, 357
Feminism, 41, 208, 350-51, 388
Feminist: historians, 208, theory, 350-51
Feminists, xii, 31, 266, 351; liberal, 266; techno-, 34
Feyerabend, Paul, 5-6, 36, 90, 283-84, 289, 332, 375
Finalization, 244
Finalizationist school, 90
Fine, Arthur, 106
Fish, Stanley, xxi, 343, 347-66
Fodor, Jerry, 144, 153, 361
Folk: psychological concepts, 79-80, 82; psychology, 368; sociology, 126; theory, 230
Forman, Paul, 141, 250
Foucault, Michel, 17, 27-28, 153, 189, 204, 250, 300
Fungibility, 296-97, 300; economists' notion of, 295; epistemic principle of, xxi, 244, 295, 299

Geertz, Clifford, 36, 43
Giddens, Anthony, 121, 292, 306
Giere, Ronald, 72, 77, 87, 92, 98, 145, 171, 173-74, 178
Global: consciousness, 280, 333, 390; critiques, 150; doubts, 178; inconsistency, 171; interpenetration of cultural boundaries, 121; epistemically appropriate ends, 324; orientation, 280; picture of knowledge, 254; principles of knowledge production, 45; psychological terms, 241; sense of epistemic legitimacy, 92; success, 88; understanding, 291
Globalization, 121, 333
Goldman, Alvin, 70-75, 79-80, 82-83, 98, 101, 267
Gorman, Michael, 22, 99, 145, 161, 218

Habermas, Jürgen, xvii, 18, 27-28, 85, 90, 209, 244, 279, 290, 292, 294, 305, 312
Hacking, Ian, 53, 96, 104, 112, 118, 241, 349
Harré, Rom, 70, 79, 80-83, 135, 153, 170, 199, 219, 253
Harvard: General Education Program, 7; School of Education, 155
Hegel, G. W. F., xvi, 18, 54, 84-94, 200, 203, 280, 282, 326-27, 341
Hirschman, Albert, 133, 291, 332
Historicism, 200-201, 203; ontologized, 204; symmetrical, 202, 204

Historicist philosophers of science, 94
Historiography, 202; under- and over-determinationist, 210-13
History and sociology of science, 106
History of physics, 5
History of psychology, 135
History of science, 4, 6-7, 9, 25, 36, 53, 81, 91-93, 102, 104, 115, 141, 144-45, 147-49, 172, 176, 179, 191, 199-200, 202, 206, 210-16, 224, 243, 266, 271, 322, 341, 379; internal, 195, 200, 219-20, 223, 227, 302; internal-external debates, 214
History of technology, 202
History of the social sciences, 115
HPS (History and Philosophy of Science), 3-4, 6-9, 63, 65, 141, 171, 186-87, 195, 218, 220-21
Humanist(s), xii, xiv, xxii, 8, 36, 43, 63, 111, 197, 201, 206-8, 300-301, 347, 363-65, 373; culture, 110; knowledge, 367; practice, 356; rhetoric, 388; theoretical, xii

Intellectual property law, 51, 242
Interdisciplinarity, 36; interpenetrative, 32; possibilities, 61; rhetoric, 85-86, 142
Interdisciplinary: alliances, 52; audience, 50; barriers, 48; coalitions, 298; differences, 43; discourses, 40; disputes, 37; exchange, 61; fields, 220;
Interpenetration, 33, 37, 44, 46, 48, 49, 60, 64, 74; interlopers, 44; logic of, 35; of opposites, 34; of psychology, 74; of science and society, 335; of theory and practice, xiv
Interpenetration(s), xx, 48, 51; mediation, 313; negotiation, xviii, 34, 316; programs, 52; project(s), 53, 311; research, xx, 33, 39, 42, 49, 123; scholarship, xx; transactions, 46
Interpenetrative, xx; rhetoric, 85-86, 142; trade, 308
Inquiry: ends of, 140, 284

Journalism, xxi, 235; scientific, 234, 236, 385; sources, 391
Journalistic: objectivity, 235; portrayal of science, 240; sources, 390
Journalists, xiii, 126, 140, 235-36
Justice, 245, 268, 290, 372; distributive, 289; epistemic, 315; social, 269

Kairos, 132, 141, 390
Kant, Immanuel, 3, 18, 71, 118-19, 161, 209, 267-68, 305; categorical imperative, 340; normative theory, 362; Kantian talk, 16; Kantianism, 21

Index

Knorr-Cetina, Karin, 10, 114, 170, 201, 325, 374
Knowledge policy, xx-xxii, 28, 33-34, 227-29, 243, 254-55, 259, 271, 275, 308, 310; research, 233
Knowledge policymaker, xxi, 26, 233, 257
Kuhn, Thomas, xiii, 7-9, 27-28, 63, 76, 103, 154, 172, 186, 190, 219, 232, 337, 376

Lakatos, Imre, 7, 36, 76, 81, 94, 169, 210, 213, 218, 238, 258, 368
Latour, Bruno, xvii, xix, 10, 13, 95, 110, 137, 150, 152, 158, 168, 176-78, 217, 243, 331, 335, 341, 353, 386; Latourian black box, 177-78
Laudan, Larry, 8, 76, 87, 92, 94, 105-7, 111, 141, 147-49, 156, 193-94, 200, 202, 204-5, 214-15, 217-18, 220-21, 223, 262, 264, 321-23, 326; Laudanian history of science, 217
Liberal(s), 35, 282, 295, 379; profession(s), xv, 23, 162, 221; and radical politics, 291
Liberal arts, 123, 222; British curriculum, 245; classical, 108
Liberalism, 220, 282, 289; equal-in-principle, 287; equal-time, 288, 294; separate-but-equal, 288
Linguistics, 161
Literary criticism, xvii, 8, 10, 100, 266, 270, 347
Literary critics, 48
Local, 50, 254, 354; autonomy, 380; changes, 212; considerations, 359; contingencies, 364; customs, 320, 328; exchanges, 199; factors, 215; historical standards, 179; improvement, 280; knowledge, 13; political contingencies, 348; localistic bias in STS literature, 275; power base, 309; social factors, 193; social realists, 326; standards, 41, 49, 181; successes of science, 88; vs. universal, 324
Logical positivism, xi, xvi, 4, 44, 162, 250, 279, 313, 343
Logical positivist(s), 72, 149, 189, 211, 310, 325, 350, 357, 359

Mannheim, Karl, 9, 266, 280, 324, 326-29, 331-32, 338, 359
Marxism, xvi, 41, 48, 105, 327, 343-44, 346, 351; critical, 386; orthodox, 34
Marxist(s), xxii, 124, 172, 228, 232, 254, 328; critique, 34; materialist, 352; traditions, 277
Mass media, 34-35, 63, 234, 300; law, xxi, 299

Media, 24, 140, 142, 259, 290, 304; advances in electronic and print, 283; icons, 290; of social relations, 228; print, 299
Merton, Robert, 9, 232, 282

Natural science(s), 26, 62, 91, 95-96, 111-13, 115-16, 219, 228, 265, 377, 381
Natural scientific: discourse, 325; knowledge, 111; research, 26
Natural scientists, xii, 56-57, 113, 117, 278, 364, 373
Norm, 32, 46, 71, 80, 101, 109, 175, 193, 218, 252, 258, 265, 267-69, 269, 295; frequency and distribution of a given, 266; unfalsifiable, 267
Normative, 5, 276, 321; agenda, 212; a posteriori claims, 19; approach to science, 12; a priori claims, 19; categories, 97; character of knowledge production, 244; claims, 218; conception of a public sphere, 294; conclusions, xvii; conditions, 187; conflicting programs, 39; considerations in a policy setting, xxi; is constitutively rhetorical, xv; constraint, 80; constructivism, 85; control, 337; correctives, 368; crossroads facing STS, 10; dimension of justification, 94; dimension of Kuhn's account of scientific revolutions, 172; epistemology, 98; inertia, 263; infraction, 209; inquiry, 20, 268; instruments, 311; intervention, 319; issues, 64; mainstream, 43; mission of epistemology, 80; mission of philosophy, xxi, 277; model of communication in the public sphere, 297; motivation of the philosophy of science, 221; naturalism, 193, 322; order, 109; normatively acceptable action, 20; normatively corrective presumptions, 278, 347; orientation, 93; perspective, 6, 131, 223; perspective of Shallow Science, 14; philosophical approaches, xiii; philosophical project, 217; positions, 210; programs, 40; project of the Frankfurt School, 329; project of STS, xx; proposal, 20; retreat, 279-80; rhetorical character of action, 32; satisfaction, 230; sensibilities, 8, 287, 310; stance, 10; standard, 98, 131; structure of science, 232, 269-70, 282; structure of work, 308; theories of action, 20; theory of scientific reasoning, 147; to full self-realization, 109; tolerance, 244; turn, 131; "versus" empirical, xiv

Index

Norms, xv, 21, 31, 101, 119, 193, 215, 263; 265-68, 279-80, 300, 331; disciplinary, 242; epistemic, 21, 98; free-standing, 215; generalizable, 216; implicit, 215, 217, 231, 275; internal 213; market, 98; methodological, 214, 218; need for, 294; philosophical, 217, 264; of physics, 242; of rationality, 71; of science, 273; scientific, xxii; unconditional, 268

Panconstructivism, 342
PC-Positivism, 146-51, 183
Philosophers of science, xxii, 6-8, 13, 20, 45, 56-57, 90, 99, 104-7, 130, 171, 186, 194, 210, 223, 228, 238, 244-45, 252, 287, 296, 308, 322, 325, 337
Philosophy and sociology of science, 378
Philosophy of science, xviii, 6-7, 26, 29, 34, 69, 79, 87, 104-6, 112, 124, 153, 216, 220, 222-23, 241, 245, 308-9, 386
Physicist(s), 13, 30, 45, 76, 112-13, 130-31, 141, 200, 215, 242, 246, 253, 273, 296, 299, 311, 358; medieval 352
Physics, 30, 45, 76, 103-4, 112, 118, 131, 147, 151, 196, 212, 215, 233, 243-44, 246-47, 250, 255, 273, 295-96, 307, 342, 377, 381, 387; high-energy, 365; nineteenth-century, 4
Pickering, Andy, xii, 10-11, 255
Plato, 15, 21, 27, 35, 58, 133, 191, 209, 282, 360-61, 377; concept of forms, 361; episteme, 27; *Meno*, 376; nous 360
Platonic Plague, 203-4, 207, 210, 216-17
Platonism, 150; micro-, 133,
Platonist(s), xv, 204, 247, 365
Polanyi, Michael, 12, 180, 201, 217, 254, 271, 283-84, 296, 367
Polanyiites, 14
Policy, 21, 26, 40, 62, 98, 126, 131-32, 169, 251, 258, 260, 262, 271-74, 290, 301, 328, 385-86, 390; consultants, 130; decisions, 276; forums, 6; potential of STS, 262; science policy, xiv, xxi, 11, 54, 56, 100, 150, 227-29, 231, 233-34, 243, 248, 251-52,254, 258, 266, 271-72, 274, 310, 334, 388; social, 75, 122, 379
Policymaker(s), xiii, xv, xxi-xxii, 8, 26, 54-55, 126, 132, 150, 229-30, 247, 252-54 , 257,259, 261, 263-64, 267-68, 271-75, 301-2, 304-7, 381, 390; in science, 246, 258, 381
Policymaking, 98, 131
Polis, 49, 291-92
Political science, xvi, 49, 53, 105, 116, 119, 122-23, 126-27, 130-31, 133; grand theory in, 123; of science, 133

Political scientists, xiii, 122, 124, 126, 130-31, 133
Political sociology, 314
Political theory, xviii; and practice, 123
Popper, Karl xxii, 6, 27-28, 37, 39, 71, 78, 89, 105, 107, 117, 180, 193, 200, 220, 238, 245, 269, 281-84, 292, 312, 333, 336, 357, 359-60, 367-68;
Popperian, 6, 7, 186, 222, 238, 239, 278, 284, 330; falsification, 22, 36, 244, 258, 335
Popperianism, 124, 288
Popperians, 223, 373
Positivism, 41, 53, 73, 97, 118, 124, 150, 189, 348, 355-56, 381; nineteenth-century, 118; philosophical, 5; versus holism, 63
Positivist(s), 4, 6-8, 23, 45, 72, 109, 117, 146-47, 179, 186, 188-89, 218, 236, 249, 254, 261, 312, 342, 348, 350, 355, 358, 359, 377, 386-87; academic rhetoric, 36; account of language, 188; conception of theory, xxii; movement, 309; philosophers, 10; philosophers of science, 253; philosophy of science, 356, 374; theory of law, 101; unity of science, 192
Positivist Constitution, xvi
Postmodern, 46, 201, 251, 304, 309, 386; geographers, 24
Postmodernism, xvii, 111, 290-91, 342
Postmodernist, xvii, 111, 244, 282, 287, 289, 291-92, 348; challenge, 345; dispute, 341; liberalism, 287-88; modernist debate, 244; science practiced by STS, 342
Postmodernists, xxii, 291-92, 294, 342-43
Pragmatism, xvi, 264-266, 356, 381
Pragmatist(s), 94, 188, 249, 262-66, 279, 377; analysis, 262; vision of science, 249
Presumption, xxii, 366-70; of guilt, 369-71; of innocence, 369, 371-72, 374; in legal matters, 368
Presumptions, 24, 28, 44, 321; incommensurable, 85; as normative instruments, 368
Progress, 8, 33, 51, 123-24, 157, 161, 178, 211, 219, 221, 232, 236, 379; collective, 270; epistemic, 187-88; of pure science, 90; in science, 196; scientific, 235, 309, 353, 375, 385; social, 3; technological, 298
Project on the Rhetoric of Inquiry, xiv
Psychological: disciplines, 5; experiments, 218; explanations, 373
Psychologist(s), 37-38, 70, 72, 74, 77, 80, 112, 128, 134-35, 150, 162, 175, 192, 195, 218, 248, 315; cognitive, 173, 339; of science, 134; social, 132, 272
Psychology, 10, 20, 38, 44, 71-76, 80, 85, 97-98, 103, 116, 125, 128, 134, 153, 160-61, 190, 198, 219, 299, 313, 358-59, 370;

420

Index

Psychology (*continued*)
  American experimental, 75; behavioral, 177, 327; developmental, 196; individual, 75; introspective, 128; moral, 21, 55, 247, of science, 133, 135; social, 75, 79, 81, 120, 141
Public policy, xvii, 130

Quine, W. V. O, xvii, 72, 79, 81, 83, 99, 148, 155, 162-63, 166, 210, 240, 319, 321, 327, 352, 359

Rational choice theory, 256, 268, 279
Rationality, 9-11, 17, 37, 71, 94, 97, 124, 131, 146, 171, 173, 183, 191, 194, 223, 255, 374; agent of, 367; arationality assumption, 149, 193, 321; bounded, 159; canons of; 278; instant, 257; instrumental, 322; models of, 20, 23; of agents, 133; of native practices, 120; of the economist, 124; philosophical models of, 159; scientific, 336; units of, 268
Reflexivity, xiii, 301, 319, 322, 341-45; in STS, 339
Relativism, 236, 319-28, 330, 332-33, 339, 364-66; epistemic, 288; realistic, 326; of STS research, 10; politics of, 333
Relativist(s), xxi, 13, 236, 319, 321-22, 325, 329-31, 333-34, 338, 366; epistemology, 334; methodological, 329; professional, 320
Replication, 258-58, 261, 267; as a norm governing science, 258-61
Representation(s), xvi, 15, 63, 174, 188, 349, 362; by classes, 50; evenhandedness of, 343; by geographical region, 50; intervention, 104; numerical, 133; political 152; privileged, 340; realistic, 327; regional 50; theoretical, 95, 99; verbal, 95; without intervention, 62, 349
Representational: approach to language, 190; function of language, 189, 322
Revolution(s), 8, 186, 212, 243; failed, 172; Kuhnian, 186-87; permanent, 269; scientific, 3, 108, 115, 152, 154, 172, 190-91, 219, 373, 375, 380
Revolutionary: interdisciplinary settings, 48; interpenetration, 47; overthrow of Newtonian mechanics, 243; periods, 4; phases, 89; scientists, 376; theorists, 44, 48; theorizing, 44, 269; theory, 44; thought, 48; scientific revolutionaries, 118
Rhetoric, xi-xii, xv, xviii, xix-xx, 3, 15, 17-18 23-24, 36, 61, 111, 115, 131, 133, 137, 343, 350; of anamnesis, 59; and argumentation, 27; of autonomous inquiry, 334; classical, xv; constructive, 255; continuum of, 313; decline as a discipline, 115; of fact-value discriminations, 257; of interpenetrability, xvii, 34, 36, 292; of interpenetration, 56; knowledge of, 108; legal, 242; of knowledge policy, 255; scientific, 174, 248, of rationality attributions, 256; of science, xii, xi, 180; of science policy, 26; of scientists, 12; of testability, 160; of truth, xvii
Rhetorical: character of disciplinary boundaries, 104; criticism, xii; function of representation, 189; impassability, 140-42; impoverishment, 102; invention, 19; knowledge is power equation, 115; rhetorically oriented view of theorizing, 347; sensibility, 22, studies of physics, 45; turn in the human sciences, xii
Rhetoricians, xii, xiv, xxii, xv, 15, 19-20, 23, 35, 60, 132, 306, 368, 390; classical, 43; distinction between contexts of invention and instruction, 118; of science, xiii
Rorty, Richard, xi, 7, 72, 92, 150, 203, 277, 280, 287, 348

Shapere, Dudley, 7, 49, 94, 103, 106, 145, 219
Shapin, Steven, 10, 81, 97, 109, 136, 211-13, 253
Simon, Herbert, 83, 139, 145-46, 148-49, 151, 154-60, 163-66, 168-69, 177-79, 198
Skinner, B. F., 113, 126, 152, 155, 161-62, 164-65, 176-78, 184
Slezak, Peter, 63, 139, 142, 144-52, 156-58, 160, 162-163, 165-67, 177-79
Social epistemology, xii-xiv, xvi-xviii, xx-xxii, 3, 5-6, 11-12, 18, 24, 26-29, 31, 34, 38-39, 43, 48, 54, 57-58, 75-76, 187, 190-91, 202, 208-9, 222-23, 229, 234, 242-43, 270, 285, 292, 310-11, 319, 337, 339-40, 344-45; policy orientation of, 277; three presumptions of, 29
Social constructivism, 253, 341
Social science(s), 52, 62-64, 74, 91-92, 96, 102-5, 112-13, 116, 118, 120-25, 129, 134, 137, 149-50, 159, 165, 186-87, 192, 195, 202, 223, 228, 243, 263, 331, 356, 373, 389; canonical histories of, 116, 133; disciplines, 192; science in, 117-18
Social scientist(s), xii, xiv, xxii, 57, 63, 110,

Index

124, 203, 296, 347, 352, 354, 373
Sociology, 9, 116, 119, 121-22, 149, 170, 172, 189, 193, 319, 387; American, 114; applied, 377; of conflict, 312; of knowledge, 9, 15, 79, 321-22, 326-28, 338, 359
Socrates, xii, 15, 58, 85, 222, 277, 319-21, 338
Socratic Conflation, 320
Sophist(s), xv, 14-15, 57, 60, 277, 319-20, 366
SSK (Sociology of Scientific Knowledge), 9, 62-64, 142, 144-46, 148, 150-51, 156-60, 162, 166-68, 171, 178, 180, 183-84, 221; theorizing, 178
Stich, Steven, 151, 163, 268, 292, 333
Strong Programme, 9, 11, 142, 221, 322-24
Structural Marxism, 202
STS (Science and Technology Studies), xii-xiv, xvii, xix-xxii, 3, 5-7, 9-11, 18-19, 22, 25-26, 38, 48, 63, 65, 96, 139, 141, 166, 168, 187, 220-21, 251, 274, 298, 301, 310, 316, 319, 336, 340-42, 345, 347, 352-53, 385, 388; disciplinization of 96; emergence of, 33; empirical base for, 95; Fundamental Mandate of, 9; High and Low Church, xiii, microsociological perspective in, 253; obscurantist tendencies in, 180; practitioner(s), xii-xiv, xx, 11, 23-24, 80, 96, 179, 221-22, 233, 253, 386; reflexive practitioners of, 85; research, xx, 10, 24, 95, 235, 274-75, 383-84, 389; researcher(s), 96, 179, 221-22, 274, 283, 301, 386, 389

Theorizing, xi, xv-xvii, 72, 94-96, 99, 171, 269-70, 343, 345, 349, 351; anti-, 362; homeopathic, 7; middle-range, 270; philosophical, xvi; role of 12; speculative, 353
Theory, 4, 17, 20, 37, 41, 53, 71, 76-77, 91-94, 96, 125, 131, 137, 141, 175-76, 178, 191, 193, 196, 221, 242-43, 245, 270, 316, 325, 338, 343-44, 347-48, 350-60, 364, 366-67, 374, 378; of actants, 183; actor-network, 13, 386; anomaly management, 258; anti-, 348, 363; choice(s), 5, 79, 141, 148, 193, 210, 212, 235, 321-22, 380; choice in science, 322; and consequences, 348-49, 350, 352-55; -driven research, 23; embedded, xxii; explanatory, 30; free-enterprisers, 269-70; and method, 63; -neutral observation, 357; organizational, 159; philosophical, xvi-xvii; positivistic, 355; of power, 266; practical applications of, 363; and practice, 14, 109, 116, 349; of presumption, 368; of reference, 162; revision, long-term rational conceptual, 377; scientific, 337, 340, of scientific rationality, xiii;; social, 121, 344; testing, 10; value-neutral, 357, theory-driven research, 23; theory-neutral observation, 358
Topoi, xv, 119
Topos, 365
Toulmin, Stephen, 23, 103, 109, 118, 390
Truth, xvi, xviii, 9, 57, 91-93, 188-89, 191, 291, 322, 334, 338, 340-41, 344, 355-56, 365, 367, 371, 378-79; disutilitarian theory of, 333; for its own sake, 334; maximizing, 30; pursuit of, 333-34; "true for me," 320; universal, 326
Truths, 83, 107, 327; demonstration of, 81; espoused by a discipline, 313; presumptive, 262; ultimate, 5 universal 327, 359; useless, 90
Tweney, Ryan, 134, 195, 197, 220

Unger, Roberto, 344, 348-50
University, 3, 49-52, 222, 289; administrators, 34; budgets, 313; consolidation, 50; departments, 48-50; employees, 52; examinations in the Middle Ages, 250; early-nineteenth-century German movement, 111; German research, 125; German system, 231; interests, 50; life, 390; politics, 49; university-oriented idealists, 128

Willard, Charles, xv, 72, 103, 285, 286
Woolgar, Steve, 10-11, 24, 95, 115, 145, 158, 168, 184, 321, 325, 341, 343

RHETORIC OF THE HUMAN SCIENCES

Lying Down Together: Law, Metaphor and Theology
Milner S. Ball

Shaping Written Knowledge: The Genre and Activity of the
Experimental Article of Science
Charles Bazerman

Textual Dynamics of the Professions: Historical and Contemporary
Studies of Writing in Professional Communities
Charles Bazerman and James Paradis

Politics and Ambiguity
William E. Connolly

Philosophy, Rhetoric and the End of Knowledge:
The Coming of Science and Technology Studies
Steve Fuller

Machiavelli and the History of Prudence
Eugene Garver

Language and Historical Representation: Getting the Story Crooked
Hans Kellner

The Rhetoric of Economics
Donald N. McCloskey

Therapeutic Discourse and Socratic Dialogue
Tullio Maranhão

The Rhetoric of the Human Sciences: Language and Argument in
Scholarship and Public Affairs
John S. Nelson, Allan Megill, and Donald N. McCloskey, eds.

What's Left? The Ecole Normale Supérieure and the Right
Diane Rubenstein

The Politics of Representation: Writing Practices in Biography, Photography and Policy Analysis
Michael J. Shapiro

The Legacy of Kenneth Burke
Herbert Simons and Trevor Melia, eds.

The Unspeakable; Discourse, Dialogue, and Rhetoric in the Postmodern World
Stephen A. Tyler

Heracles' Bow: Essays on the Rhetoric and the Poetics of the Law
James Boyd White